Fourth Edition

Partial Solutions Guide

Chemistry

Steven S. Zumdahl

Thomas J. Hummel
Susan Arena Zumdahl
Steven S. Zumdahl

University of Illinois at Urbana-Champaign

HOUGHTON MIFFLIN COMPANY Boston New York

Senior Sponsoring Editor:	Richard Stratton
Assistant Editor:	Marianne Stepanian
Director of Manufacturing:	Michael O'Dea
Executive Marketing Manager:	Karen Natale

Cover design:	Harold Burch, Harold Burch Design, New York City.
Cover image:	Pete Turner, The Image Bank.

Printed in the U.S.A.

ISBN: 0-669-41798-X

23456789-VG-00 99 98 97

TABLE OF CONTENTS

TO THE STUDENT: HOW TO USE THIS GUIDE

Solutions to all of the odd numbered end of chapter questions and exercises are in this manual. This "Solutions Guide" can be very valuable if you use it properly. The way NOT to use it is to look at an exercise in the book and then immediately check the solution, often saying to yourself, "That's easy, I can do it." Chemistry is easy once you get the hang of it, but it takes practice. Don't look up a solution to a problem until you have tried to work it on your own. If you are completely stuck, see if you can find a similar problem in the Sample Exercises in the chapter. Only look up the solution as a last resort. If you do this for a problem, look for a similar problem in the end of chapter exercises and try working it. The more problems you do, the easier chemistry becomes. It is also in your self interest to try to work as many problems as possible. Most exams that you will take in chemistry will involve a lot of problem solving. If you have worked several problems similar to the ones on an exam, you will do much better than if the exam is the first time you try to solve a particular type of problem. No matter how much you read and study the text, or how well you think you understand the material, you don't really understand it until you have taken the information in the text and applied the principles to problem solving. You will make mistakes, but the good students learn from their mistakes.

In this manual we have worked problems as in the textbook. We have shown intermediate answers to the correct number of significant figures and used the rounded answer in later calculations. Thus, some of your answers may differ slightly from ours. When we have not followed this convention, we have usually noted this in the solution.

We are grateful to Delores Wyatt for her outstanding effort in preparing the manuscript of this manual. We also thank Robert Pfaff for his careful and thorough accuracy review of the Solutions Guide.

TJH
SAZ
SSZ

CHAPTER ONE

CHEMICAL FOUNDATIONS

Questions

13. No, it is useful whenever a systematic approach of observation and hypothesis testing can be used.

15. Precision is related to how many significant figures one can associate with a measurement. Consider weighing an object on three different balances with the following results: 11 g; 11.25 g; 11.2456 g. Since the assumed uncertainty in all measurements is ± 1 in the last digit, then the uncertainty of the three balances are: ± 1 g; ± 0.01 g; ± 0.0001 g, respectively. The balance with the smallest uncertainty is the balance with the most significant figures associated with the measurement, which is also the balance that is assumed most precise.

17. Chemical changes involve the making and breaking of chemical forces (bonds). Physical changes do not. The identity of a substance changes after a chemical change, but not after a physical change.

Exercises

Significant Figures and Unit Conversions

19. a. inexact b. exact c. exact

For c, $\frac{36 \text{ in}}{\text{yd}} \times \frac{2.54 \text{ cm}}{\text{in}} \times \frac{1 \text{ m}}{100 \text{ cm}} = \frac{0.9144 \text{ m}}{\text{yd}}$ (All conversion factors used are exact.)

d. inexact; Although this number appears to be exact, it probably isn't. The announced attendance may be tickets sold but not the number who were actually in the stadium. Some people who paid may not have gone, some may leave early, or arrive late, some may sneak in without paying, etc.

e. exact f. inexact

21. a. $0.00\underline{12}$; 2 S.F., 1.2×10^{-3} b. $\underline{437}{,}000$; 3 S.F., 4.37×10^5

c. $\underline{900.0}$; 4 S.F., 9.000×10^2 d. $\underline{106}$; 3 S.F.; 1.06×10^2

e. $\underline{125{,}904}{,}000$; 6 S.F., 1.25904×10^8 f. $\underline{1.0012}$; 5 S.F., 1.0012×10^0

g. 2006; 4 S.F., 2.006×10^3

h. 3050; 3 S.F., 3.05×10^3

i. 0.001060; 4 S.F., 1.060×10^{-3}

23. a. 6×10^8 b. 5.8×10^8 c. 5.82×10^8

d. 5.8200×10^8 e. 5.820000×10^8

25. a. 467; The difference of 25.27 - 24.16 = 1.11 has only three significant figures, so the answer will only have 3 significant figures. For this problem and for subsequent problems, the addition/ subtraction rule is applied separately from the multiplication/division rule.

b. 0.24; The difference of 8.925 - 8.904 = 0.021 has only 2 significant figures.

c. $(9.04 - 8.23 + 21.954 + 81.0) \div 3.1416 = 103.8 \div 3.1416 = 33.04$

We will generally round off at intermediate steps in order to show the correct number of significant figures. However, you should round off at the end of all the mathematical operations in order to avoid round off error. Make sure you keep track of the correct number of significant figures during intermediate steps, but round off at the end.

d. $\frac{9.2 \times 100.65}{8.321 + 4.026} = \frac{9.2 \times 100.65}{12.347} = 75$

e. $0.1654 + 2.07 - 2.114 = 0.12$

Uncertainty begins to appear in the second decimal place. Numbers were added as written and the answer was rounded off to 2 decimal places at the end. If you round to 2 decimal places and add you get 0.13. Always round off at the end of the operation to avoid round off error.

f. $8.27(4.987 - 4.962) = 8.27(0.025) = 0.21$

g. $\frac{9.5 + 4.1 + 2.8 + 3.175}{4} = \frac{19.6}{4} = 4.90 = 4.9$

Uncertainty appears in the first decimal place. The average of several numbers can only be as precise as the least precise number. Averages can be exceptions to the significant figure rules.

h. $\frac{9.025 - 9.024}{9.025} \times 100 = \frac{0.001}{9.025} \times 100 = 0.01$

27. a. $1\text{ km} = 10^3\text{ m} = 10^6\text{ mm} = 10^{15}\text{ pm}$

b. $1\text{ g} = 10^{-3}\text{ kg} = 10^3\text{ mg} = 10^9\text{ ng}$;

c. $1\text{ mL} = 10^{-3}\text{ L} = 10^{-3}\text{ dm}^3 = 1\text{ cm}^3$

d. $1\text{ mg} = 10^{-6}\text{ kg} = 10^{-3}\text{ g} = 10^3\ \mu\text{g} = 10^6\text{ ng} = 10^9\text{ pg} = 10^{12}\text{ fg}$

e. $1\text{ s} = 10^3\text{ ms} = 10^9\text{ ns}$

29. $1\ \text{Å} \times \frac{10^{-8}\ \text{cm}}{\text{Å}} \times \frac{1\ \text{m}}{100\ \text{cm}} \times \frac{10^{9}\ \text{nm}}{\text{m}} = 1 \times 10^{-1}\ \text{nm}$

$1\ \text{Å} \times \frac{1 \times 10^{-1}\ \text{nm}}{\text{Å}} \times \frac{1\ \text{m}}{10^{9}\ \text{nm}} \times \frac{10^{12}\ \text{pm}}{\text{m}} = 1 \times 10^{2}\ \text{pm}$

31. a. Appropriate conversion factors are found on the back cover of the text. In general, the number of significant figures we use in the conversion factors will be one more than the number of significant figures from the numbers given in the problem. This is usually sufficient to avoid round off error.

$3.91\ \text{kg} \times \frac{1\ \text{lb}}{0.4536\ \text{kg}} = 8.62\ \text{lb};\ \ 0.62\ \text{lb} \times \frac{16\ \text{oz}}{\text{lb}} = 9.9\ \text{oz}$

Baby's weight = 8 lb and 9.9 oz or to the nearest ounce, 8 lb and 10 oz.

$51.4\ \text{cm} \times \frac{1\ \text{in}}{2.54\ \text{cm}} = 20.2\ \text{in} \approx 20\ 1/4\ \text{in} = \text{baby's height}$

b. $25{,}000\ \text{mi} \times \frac{1.61\ \text{km}}{\text{mi}} = 4.0 \times 10^{4}\ \text{km};\ \ 4.0 \times 10^{4}\ \text{km} \times \frac{1000\ \text{m}}{\text{km}} = 4.0 \times 10^{7}\ \text{m}$

c. $V = 1 \times w \times h = 1.0\ \text{m} \times \left(5.6\ \text{cm} \times \frac{1\ \text{m}}{100\ \text{cm}}\right) \times \left(2.1\ \text{dm} \times \frac{1\ \text{m}}{10\ \text{dm}}\right) = 1.2 \times 10^{-2}\ \text{m}^3$

$1.2 \times 10^{-2}\ \text{m}^3 \times \left(\frac{10\ \text{dm}}{\text{m}}\right)^3 \times \frac{1\ \text{L}}{\text{dm}^3} = 12\ \text{L}$

$12\ \text{L} \times \frac{1000\ \text{cm}^3}{\text{L}} \times \left(\frac{1\ \text{in}}{2.54\ \text{cm}}\right)^3 = 730\ \text{in}^3;\ \ 730\ \text{in}^3 \times \left(\frac{1\ \text{ft}}{12\ \text{in}}\right)^3 = 0.42\ \text{ft}^3$

33. a. $928\ \text{mi} \times \frac{5280\ \text{ft}}{\text{mi}} \times \frac{1\ \text{fathom}}{6\ \text{ft}} \times \frac{1\ \text{cable length}}{100\ \text{fathoms}} \times \frac{1\ \text{nautical mi}}{10\ \text{cable lengths}}$

$\times \frac{1\ \text{league}}{3\ \text{nautical miles}} = 272\ \text{leagues}$

$928\ \text{mi} \times \frac{1.609\ \text{km}}{1\ \text{mi}} = 1.49 \times 10^{3}\ \text{km}$

b. $1.0\ \text{cable length} \times \frac{100\ \text{fathom}}{\text{cable length}} \times \frac{6\ \text{ft}}{\text{fathom}} \times \frac{1\ \text{yd}}{3\ \text{ft}} \times \frac{1\ \text{m}}{1.09\ \text{yd}} \times \frac{1\ \text{km}}{1000\ \text{m}} = 0.18\ \text{km}$

$1.0\ \text{cable length} = 0.18\ \text{km} \times \frac{1000\ \text{m}}{\text{km}} \times \frac{100\ \text{cm}}{\text{m}} = 1.8 \times 10^{4}\ \text{cm}$

c. $315\ \text{ft} \times \frac{12\ \text{in}}{\text{ft}} \times \frac{2.54\ \text{cm}}{\text{in}} \times \frac{1\ \text{m}}{100\ \text{cm}} = 96.0\ \text{m}$

$$37 \text{ ft} \times \frac{12 \text{ in}}{\text{ft}} \times \frac{2.54 \text{ cm}}{\text{in}} \times \frac{1 \text{ m}}{100 \text{ cm}} = 11 \text{ m}$$

$$315 \text{ ft} \times \frac{1 \text{ fathom}}{6 \text{ ft}} \times \frac{1 \text{ cable length}}{100 \text{ fathoms}} = 5.25 \times 10^{-1} \text{ cable lengths}$$

$$37 \text{ ft} \times \frac{1 \text{ fathom}}{6 \text{ ft}} = 6.2 \text{ fathoms}$$

35. a. $1 \text{ troy lb} \times \frac{12 \text{ troy oz}}{\text{troy lb}} \times \frac{20 \text{ pw}}{\text{troy oz}} \times \frac{24 \text{ grains}}{\text{pw}} \times \frac{0.0648 \text{ g}}{\text{grain}} \times \frac{1 \text{ kg}}{1000 \text{ g}} = 0.373 \text{ kg}$

$1 \text{ troy lb} = 0.373 \text{ kg} \times \frac{2.205 \text{ lb}}{\text{kg}} = 0.822 \text{ lb}$

b. $1 \text{ troy oz} \times \frac{20 \text{ pw}}{\text{troy oz}} \times \frac{24 \text{ grains}}{\text{pw}} \times \frac{0.0648 \text{ g}}{\text{grain}} = 31.1 \text{ g}$

$1 \text{ troy oz} = 31.1 \text{ g} \times \frac{1 \text{ carat}}{0.200 \text{ g}} = 156 \text{ carats}$

c. $1 \text{ troy lb} = 0.373 \text{ kg};\ 0.373 \text{ kg} \times \frac{1000 \text{ g}}{\text{kg}} \times \frac{1 \text{ cm}^3}{19.3 \text{ g}} = 19.3 \text{ cm}^3$

37. $\frac{4.4 \times 10^9 \text{ mi}}{2.0 \text{ yr}} \times \frac{1.61 \text{ km}}{\text{mi}} \times \frac{1 \text{ yr}}{365 \text{ d}} \times \frac{1 \text{ d}}{24 \text{ hr}} \times \frac{1 \text{ hr}}{60 \text{ min}} \times \frac{1 \text{ min}}{60 \text{ s}} = 110 \text{ km/s}$

39. $\frac{14 \text{ km}}{\text{L}} \times \frac{1 \text{ mi}}{1.61 \text{ km}} \times \frac{3.79 \text{ L}}{\text{gal}} = 33 \text{ mi/gal}$; The spouse's car has the better gas mileage.

Temperature

41. $T_C = \frac{5}{9}(T_F - 32) = \frac{5}{9}(102.5 - 32) = 39.2°C$; $T_K = T_C + 273.2 = 312.4$ K (Note: 32 is exact)

43. $T_F = \frac{9}{5} \times T_C + 32 = \frac{9}{5} \times 25 + 32 = 77°F$; $T_K = 25 + 273 = 298$ K

45. We can do this two ways. First, we calculate the high and low temperature and get the uncertainty from the range. 20.6°C ± 0.1°C means the temperature can range from 20.5°C to 20.7°C.

$T_F = \frac{9}{5} \times T_c + 32 \leftarrow$ (exact); $T_F = \frac{9}{5} \times 20.6 + 32 = 69.1°F$

$T_F(\text{min}) = \frac{9}{5} \times 20.5 + 32 = 68.9°F$; $T_F(\text{max}) = \frac{9}{5} \times 20.7 + 32 = 69.3°F$

So the temperature ranges from 68.9°F to 69.3°F which we can express as 69.1 ± 0.2°F.

An alternative way is to treat the uncertainty and the temperature in °C separately.

$$T_F = \frac{9}{5} \times T_C + 32 = \frac{9}{5} \times 20.6 + 32 = 69.1°F; \quad \pm 0.1°C \times \frac{9°F}{5°C} = \pm 0.18°F \approx \pm 0.2°F$$

Combining the two calculations: $T_F = 69.1 \pm 0.2°F$

Density

47. $$\frac{2.70 \text{ g}}{\text{cm}^3} \times \frac{1 \text{ kg}}{1000 \text{ g}} \times \left(\frac{100 \text{ cm}}{\text{m}}\right)^3 = \frac{2.70 \times 10^3 \text{ kg}}{\text{m}^3}$$

$$\frac{2.70 \text{ g}}{\text{cm}^3} \times \frac{1 \text{ lb}}{453.6 \text{ g}} \times \left(\frac{2.54 \text{ cm}}{\text{in}}\right)^3 \times \left(\frac{12 \text{ in}}{\text{ft}}\right)^3 = \frac{169 \text{ lb}}{\text{ft}^3}$$

49. $d = \text{density} = \frac{\text{mass}}{\text{volume}}$; mass $= 1.67 \times 10^{-24}$ g; radius = diameter/2 $= 5.0 \times 10^{-4}$ pm

$$V = \frac{4}{3}\pi r^3 = \frac{4}{3} \times 3.14 \times \left(5.0 \times 10^{-4} \text{ pm} \times \frac{1 \text{ m}}{10^{12} \text{ pm}} \times \frac{100 \text{ cm}}{\text{m}}\right)^3 = 5.2 \times 10^{-40} \text{ cm}^3$$

$$d = \frac{1.67 \times 10^{-24} \text{ g}}{5.2 \times 10^{-40} \text{ cm}^3} = \frac{3.2 \times 10^{15} \text{ g}}{\text{cm}^3}$$

51. $$5.0 \text{ carat} \times \frac{0.200 \text{ g}}{\text{carat}} \times \frac{1 \text{ cm}^3}{3.51 \text{ g}} = 0.28 \text{ cm}^3$$

53. a. Both have the same mass.

b. 1.0 mL of mercury; Mercury has a larger density than water.

$$1.0 \text{ mL} \times \frac{13.6 \text{ g}}{\text{mL}} = 13.6 \text{ g of mercury}; \quad 1.0 \text{ mL} \times \frac{0.998 \text{ g}}{\text{mL}} = 1.0 \text{ g of water}$$

c. Same; Both represent 19.3 g of substance.

d. 1.0 L of benzene (880 g vs 670 g)

Classification and Separation of Matter

55. Solid: own volume, own shape, does not flow; Liquid: own volume, takes shape of container, flows; Gas: takes volume and shape of container, flows

57. a. pure b. mixture c. mixture d. pure e. mixture (copper and zinc)

f. pure g. mixture h. mixture i. pure

Iron and uranium are elements. Water and table salt are compounds. Water is H_2O and table salt is NaCl. Compounds are composed of two or more elements.

Additional Exercises

59. a. 8.41 (2.16 has only three significant figures.)

b. 16.1 (Uncertainty appears in the first decimal place for 8.1.)

c. 52.5 (All numbers have 3 significant figures.)

d. 5 (2 contains one significant figure.) e. 0.009

f. 429.59 (Uncertainty appears in 2nd decimal place for 2.17 and 4.32.)

61. $1.5 \text{ teaspoons} \times \frac{80.\text{ mg acet}}{0.50 \text{ teaspoons}} = 240 \text{ mg acetaminophen}$

$$\frac{240 \text{ mg acet}}{24 \text{ lb}} \times \frac{1 \text{ lb}}{0.454 \text{ kg}} = 22 \text{ mg acetaminophen/kg}$$

$$\frac{240 \text{ mg acet}}{35 \text{ lb}} \times \frac{1 \text{ lb}}{0.454 \text{ kg}} = 15 \text{ mg acetaminophen/kg}$$

The range is from 15 mg to 22 mg acetaminophen per kg of body weight.

63. $126 \text{ gal} \times \frac{4 \text{ qt}}{\text{gal}} \times \frac{1 \text{ L}}{1.057 \text{ qt}} = 477 \text{ L}$

65. $\text{Total volume} = \left(200.\text{ m} \times \frac{100 \text{ cm}}{\text{m}}\right) \times \left(300.\text{ m} \times \frac{100 \text{ cm}}{\text{m}}\right) \times 4.0 \text{ cm} = 2.4 \times 10^9 \text{ cm}^3$

$$\text{Vol. covered by 1 bag of topsoil} = \left[10 \text{ ft}^2 \times \left(\frac{12 \text{ in}}{\text{ft}}\right)^2 \times \left(\frac{2.54 \text{ cm}}{\text{in}}\right)^2\right] \times \left(1.0 \text{ in} \times \frac{2.54 \text{ cm}}{\text{in}}\right) = 2.4 \times 10^4 \text{ cm}^3$$

$$2.4 \times 10^9 \text{ cm}^3 \times \frac{1 \text{ bag}}{2.4 \times 10^4 \text{ cm}^3} = 1.0 \times 10^5 \text{ bags topsoil}$$

67. $T_c = (68°\text{F} - 32°\text{F}) \times \frac{5°\text{C}}{9°\text{F}} = 20.°\text{C}$

Gallium is a solid at 20.°C. From the melting point, gallium doesn't convert to the liquid state until 29.8°C.

69. $\text{Volume of lake} = 100 \text{ mi}^2 \times \left(\frac{5280 \text{ ft}}{\text{mi}}\right)^2 \times 20 \text{ ft} = 6 \times 10^{10} \text{ ft}^3$

$$6 \times 10^{10} \text{ ft}^3 \times \left(\frac{12 \text{ in}}{\text{ft}} \times \frac{2.54 \text{ cm}}{\text{in}}\right)^3 \times \frac{1 \text{ mL}}{\text{cm}^3} \times \frac{0.4 \text{ µg}}{\text{mL}} = 7 \times 10^{14} \text{ µg}$$

$$7 \times 10^{14}\ \mu g \times \frac{1\ g}{10^6\ \mu g} \times \frac{1\ kg}{10^3\ g} = 7 \times 10^5 \text{ kg of mercury}$$

71. $V = 5.00\ cm \times 4.00\ cm \times 2.50\ cm = 50.0\ cm^3$; $50.0\ cm^3 \times \frac{22.57\ g}{cm^3} = 1130\ g$

$$1.00\ kg \times \frac{1000\ g}{kg} \times \frac{1\ cm^3}{22.57\ g} = 44.3\ cm^3$$

73. Circumference = c = $2\pi r$; $V = \frac{4\pi r^3}{3} = \frac{4\pi}{3}\left(\frac{c}{2\pi}\right)^3 = \frac{c^3}{6\pi^2}$

$$\text{Largest density} = \frac{5.25\ oz}{\frac{(9.00\ in)^3}{6\pi^2}} = \frac{5.25\ oz}{12.3\ in^3} = \frac{0.427\ oz}{in^3}$$

$$\text{Smallest density} = \frac{5.00\ oz}{\frac{(9.25\ in)^3}{6\pi^2}} = \frac{5.00\ oz}{13.4\ in^3} = \frac{0.373\ oz}{in^3}$$

Maximum range is: $\frac{(0.373 - 0.427)\ oz}{in^3}$ or $0.40 \pm 0.03\ oz/in^3$ (Uncertainty in 2nd decimal place.)

75. We need to calculate the maximum and minimum values of the density, given the uncertainty in each measurement. The maximum value is:

$$d_{max} = \frac{19.625\ g + 0.002\ g}{25.00\ cm^3 - 0.03\ cm^3} = \frac{19.627\ g}{24.97\ cm^3} = 0.7860\ g/cm^3$$

The minimum value of the density is:

$$d_{min} = \frac{19.625\ g - 0.002\ g}{25.00\ cm^3 + 0.03\ cm^3} = \frac{19.623\ g}{25.03\ cm^3} = 0.7840\ g/cm^3$$

The density of the liquid is between $0.7840\ g/cm^3$ and $0.7860\ g/cm^3$. These measurements are sufficiently precise to distinguish between ethanol ($d = 0.789\ g/cm^3$) and isopropyl alcohol ($d = 0.785\ g/cm^3$).

77. $V = V(\text{final}) - V(\text{initial})$; $d = \frac{28.90\ g}{9.8\ cm^3 - 6.4\ cm^3} = \frac{28.90\ g}{3.4\ cm^3} = 8.5\ g/cm^3$

$d_{max} = \frac{mass_{max}}{V_{min}}$, We get V_{min} from $9.7\ cm^3 - 6.5\ cm^3 = 3.2\ cm^3$

$$d_{max} = \frac{28.93\ g}{3.2\ cm^3} = \frac{9.0\ g}{cm^3};\ d_{min} = \frac{mass_{min}}{V_{max}} = \frac{28.87\ g}{9.9\ cm^3 - 6.3\ cm^3} = \frac{8.0\ g}{cm^3}$$

The density is: $8.5 \pm 0.5\ g/cm^3$.

Challenge Problems

79. In a subtraction, the result gets smaller but the uncertainties add. If the two numbers are very close together, the uncertainty may be larger than the result. For example, let us assume we want to take the difference of the following two measured quantities, 999,999 ± 2 and 999,996 ± 2. The difference is 3 ± 4. Because of the large uncertainty, subtracting two similar numbers is bad practice.

81. Heavy pennies (old): mean mass = 3.08 ± 0.05 g

Light pennies (new): mean mass $= \frac{(2.467 + 2.545 + 2.518)}{3} = 2.51 \pm 0.04$ g

Average density of old pennies: $d_{old} = \dfrac{\dfrac{95 \times 8.96\ g}{cm^3} + \dfrac{5 \times 7.14\ g}{cm^3}}{100} = \dfrac{8.9\ g}{cm^3}$

Average density of new pennies: $d_{new} = \dfrac{\dfrac{2.4 \times 8.96\ g}{cm^3} + \dfrac{97.6 \times 7.14\ g}{cm^3}}{100} = \dfrac{7.18\ g}{cm^3}$

Since $d = \frac{mass}{volume}$ and the volume of old and new pennies are the same, then:

$$\frac{d_{new}}{d_{old}} = \frac{mass_{new}}{mass_{old}};\quad \frac{d_{new}}{d_{old}} = \frac{7.18}{8.9} = 0.81;\quad \frac{mass_{new}}{mass_{old}} = \frac{2.51}{3.08} = 0.815$$

To the first two decimal places, the ratios are the same. We can reasonably conclude that yes, the difference in mass is accounted for by the difference in the alloy used.

83. a. One possibility is that rope B is not attached to anything and rope A and rope C are connected via a pair of pulleys and/or gears.

b. Try to pull rope B out of the box. Measure the distance moved by C for a given movement of A. Hold either A or C firmly while pulling on the other.

CHAPTER TWO

ATOMS, MOLECULES AND IONS

Questions

11. a. Atoms have mass and are neither destroyed nor created by chemical reactions. Therefore, mass is neither created nor destroyed by chemical reactions. Mass is conserved.

    b. The composition of a substance depends on the number and kinds of atoms that form it.

    c. Compounds of the same elements differ only in the numbers of atoms of the elements forming them, i.e., NO, N_2O, NO_2.

13. Deflection of cathode rays by magnetic and electric fields led to the conclusion that they were negatively charged. The cathode ray was produced at the negative electrode and repelled by the negative pole of the applied electric field.

15. The atomic number of an element is equal to the number of protons in the nucleus of an atom of that element. The mass number is the sum of the number of protons plus neutrons in the nucleus. The atomic mass is the actual mass of a particular isotope (including electrons). As we will see in chapter three, the average mass of an atom is taken from a measurement made on a large number of atoms. The average atomic mass value is listed in the periodic table.

17. A compound will always contain the same numbers (and types) of atoms. A given amount of hydrogen will react only with a specific amount of oxygen. Any excess oxygen will remain unreacted.

Exercises

Development of the Atomic Theory

19. a. The composition of a substance depends on the numbers of atoms of each element making up the compound (i.e., the formula of the compound) and not on the composition of the mixture from which it was formed.

    b. Avogadro's hypothesis implies that volume ratios are equal to molecule ratios at constant temperature and pressure. $H_2 + Cl_2 \rightarrow 2\ HCl$. From the balanced equation, the volume of HCl produced will be twice the volume of H_2 (or Cl_2) reacted.

21. $\frac{1.188}{1.188} = 1.000$; $\frac{2.375}{1.188} = 1.999$; $\frac{3.563}{1.188} = 2.999$

The masses of fluorine are simple ratios of whole numbers to each other, 1:2:3.

23. To get the atomic mass of H to be 1.00, we divide the mass of hydrogen that reacts with 1.00 g of oxygen by 0.126, i.e., $\frac{0.126}{0.126} = 1.00$. To get Na, Mg and O on the same scale, we do the same division.

Na: $\frac{2.875}{0.126} = 22.8$; Mg: $\frac{1.500}{0.126} = 11.9$; O: $\frac{1.00}{0.126} = 7.94$

	H	O	Na	Mg
Relative Value	1.00	7.94	22.8	11.9
Accepted Value	1.008	16.00	22.99	24.31

The atomic masses of O and Mg are incorrect. The atomic masses of H and Na are close. Something must be wrong about the assumed formulas of the compounds. It turns out the correct formulas are H_2O, Na_2O, and MgO. The smaller discrepancies result from the error in the atomic mass of H.

The Nature of the Atom

25. Density of hydrogen nucleus (contains one proton only):

$$V_{nucleus} = \frac{4}{3}\pi r^3 = \frac{4}{3}(3.14)(5 \times 10^{-14}\ \text{cm})^3 = 5 \times 10^{-40}\ \text{cm}^3$$

$$d = \frac{1.67 \times 10^{-24}\ \text{g}}{5 \times 10^{-40}\ \text{cm}^3} = 3 \times 10^{15}\ \text{g/cm}^3$$

Density of H-atom (contains one proton and one electron):

$$V_{atom} = \frac{4}{3}(3.14)(1 \times 10^{-8}\ \text{cm})^3 = 4 \times 10^{-24}\ \text{cm}^3$$

$$d = \frac{1.67 \times 10^{-24} + 9 \times 10^{-28}\ \text{g}}{4 \times 10^{-24}\ \text{cm}^3} = 0.4\ \text{g/cm}^3$$

27. $5.93 \times 10^{-18}\ \text{C} \times \frac{1\ \text{electron charge}}{1.602 \times 10^{-19}\ \text{C}}$ = 37 negative (electron) charges on the oil drop

29. gold - Au; silver - Ag; mercury - Hg; potassium - K; iron - Fe; antimony - Sb; tungsten - W

31. fluorine - F; chlorine - Cl; bromine - Br; sulfur - S; oxygen - O; phosphorus - P

33. Sn - tin; Pt - platinum; Co - cobalt; Ni - nickel; Mg - magnesium; Ba - barium; K - potassium

35. The noble gases are He, Ne, Ar, Kr, Xe, and Rn (helium, neon, argon, krypton, xenon, and radon). Radon has only radioactive isotopes. In the periodic table the whole number enclosed in parenthesis is the mass number of the longest lived isotope of the element.

37. a. Eight; Li to Ne
b. Eight; Na to Ar
c. Eighteen; K to Kr
d. Five; N, P, As, Sb, Bi

39. a. $^{238}_{94}Pu$; 94 protons, 238 - 94 = 144 neutrons
b. $^{65}_{29}Cu$; 29 protons, 65 - 29 = 36 neutrons
c. $^{52}_{24}Cr$; 24 protons, 28 neutrons
d. $^{4}_{2}He$; 2 protons, 2 neutrons
e. $^{60}_{27}Co$; 27 protons, 33 neutrons
f. $^{54}_{24}Cr$; 24 protons, 30 neutrons

41. 9 protons means the atomic number is 9. The mass number is 9 + 10 = 19; Symbol: $^{19}_{9}F$

43. Atomic number = 63 (Eu); Charge = +63 - 60 = +3; Mass number = 63 + 88 = 151;
Symbol: $^{151}_{63}Eu^{3+}$

45. Atomic number = 16 (S); Charge = +16 - 18 = -2; Mass number = 16 + 18 = 34; Symbol: $^{34}_{16}S^{2-}$

47.

Symbol	Number of protons in nucleus	Number of neutrons in nucleus	Number of electrons	Net charge
$^{75}_{33}As^{3+}$	33	42	30	3+
$^{128}_{52}Te^{2-}$	52	76	54	2-
$^{32}_{16}S$	16	16	16	0
$^{204}_{81}Tl^{+}$	81	123	80	1+
$^{195}_{78}Pt$	78	117	78	0

49. Metals: Mg, Ti, Au, Bi, Ge, Eu, Am; Nonmetals: Si, B, At, Rn, Br

51. a and d; A group is a vertical column of elements in the periodic table. Elements in the same family (group) have similar chemical properties.

53. Carbon is a nonmetal. Silicon and germanium are metalloids. Tin and lead are metals. Thus, metallic character increases as one goes down a family in the periodic table.

55. Metals lose electrons to form cations and nonmetals gain electrons to form anions. Group IA, IIA and IIIA metals form stable +1,+2 and +3 charged cations, repectively. Group VA, VIA and VIIA nonmetals form -3, -2 and -1 charged anions, respectively.

a. Lose 1 e^- to form Na^+. b. Lose 2 e^- to form Sr^{2+}. c. Lose two e^- to form Ba^{2+}.

d. Gain 1 e^- to form I^-. e. Lose 3 e^- to form Al^{3+}. f. Gain 2 e^- to form S^{2-}.

Nomenclature

57. a. sodium chloride b. rubidium oxide
c. calcium sulfide d. aluminum iodide

59. a. chromium(VI) oxide b. chromium(III) oxide c. aluminum oxide
d. sodium hydride e. calcium bromide
f. zinc chloride (Zinc only forms +2 ions so no roman numerals are needed for zinc compounds.)

61. a. potassium perchlorate b. calcium phosphate
c. aluminum sulfate d. lead(II) nitrate

63. a. nitrogen triiodide b. phosphorus trichloride
c. sulfur difluoride d. dinitrogen tetrafluoride

65. a. copper(I) iodide b. copper(II) iodide c. cobalt(II) iodide
d. sodium carbonate e. sodium hydrogen carbonate or sodium bicarbonate
f. tetrasulfur tetranitride g. sulfur hexafluoride h. sodium hypochlorite
i. barium chromate j. ammonium nitrate

67. a. CsBr b. $BaSO_4$ c. NH_4Cl d. ClO
e. $SiCl_4$ f. ClF_3 g. BeO h. MgF_2

69. a. NaOH b. $Al(OH)_3$ c. HCN
d. Na_2O_2 e. $Cu(C_2H_3O_2)_2$ f. CF_4
g. PbO h. PbO_2 i. $HC_2H_3O_2$
j. CuBr k. H_2SO_3 l. GaAs (Ga^{3+} and As^{3-} ions)

Additional Exercises

71. There should be no difference. The composition of insulin from both sources will be the same and therefore, it will have the same activity regardless of the source. As a practical note, trace contaminants in the two types of insulin may be different. These trace contaminates may be important.

73. a. $Pb(C_2H_3O_2)_2$: lead(II) acetate
b. $CuSO_4$: copper(II) sulfate
c. CaO: calcium oxide
d. $MgSO_4$: magnesium sulfate
e. $Mg(OH)_2$: magnesium hydroxide
f. $CaSO_4$: calcium sulfate
g. N_2O: dinitrogen monoxide or nitrous oxide

75. A chemical formula gives the actual number and kind of atoms in a compound. In all cases, 12 hydrogen atoms are present. For example:

$$4 \text{ molecules } H_3PO_4 \times \frac{3 \text{ atoms H}}{\text{molecule } H_3PO_4} = 12 \text{ atoms H}$$

Challenge Problems

77. Copper(Cu), silver(Ag) and gold(Au) make up the coinage metals.

79. Avogadro proposed that equal volumes of gases (at constant temperature and pressure) contains equal numbers of molecules. In terms of balanced equations, Avogadro's hypothesis implies that volume ratios will be identical to molecule ratios. Assuming one molecule of octane reacting, then 1 molecule of C_xH_y produces 8 molecules of CO_2 and 9 molecules of H_2O. $C_xH_y + O_2 \rightarrow 8\ CO_2 + 9\ H_2O$. Since all the carbon in octane ends up as carbon in CO_2, then octane contains 8 atoms of C. Similarly, all hydrogen in octane ends up as hydrogen in H_2O, so one molecule of octane contains $9 \times 2 = 18$ atoms of H. Octane formula = C_8H_{18} and the ratio of C:H = 8:18 or 4:9.

81. Compound I: $\frac{14.0 \text{ g R}}{3.00 \text{ g Q}} = \frac{4.67 \text{ g R}}{1.00 \text{ g Q}}$; Compound II: $\frac{7.00 \text{ g R}}{4.50 \text{ g Q}} = \frac{1.56 \text{ g R}}{1.00 \text{ g Q}}$

The ratio of the masses of R that combines with 1.00 g Q is: $\frac{4.67}{1.56} = 2.99 \approx 3$

As expected from the law of multiple proportions, this ratio is a small whole number.

Since Compound I contains three times the mass of R per gram of Q as compared to Compound II (RQ), then the formula of Compound I should be R_3Q.

83. a. The compounds have the same number and types of atoms (some formula), but the atoms in the molecules are bonded together differently. Therefore, the two compounds are different compounds with different properties. The compounds are called isomers of each other.

b. When wood burns, most of the solid material in wood is converted to gases, which escape. The gases produced are most likely CO_2 and H_2O.

c. The atom is not an indivisible particle, but is instead composed of other smaller particles, e.g., electrons, neutrons, protons.

d. The two hydride samples contain different isotopes of either hydrogen and/or lithium. Although the compounds are composed of different isotopes, their properties are similar because different isotopes of the same element have similar properties (except, of course, their mass).

CHAPTER THREE

STOICHIOMETRY

Questions

13. The molecular formula tells us the actual number of atoms of each element in a molecule (or formula unit) of a compound. The empirical formula tells only the simplest whole number ratio of atoms of each element in a molecule. The molecular formula is a whole number multiple of the empirical formula. If that multiplier is one, the molecular and empirical formulas are the same. For example, both the molecular and empirical formulas of water are H_2O. They are the same. For hydrogen peroxide, the empirical formula is OH; the molecular formula is H_2O_2.

Exercises

Atomic Masses and the Mass Spectrometer

15. A = atomic mass = 0.7899(23.9850 amu) + 0.1000(24.9858 amu) + 0.1101(25.9826 amu)

    A = 18.95 amu + 2.499 amu + 2.861 amu = 24.31 amu

17. A = atomic mass = 0.7553 (34.96885 amu) + 0.2447 (36.96590 amu)

    A = 26.41 amu + 9.046 = 35.46 amu; From atomic masses in the period table, this is chlorine.

19. Let x = % of ^{151}Eu and y = % of ^{153}Eu, then $x + y = 100$ and $y = 100 - x$.

$$151.96 = \frac{x(150.9196) + (100 - x)(152.9209)}{100}$$

$15196 = 150.9196\,x + 15292.09 - 152.9209\,x$, $-96 = -2.0013\,x$

x = 48%; 48% ^{151}Eu and 100 - 48 = 52% ^{153}Eu

21. There are three peaks in the mass spectrum, each 2 mass units apart. This is consistent with two isotopes, differing in mass by two mass units. The peak at 157.84 corresponds to a Br_2 molecule composed of two atoms of the lighter isotope. This isotope has mass equal to 157.84/2 or 78.92. This corresponds to ^{79}Br. The second isotope is ^{81}Br with mass equal to 161.84/2 = 80.92. The peaks in the mass spectrum correspond to $^{79}Br_2$, $^{79}Br^{81}Br$ and $^{81}Br_2$ in order of increasing mass. The intensities of the highest and lowest mass tell us the two isotopes are present at about equal abundance. The actual abundance is 50.69% ^{79}Br and 49.31% ^{81}Br. The calculation of the abundance from the mass spectrum is beyond the scope of this text.

Moles and Molar Masses

23. When more than one conversion factor is necessary to determine the answer, we will apply the conversion factors into one calculation instead of determining intermediate answers. This method reduces round-off error and is a time saver.

$$500.\ \text{atoms Fe} \times \frac{1\ \text{mol Fe}}{6.022 \times 10^{23}\ \text{atoms Fe}} \times \frac{55.85\ \text{g Fe}}{\text{mol Fe}} = 4.64 \times 10^{-20}\ \text{g Fe}$$

25. $1.00\ \text{carat} \times \frac{0.200\ \text{g C}}{\text{carat}} \times \frac{1\ \text{mol C}}{12.01\ \text{g C}} \times \frac{6.022 \times 10^{23}\ \text{atoms C}}{\text{mol C}} = 1.00 \times 10^{22}\ \text{atoms C}$

27. Al_2O_3: 2(26.98) + 3(16.00) = 101.96 g/mol

Na_3AlF_6: 3(22.99) + 1(26.98) + 6(19.00) = 209.95 g/mol

29. a. NH_3: 14.01 g/mol + 3(1.008 g/mol) = 17.03 g/mol

b. N_2H_4: 2(14.01) + 4(1.008) = 32.05 g/mol

c. $(NH_4)_2Cr_2O_7$: 2(14.01) + 8(1.008) + 2(52.00) + 7(16.00) = 252.08 g/mol

31. a. $1.00\ \text{g}\ NH_3 \times \frac{1\ \text{mol}\ NH_3}{17.03\ \text{g}\ NH_3} = 0.0587\ \text{mol}\ NH_3$

b. $1.00\ \text{g}\ N_2H_4 \times \frac{1\ \text{mol}\ N_2H_4}{32.05\ \text{g}\ N_2H_4} = 0.0312\ \text{mol}\ N_2H_4$

c. $1.00\ \text{g}\ (NH_4)_2Cr_2O_7 \times \frac{1\ \text{mol}\ (NH_4)_2Cr_2O_7}{252.08\ \text{g}\ (NH_4)_2Cr_2O_7} = 3.97 \times 10^{-3}\ \text{mol}\ (NH_4)_2Cr_2O_7$

33. a. $5.00\ \text{mol}\ NH_3 \times \frac{17.03\ \text{g}\ NH_4}{\text{mol}\ NH_3} = 85.2\ \text{g}\ NH_3$

b. $5.00\ \text{mol}\ N_2H_4 \times \frac{32.05\ \text{g}\ N_2H_4}{\text{mol}\ N_2H_4} = 160.\ \text{g}\ N_2H_4$

c. $5.00\ \text{mol}\ (NH_4)_2Cr_2O_7 \times \frac{252.08\ \text{g}\ (NH_4)_2Cr_2O_7}{\text{mol}\ (NH_4)_2Cr_2O_7} = 1260\ \text{g}\ (NH_4)_2Cr_2O_7$

35. Chemical formulas give atom ratios as well as mol ratios.

a. $5.00\ \text{mol}\ NH_3 \times \frac{1\ \text{mol N}}{\text{mol}\ NH_3} \times \frac{14.01\ \text{g N}}{\text{mol N}} = 70.1\ \text{g N}$

b. $5.00\ \text{mol}\ N_2H_4 \times \frac{2\ \text{mol N}}{\text{mol}\ N_2H_4} \times \frac{14.01\ \text{g N}}{\text{mol N}} = 140.\ \text{g N}$

c. $5.00\ \text{mol}\ (NH_4)_2Cr_2O_7 \times \frac{2\ \text{mol N}}{\text{mol}\ (NH_4)_2Cr_2O_7} \times \frac{14.01\ \text{g N}}{\text{mol N}} = 140.\ \text{g N}$

37. a. $1.00 \text{ g } NH_3 \times \frac{1 \text{ mol } NH_3}{17.03 \text{ g } NH_3} \times \frac{6.022 \times 10^{23} \text{ molecules } NH_3}{\text{mol } NH_3} = 3.54 \times 10^{22} \text{ molecules } NH_3$

b. $1.00 \text{ g } N_2H_4 \times \frac{1 \text{ mol } N_2H_4}{32.05 \text{ g } N_2H_4} \times \frac{6.022 \times 10^{23} \text{ molecules } N_2H_4}{\text{mol } N_2H_4} = 1.88 \times 10^{22} \text{ molecules } N_2H_4$

c. $1.00 \text{ g } (NH_4)Cr_2O_7 \times \frac{1 \text{ mol } (NH_4)_2Cr_2O_7}{252.08 \text{ g } (NH_4)_2Cr_2O_7} \times \frac{6.022 \times 10^{23} \text{ molecules } (NH_4)_2Cr_2O_7}{\text{mol } (NH_4)_2Cr_2O_7}$

$= 2.39 \times 10^{21} \text{ molecules } (NH_4)_2Cr_2O_7$

39. Using answers from Exercise 3.37:

a. $3.54 \times 10^{22} \text{ molecules } NH_3 \times \frac{1 \text{ atom N}}{\text{molecule } NH_3} = 3.54 \times 10^{22} \text{ atoms N}$

b. $1.88 \times 10^{22} \text{ molecules } N_2H_4 \times \frac{2 \text{ atoms N}}{\text{molecule } N_2H_4} = 3.76 \times 10^{22} \text{ atoms N}$

c. $2.39 \times 10^{21} \text{ molecules } (NH_4)_2Cr_2O_7 \times \frac{2 \text{ atoms N}}{\text{molecule } (NH_4)_2Cr_2O_7} = 4.78 \times 10^{21} \text{ atoms N}$

41. Molar mass of $C_6H_8O_6 = 6(12.01) + 8(1.008) + 6(16.00) = 176.12$ g/mol

$500.0 \text{ mg} \times \frac{1 \text{ g}}{1000 \text{ mg}} \times \frac{1 \text{ mol}}{176.12 \text{ g}} = 2.839 \times 10^{-3} \text{ mol}$

$2.839 \times 10^{-3} \text{ mol} \times \frac{6.022 \times 10^{23} \text{ molecules}}{\text{mol}} = 1.710 \times 10^{21} \text{ molecules}$

43. a. $100 \text{ molecules } H_2O \times \frac{1 \text{ mol } H_2O}{6.022 \times 10^{23} \text{ molecules } H_2O} = 1.661 \times 10^{-22} \text{ mol } H_2O$

b. $100.0 \text{ g } H_2O \times \frac{1 \text{ mol } H_2O}{18.02 \text{ g } H_2O} = 5.549 \text{ mol } H_2O$

c. $150 \text{ molecules } O_2 \times \frac{1 \text{ mol } O_2}{6.022 \times 10^{23} \text{ molecules } O_2} = 2.491 \times 10^{-22} \text{ mol } O_2$

45. a. $3.00 \times 10^{20} \text{ molecules } N_2 \times \frac{1 \text{ mol } N_2}{6.02 \times 10^{23} \text{ molecules}} \times \frac{28.02 \text{ g } N_2}{\text{mol } N_2} = 1.40 \times 10^{-2} \text{ g } N_2$

b. $3.00 \times 10^{-3} \text{ mol } N_2 \times \frac{28.02 \text{ g } N_2}{\text{mol } N_2} = 8.41 \times 10^{-2} \text{ g } N_2$

c. $1.5 \times 10^{2} \text{ mol } N_2 \times \frac{28.02 \text{ g } N_2}{\text{mol } N_2} = 4.2 \times 10^{3} \text{ g } N_2$

d. $1 \text{ molecule } N_2 \times \frac{1 \text{ mol } N_2}{6.022 \times 10^{23} \text{ molecules } N_2} \times \frac{28.02 \text{ g } N_2}{\text{mol } N_2} = 4.653 \times 10^{-23} \text{ g } N_2$

e. 2.00×10^{-15} mol $N_2 \times \dfrac{28.02 \text{ g } N_2}{\text{mol } N_2} = 5.60 \times 10^{-14}$ g N_2 = 56.0 fg N_2

f. 18.0 pmol $N_2 \times \dfrac{1 \text{ mol } N_2}{10^{12} \text{ pmol}} \times \dfrac{28.02 \text{ g } N_2}{\text{mol } N_2} = 5.04 \times 10^{-10}$ g N_2 = 504 pg N_2

g. 5.0 nmol $N_2 \times \dfrac{1 \text{ mol } N_2}{10^{9} \text{ nmol}} \times \dfrac{28.02 \text{ g } N_2}{\text{mol } N_2} = 1.4 \times 10^{-7}$ g N_2 = 140 ng

47. a. $14 \text{ mol C}\left(\dfrac{12.01 \text{ g}}{\text{mol C}}\right) + 18 \text{ mol H}\left(\dfrac{1.008 \text{ g}}{\text{mol H}}\right) + 2 \text{ mol N}\left(\dfrac{14.01 \text{ g}}{\text{mol N}}\right) + 5 \text{ mol O}\left(\dfrac{16.00 \text{ g}}{\text{mol O}}\right)$

= 294.30 g/mol

b. 10.0 g aspartame $\times \dfrac{1 \text{ mol}}{294.30 \text{ g}} = 3.40 \times 10^{-2}$ mol

c. 1.56 mol $\times \dfrac{294.30 \text{ g}}{\text{mol}}$ = 459 g

d. 5.0 mg $\times \dfrac{1 \text{ g}}{1000 \text{ mg}} \times \dfrac{1 \text{ mol}}{294.30 \text{ g}} \times \dfrac{6.02 \times 10^{23} \text{ molecules}}{\text{mol}} = 1.0 \times 10^{19}$ molecules

e. The chemical formula tells us that 1 molecule of aspartame contains two atoms of N. The chemical formula also says that 1 mol of aspartame contains two mol of N.

1.2 g aspartame $\times \dfrac{1 \text{ mol aspartame}}{294.30 \text{ g aspartame}} \times \dfrac{2 \text{ mol N}}{\text{mol aspartame}} \times \dfrac{6.02 \times 10^{23} \text{ atoms N}}{\text{mol N}}$

$= 4.9 \times 10^{21}$ atoms of nitrogen

f. 1.0×10^{9} molecules $\times \dfrac{1 \text{ mol}}{6.02 \times 10^{23} \text{ molecules}} \times \dfrac{294.30 \text{ g}}{\text{mol}} = 4.9 \times 10^{-13}$ g or 490 fg

g. 1 molecule aspartame $\times \dfrac{1 \text{ mol}}{6.022 \times 10^{23} \text{ molecules}} \times \dfrac{294.30 \text{ g}}{\text{mol}} = 4.887 \times 10^{-22}$ g

Percent Composition

49. mass %Cd $= \dfrac{\text{mass of Cd in 1 mol compound}}{\text{molar mass of compound}} \times 100$

CdS: %Cd $= \dfrac{112.4 \text{ g Cd}}{144.5 \text{ g CdS}} \times 100 = 77.79\%$ Cd

CdSe: %Cd $= \dfrac{112.4 \text{ g}}{191.4 \text{ g}} \times 100 = 58.73\%$ Cd; CdTe: %Cd $= \dfrac{112.4 \text{ g}}{240.0 \text{ g}} \times 100 = 46.83\%$ Cd

51. In 1 mole of $YBa_2Cu_3O_7$, there are 1 mole of Y, 2 moles of Ba, 3 moles of Cu, and 7 moles of O.

$$\text{Molar mass} = 1 \text{ mol Y}\left(\frac{88.91 \text{ g Y}}{\text{mol Y}}\right) + 2 \text{ mol Ba}\left(\frac{137.3 \text{ g Ba}}{\text{mol Ba}}\right)$$

$$+ 3 \text{ mol Cu}\left(\frac{63.55 \text{ g Cu}}{\text{mol Cu}}\right) + 7 \text{ mol O}\left(\frac{16.00 \text{ g O}}{\text{mol O}}\right)$$

Molar mass = 88.91 + 274.6 + 190.65 + 112.00 = 666.2 g/mol

$$\%Y = \frac{88.91 \text{ g}}{666.2 \text{ g}} \times 100 = 13.35\% \text{ Y}; \ \%Ba = \frac{274.6 \text{ g}}{666.2 \text{ g}} \times 100 = 41.22\% \text{ Ba}$$

$$\%Cu = \frac{190.65 \text{ g}}{666.2 \text{ g}} \times 100 = 28.62\% \text{ Cu}; \ \%O = \frac{112.0 \text{ g}}{666.2 \text{ g}} \times 100 = 16.81\% \text{ O}$$

53. Na_3PO_4: molar mass = 3(22.99) + 30.97 + 4(16.00) = 163.94 g/mol

$$\%P = \frac{30.97 \text{ g}}{163.94 \text{ g}} \times 100 = 18.89\% \text{ P}$$

PH_3: molar mass = 30.97 + 3(1.008) = 33.99 g/mol

$$\%P = \frac{30.97 \text{ g}}{33.99 \text{ g}} \times 100 = 91.12\% \text{ P}$$

P_4O_{10}: molar mass = 4(30.97) + 10(16.00) = 283.88 g/mol

$$\%P = \frac{123.88 \text{ g}}{283.88 \text{ g}} \times 100 = 43.638\% \text{ P}$$

$(NPCl_2)_3$: molar mass = 3(14.01) + 3(30.97) + 6(35.45) = 347.64 g/mol

$$\%P = \frac{92.91 \text{ g}}{347.64 \text{ g}} \times 100 = 26.73\% \text{ P}$$

The order from lowest to highest percentage of phosphorus is:

$$Na_3PO_4 < (NPCl_2)_3 < P_4O_{10} < PH_3$$

55. There are many valid methods to solve this problem. We will assume 100.00 g of compound, then determine from the information in the problem how many mol of compound equals 100.00 g of compound. From this information, we can determine the mass of one mol of compound (the molar mass) by setting up a ratio. Assuming 100.00 g cyanocobalamin:

$$\text{mol cyanocobalamin} = 4.34 \text{ g Co} \times \frac{1 \text{ mol Co}}{58.93 \text{ g Co}} \times \frac{1 \text{ mol cyanocobalamin}}{\text{mol Co}}$$

$$= 7.36 \times 10^{-2} \text{ mol cyanocobalamin}$$

$$\frac{x \text{ g cyanocobalamin}}{1 \text{ mol cyanocobalamin}} = \frac{100.00 \text{ g}}{7.36 \times 10^{-2} \text{ mol}}, \ x = \text{molar mass} = 1360 \text{ g/mol}$$

Empirical and Molecular Formulas

57. a. Molar mass of CH_2O = $1 \text{ mol C}\left(\frac{12.01 \text{ g}}{\text{mol C}}\right) + 2 \text{ mol H}\left(\frac{1.008 \text{ g H}}{\text{mol H}}\right)$

$$+ 1 \text{ mol O}\left(\frac{16.00 \text{ g}}{\text{mol O}}\right) = 30.03 \text{ g/mol}$$

$$\%C = \frac{12.01 \text{ g C}}{30.03 \text{ g } CH_2O} \times 100 = 39.99\% \text{ C}; \ \%H = \frac{2.016 \text{ g H}}{30.03 \text{ g } CH_2O} \times 100 = 6.713\% \text{ H}$$

$$\%O = \frac{16.00 \text{ g O}}{30.03 \text{ g } CH_2O} \times 100 = 53.28\% \text{ O} \qquad \text{or } \%O = 100.00 - (39.99 + 6.713) = 53.30\%$$

b. Molar Mass of $C_6H_{12}O_6$ = 6(12.01) + 12(1.008) + 6(16.00) = 180.16 g/mol

$$\%C = \frac{72.06 \text{ g C}}{180.16 \text{ g } C_6H_{12}O_6} \times 100 = 40.00\%; \quad \%H = \frac{12.096 \text{ g}}{180.16 \text{ g}} \times 100 = 6.7140\%$$

%O = 100.00 - (40.00 + 6.714) = 53.29%

c. Molar mass of $HC_2H_3O_2$ = 2(12.01) + 4(1.008) + 2(16.00) = 60.05 g/mol

$$\%C = \frac{24.02 \text{ g}}{60.05 \text{ g}} \times 100 = 40.00\%; \quad \%H = \frac{4.032 \text{ g}}{60.05 \text{ g}} \times 100 = 6.714\%$$

%O = 100.00 - (40.00 + 6.714) = 53.29%

59. a. $C_3H_4O_3$ b. CH c. CH d. P_2O_5 e. CH_2O f. CH_2O

61. Out of 100.00 g of compound, there are:

$$48.64 \text{ g C} \times \frac{1 \text{ mol C}}{12.01 \text{ g C}} = 4.050 \text{ mol C}; \ 8.16 \text{ g H} \times \frac{1 \text{ mol H}}{1.008 \text{ g H}} = 8.10 \text{ mol H}$$

$$\%O = 100.00 - 48.64 - 8.16 = 43.20\%; \ 43.20 \text{ g O} \times \frac{1 \text{ mol O}}{16.00 \text{ g O}} = 2.700 \text{ mol O}$$

Dividing each mol value by the smallest number:

$$\frac{4.050}{2.700} = 1.500, \ \frac{8.10}{2.700} = 3.00, \ \frac{2.700}{2.700} = 1.000$$

Since a whole number ratio is required, the C:H:O ratio is 1.5:3:1 or 3:6:2. So the empirical formula is $C_3H_6O_2$.

63. $0.979 \text{ g Na} \times \frac{1 \text{ mol Na}}{22.99 \text{ g Na}} = 4.26 \times 10^{-2} \text{ mol Na}$

$$1.365 \text{ g S} \times \frac{1 \text{ mol S}}{32.07 \text{ g S}} = 4.256 \times 10^{-2} \text{ mol S}$$

$$1.021 \text{ g O} \times \frac{1 \text{ mol O}}{16.00 \text{ g O}} = 6.381 \times 10^{-2} \text{ mol O}$$

Determine the mol ratios by dividing by mol S (the smallest number):

$$\frac{6.381 \times 10^{-2} \text{ mol O}}{4.256 \times 10^{-2} \text{ mol S}} = 1.499 \approx \frac{1.5 \text{ mol O}}{\text{mol S}} = \frac{3 \text{ mol O}}{2 \text{ mol S}}; \quad \frac{4.26 \times 10^{-2} \text{ mol Na}}{4.256 \times 10^{-2} \text{ mol S}} = 1.00$$

From mol ratios, the empirical formula = $Na_2S_2O_3$.

65. Out of 100.0 g compound: $30.4 \text{ g N} \times \frac{1 \text{ mol N}}{14.01 \text{ g N}} = 2.17 \text{ mol N}$

$\%O = 100.0 - 30.4 = 69.6\%$ O; $69.6 \text{ g O} \times \frac{1 \text{ mol O}}{16.00 \text{ g O}} = 4.35 \text{ mol O}$

$\frac{2.17}{2.17} = 1.00$; $\frac{4.35}{2.17} = 2.00$; Empirical formula is NO_2.

The empirical formula mass of $NO_2 \approx 14 + 2(16) = 46$ g/mol.

$\frac{92 \text{ g}}{46 \text{ g}} = 2.0$; Therefore, the molecular formula is N_2O_4.

67. Assuming 100.00 g of compound:

$$7.74 \text{ g H} \times \frac{1 \text{ mol H}}{1.008 \text{ g H}} = 7.68 \text{ mol H}; \quad 92.26 \text{ g C} \times \frac{1 \text{ mol C}}{12.01 \text{ g C}} = 7.682 \text{ mol C}$$

The mole ratio between C and H is 1:1 so the empirical formula is CH.

empirical formula mass = 12.01 + 1.008 = 13.02 g/mol

$$\frac{\text{molar mass}}{\text{empirical formula mass}} = \frac{78.1}{13.02} = 6.00$$

molecular formula = $(CH)_6 = C_6H_6$

69. When combustion data is given, it is assumed that all the carbon in the compound ends up as carbon in CO_2 and all the hydrogen in the compound ends up a hydrogen in H_2O. In the sample of propane combusted, the moles of C and H are:

$$\text{mol C} = 2.641 \text{ g } CO_2 \times \frac{1 \text{ mol } CO_2}{44.01 \text{ g } CO_2} \times \frac{1 \text{ mol C}}{\text{mol } CO_2} = 0.06001 \text{ mol C}$$

$$\text{mol H} = 1.442 \text{ g } H_2O \times \frac{1 \text{ mol } H_2O}{18.02 \text{ g } H_2O} \times \frac{2 \text{ mol H}}{\text{mol } H_2O} = 0.1600 \text{ mol H}$$

$$\frac{\text{mol H}}{\text{mol C}} = \frac{0.1600}{0.06001} = 2.666$$

Multiplying this ratio by three gives the empirical formula of C_3H_8.

71. Since the compound contains only carbon and hydrogen, we could solve this problem using the same procedure as in Exercise 3.69. Instead, we will solve this problem using the procedure outlined in the text, i.e., we will determine the composition by mass percent and then solve for the empirical formula.

$$156.8 \text{ mg } CO_2 \times \frac{1 \text{ g}}{1000 \text{ mg}} \times \frac{12.01 \text{ g C}}{44.01 \text{ g } CO_2} \times \frac{1000 \text{ mg}}{\text{g}} = 42.79 \text{ mg C}$$

$$42.8 \text{ mg } H_2O \times \frac{1 \text{ g}}{1000 \text{ mg}} \times \frac{2.016 \text{ g H}}{18.02 \text{ g } H_2O} \times \frac{1000 \text{ mg}}{\text{g}} = 4.79 \text{ mg H}$$

$$\%C = \frac{42.79 \text{ mg}}{47.6 \text{ mg}} \times 100 = 89.9\% \text{ C}; \ \%H = 100.0 - 89.9 = 10.1\% \text{ H}$$

Out of 100.0 g Cumene, we have:

$$89.9 \text{ g C} \times \frac{1 \text{ mol C}}{12.01 \text{ g C}} = 7.49 \text{ mol C}; \ 10.1 \text{ g H} \times \frac{1 \text{ mol H}}{1.008 \text{ g H}} = 10.0 \text{ mol H}$$

$\frac{10.0}{7.49} = 1.34 \approx \frac{4}{3}$, i.e., mol H to mol C are in a 4:3 ratio. Empirical formula = C_3H_4

Empirical formula mass $\approx 3(12) + 4(1) = 40$ g/mol

The molecular formula is $(C_3H_4)_3$ or C_9H_{12} since the molar mass will be between 115 and 125 g/mol.

Balancing Chemical Equations

73. a. $In + O_2 \rightarrow In_2O_3$, $4\ In(s) + 3\ O_2(g) \rightarrow 2\ In_2O_3(s)$

b. $C_6H_{12}O_6 \rightarrow C_2H_6O + CO_2$

Balance C-atoms first. 6-C on left, 3-C on right.

First, try multiplying both products by 2:

$$C_6H_{12}O_6 \rightarrow 2\ C_2H_6O + 2\ CO_2$$

O and H are also balanced, and the balanced equation is:

$$C_6H_{12}O_6(aq) \rightarrow 2\ C_2H_6O(aq) + 2\ CO_2(g)$$

c. $K + H_2O \rightarrow KOH + H_2$, $2\ K(s) + 2\ H_2O(l) \rightarrow 2\ KOH(aq) + H_2(g)$

75. a. $Cu(s) + 2\ AgNO_3(aq) \rightarrow 2\ Ag(s) + Cu(NO_3)_2(aq)$

b. $Zn(s) + 2\ HCl(aq) \rightarrow ZnCl_2(aq) + H_2(g)$

c. $Au_2S_3(s) + 3\ H_2(g) \rightarrow 2\ Au(s) + 3\ H_2S(g)$; d. $Ca(s) + 2\ H_2O(l) \rightarrow Ca(OH)_2(aq) + H_2(g)$

77. a. $C_{12}H_{22}O_{11}(s) + 12\ O_2(g) \rightarrow 12\ CO_2(g) + 11\ H_2O(g)$

b. $C_6H_6(l) + \frac{15}{2}O_2(g) \rightarrow 6\ CO_2(g) + 3\ H_2O(g)$; Multiply by two to give whole numbers.

$2\ C_6H_6(l) + 15\ O_2(g) \rightarrow 12\ CO_2(g) + 6\ H_2O(g)$

c. $2\ Fe + \frac{3}{2}O_2 \rightarrow Fe_2O_3$; For whole numbers: $4\ Fe(s) + 3\ O_2(g) \rightarrow 2\ Fe_2O_3(s)$

d. $C_4H_{10} + \frac{13}{2}O_2 \rightarrow 4\ CO_2 + 5\ H_2O$; Multiply by two to give whole numbers.

$2\ C_4H_{10}(g) + 13\ O_2(g) \rightarrow 8\ CO_2(g) + 10\ H_2O(g)$

e. $2\ FeO(s) + \frac{1}{2}O_2(g) \rightarrow Fe_2O_3(s)$; For whole numbers, multiply by two.

$4\ FeO(s) + O_2(g) \rightarrow 2\ Fe_2O_3(s)$

79. a. $SiO_2(s) + C(s) \rightarrow Si(s) + CO(g)$

Balance oxygen atoms: $SiO_2 + C \rightarrow Si + 2\ CO$

Balance carbon atoms: $SiO_2(s) + 2\ C(s) \rightarrow Si(s) + 2\ CO(g)$

b. $SiCl_4(l) + Mg(s) \rightarrow Si(s) + MgCl_2(s)$

Balance Cl atoms: $SiCl_4 + Mg \rightarrow Si + 2\ MgCl_2$

Balance Mg atoms: $SiCl_4(l) + 2\ Mg(s) \rightarrow Si(s) + 2\ MgCl_2(s)$

c. $Na_2SiF_6(s) + Na(s) \rightarrow Si(s) + NaF(s)$

Balance F atoms: $Na_2SiF_6 + Na \rightarrow Si + 6\ NaF$

Balance Na atoms: $Na_2SiF_6(s) + 4\ Na(s) \rightarrow Si(s) + 6\ NaF(s)$

81. $Pb(NO_3)_2(aq) + H_3AsO_4(aq) \rightarrow PbHAsO_4(s) + 2\ HNO_3(aq)$

Note: The insecticide used is $PbHAsO_4$ and is commonly called lead arsenate. This is not the correct name, however. Correctly, lead arsenate would be $Pb_3(AsO_4)_2$ and $PbHAsO_4$ should be named lead hydrogen arsenate.

Reaction Stoichiometry

83. The stepwise method to solve stoichiometry problems is outlined in the text. Instead of calculating intermediate answers for each step, we will combine conversion factors into one calculation. This practice reduces round off error and saves time.

The balanced reaction is: $(NH_4)_2Cr_2O_7(s) \rightarrow Cr_2O_3(s) + N_2(g) + 4\ H_2O(g)$

$$10.8\text{ g }(NH_4)_2Cr_2O_7 \times \frac{1\text{ mol }(NH_4)_2Cr_2O_7}{252.08\text{ g}} = 4.28 \times 10^{-2}\text{ mol }(NH_4)_2Cr_2O_7$$

$$4.28 \times 10^{-2}\text{ mol }(NH_4)_2Cr_2O_7 \times \frac{1\text{ mol }Cr_2O_3}{\text{mol }(NH_4)_2Cr_2O_7} \times \frac{152.00\text{ g }Cr_2O_3}{\text{mol }Cr_2O_3} = 6.51\text{ g }Cr_2O_3$$

$$4.28 \times 10^{-2}\text{ mol }(NH_4)_2Cr_2O_7 \times \frac{1\text{ mol }N_2}{\text{mol }(NH_4)_2Cr_2O_7} \times \frac{28.02\text{ g }N_2}{\text{mol }N_2} = 1.20\text{ g }N_2$$

$$4.28 \times 10^{-2}\text{ mol }(NH_4)_2Cr_2O_7 \times \frac{4\text{ mol }H_2O}{\text{mol }(NH_4)_2Cr_2O_7} \times \frac{18.02\text{ g }H_2O}{\text{mol }H_2O} = 3.09\text{ g }H_2O$$

85. $$1.000\text{ kg Al} \times \frac{1000\text{ g Al}}{\text{kg Al}} \times \frac{1\text{ mol Al}}{26.98\text{ g Al}} \times \frac{3\text{ mol }NH_4ClO_4}{3\text{ mol Al}} \times \frac{117.49\text{ g }NH_4ClO_4}{\text{mol }NH_4ClO_4} = 4355\text{ g}$$

87. $$1.0 \times 10^2\text{ g }Ca_3(PO_4)_2 \times \frac{1\text{ mol }Ca_3(PO_4)_2}{310.2\text{ g }Ca_3(PO_4)_2} \times \frac{3\text{ mol }H_2SO_4}{\text{mol }Ca_3(PO_4)_2} \times \frac{98.09\text{ g }H_2SO_4}{\text{mol }H_2SO_4}$$

$= 95$ g H_2SO_4 are needed

$$95\text{ g }H_2SO_4 \times \frac{100\text{ g concentrated reagent}}{98\text{ g }H_2SO_4} = 97\text{ g of concentrated sulfuric acid}$$

89. a. $$1.0 \times 10^2\text{ mg }NaHCO_3 \times \frac{1\text{ g}}{1000\text{ mg}} \times \frac{1\text{ mol }NaHCO_3}{84.01\text{ g }NaHCO_3} \times \frac{1\text{ mol }C_6H_8O_7}{3\text{ mol }NaHCO_3} \times \frac{192.12\text{ g }C_6H_8O_7}{\text{mol }C_6H_8O_7}$$

$= 0.076$ g or 76 mg $C_6H_8O_7$

b. $$0.10\text{ g }NaHCO_3 \times \frac{1\text{ mol }NaHCO_3}{84.01\text{ g }NaHCO_3} \times \frac{3\text{ mol }CO_2}{3\text{ mol }NaHCO_2} \times \frac{44.01\text{ g }CO_2}{\text{mol }CO_2} = 0.052\text{ g or 52 mg }CO_2$$

Limiting Reactants and Percent Yield

91. a. $Mg(s) + I_2(s) \rightarrow MgI_2(s)$

From the balanced equation, 100 molecules of I_2 reacts completely with 100 atoms of Mg. We have a stoichiometric mixture. Neither is limiting.

b. $$150\text{ atoms Mg} \times \frac{1\text{ molecule }I_2}{1\text{ atom Mg}} = 150\text{ molecules }I_2\text{ needed}$$

We need 150 molecules I_2 to react completely with 150 atoms Mg; we only have 100 molecules I_2. Therefore, I_2 is limiting.

c. $200 \text{ atoms Mg} \times \frac{1 \text{ molecule } I_2}{1 \text{ atom Mg}} = 200 \text{ molecules } I_2$; Mg is limiting since 300 molecules I_2 are present.

d. $0.16 \text{ mol Mg} \times \frac{1 \text{ mol } I_2}{1 \text{ mol Mg}} = 0.16 \text{ mol } I_2$; Mg is limiting since 0.25 mol I_2 are present.

e. $0.14 \text{ mol Mg} \times \frac{1 \text{ mol } I_2}{1 \text{ mol Mg}} = 0.14 \text{ mol } I_2$ needed; Stoichiometric mixture. Neither is limiting.

f. $0.12 \text{ mol Mg} \times \frac{1 \text{ mol } I_2}{1 \text{ mol Mg}} = 0.12 \text{ mol } I_2$ needed; I_2 is limiting since only 0.08 mol I_2 are present.

g. $6.078 \text{ g Mg} \times \frac{1 \text{ mol Mg}}{24.31 \text{ g Mg}} \times \frac{1 \text{ mol } I_2}{1 \text{ mol Mg}} \times \frac{253.8 \text{ g } I_2}{\text{mol } I_2} = 63.46 \text{ g } I_2$

Stoichiometric mixture. Neither is limiting.

h. $1.00 \text{ g Mg} \times \frac{1 \text{ mol Mg}}{24.31 \text{ g Mg}} \times \frac{1 \text{ mol } I_2}{1 \text{ mol Mg}} \times \frac{253.8 \text{ g } I_2}{\text{mol } I_2} = 10.4 \text{ g } I_2$

10.4 g I_2 needed, but we only have 2.00 g. I_2 is limiting.

i. From h above, we calculated that 10.4 g I_2 will react completely with 1.00 g Mg. We have 20.00 g I_2. I_2 is in excess. Mg is limiting.

93. a. $\text{mol Ag} = 2.0 \text{ g Ag} \times \frac{1 \text{ mol Ag}}{107.9 \text{ g Ag}} = 1.9 \times 10^{-2} \text{ mol Ag}$

$\text{mol } S_8 = 2.0 \text{ g } S_8 \times \frac{1 \text{ mol } S_8}{256.56 \text{ g } S_8} = 7.8 \times 10^{-3} \text{ mol } S_8$

From the balanced equation the required mol Ag to mol S_8 ratio is 16:1. The actual mol ratio is:

$$\frac{1.9 \times 10^{-2} \text{ mol Ag}}{7.8 \times 10^{-3} \text{ mol } S_8} = 2.4$$

This is well below the required ratio so Ag is the limiting reagent.

$$1.9 \times 10^{-2} \text{ mol Ag} \times \frac{8 \text{ mol } Ag_2S}{16 \text{ mol Ag}} \times \frac{247.9 \text{ g } Ag_2S}{\text{mol } Ag_2S} = 2.4 \text{ g } Ag_2S$$

b. $1.9 \times 10^{-2} \text{ mol Ag} \times \frac{1 \text{ mol } S_8}{16 \text{ mol Ag}} \times \frac{256.56 \text{ g } S_8}{\text{mol } S_8} = 0.30 \text{ g } S_8$

0.30 g S_8 are required to react with all of the Ag present.

S_8 in excess = 2.0 g S_8 - 0.30 g S_8 = 1.7 g S_8 in excess

95. $Ca_3(PO_4)_2 + 3\ H_2SO_4 \rightarrow 3\ CaSO_4 + 2\ H_3PO_4$

$$1.0 \times 10^3\ g\ Ca_3(PO_4)_2 \times \frac{1\ mol\ Ca_3(PO_4)_2}{310.18\ g\ Ca_3(PO_4)_2} = 3.2\ mol\ Ca_3(PO_4)_2$$

$$1.0 \times 10^3\ g\ conc.\ H_2SO_4 \times \frac{98\ g\ H_2SO_4}{100\ g\ conc.\ H_2SO_4} \times \frac{1\ mol\ H_2SO_4}{98.09\ g\ H_2SO_4} = 10.\ mol\ H_2SO_4$$

The required mol ratio from the balanced equation is 3 mol H_2SO_4 to 1 mol $Ca_3(PO_4)_2$. The actual ratio is: $\frac{10.\ mol\ H_2SO_4}{3.2\ mol\ Ca_3(PO_4)_2} = 3.1$

This is higher than the required mol ratio so $Ca_3(PO_4)_2$ is the limiting reagent.

$$3.2\ mol\ Ca_3(PO_4)_2 \times \frac{3\ mol\ CaSO_4}{mol\ Ca_3(PO_4)_2} \times \frac{136.15\ g\ CaSO_4}{mol\ CaSO_4} = 1300\ g\ CaSO_4\ produced$$

$$3.2\ mol\ Ca_3(PO_4)_2 \times \frac{2\ mol\ H_3PO_4}{mol\ Ca_3(PO_4)_2} \times \frac{97.99\ g\ H_3PO_4}{mol\ H_3PO_4} = 630\ g\ H_3PO_4\ produced$$

97. $2\ Cu(s) + S(s) \rightarrow Cu_2S(s)$ or $16\ Cu(s) + S_8(s) \rightarrow 8\ Cu_2S(s)$

$$1.50\ g\ Cu \times \frac{1\ mol\ Cu}{63.55\ g\ Cu} \times \frac{1\ mol\ Cu_2S}{2\ mol\ Cu} \times \frac{159.17\ g\ Cu_2S}{mol\ Cu_2S} = 1.88\ g\ Cu_2S \text{ is theoretical yield.}$$

$$\%\ yield = \frac{actual\ yield}{theoretical\ yield} \times 100 = \frac{1.76\ g}{1.88\ g} \times 100 = 93.6\%$$

99. $C_2H_6 + Cl_2 \rightarrow C_2H_5Cl + HCl$

$$300.\ g\ C_2H_6 \times \frac{1\ mol\ C_2H_6}{30.07\ g\ C_2H_6} = 9.98\ mol\ C_2H_6;\ \ 650.\ g\ Cl_2 \times \frac{1\ mol\ Cl_2}{70.90\ g\ Cl_2} = 9.17\ mol\ Cl_2$$

The balanced equation requires a 1:1 mol ratio between reactants. 9.17 mol of C_2H_6 will react with all of the Cl_2 present (9.17 mol). Since 9.98 mol C_2H_6 is present, then Cl_2 is the limiting reagent.

The theoretical yield of C_2H_5Cl is:

$$9.17\ mol\ Cl_2 \times \frac{1\ mol\ C_2H_5Cl}{mol\ Cl_2} \times \frac{64.51\ g\ C_2H_5Cl}{mol\ C_2H_5Cl} = 592\ g\ C_2H_5Cl$$

$$Percent\ yield = \frac{actual}{theoretical} \times 100 = \frac{490.\ g}{592\ g} \times 100 = 82.8\%$$

Additional Exercises

101. $\frac{9.123 \times 10^{-23}\ \text{g}}{\text{atom}} \times \frac{6.022 \times 10^{23}\ \text{atom}}{\text{mol}} = \frac{54.94\ \text{g}}{\text{mol}}$

The atomic mass is 54.94. From the periodic table, the element is maganese (Mn).

103. Out of 100.00 g of compound there are:

$$83.53\ \text{g Sb} \times \frac{1\ \text{mol Sb}}{121.8\ \text{g Sb}} = 0.6858\ \text{mol Sb};\quad 16.47\ \text{g O} \times \frac{1\ \text{mol O}}{16.00\ \text{g O}} = 1.029\ \text{mol O}$$

$\frac{1.029}{0.6858} = 1.500 = \frac{3}{2}$; Empirical formula: Sb_2O_3; Empirical formula mass = 291.6 g/mol

Mass of Sb_4O_6 is 583.2, which is in the correct range. Molecular formula: Sb_4O_6.

105. Compound I: mass O = 0.6498 g Hg_xO_y - 0.6018 g Hg = 0.0480 g O

$$0.6018\ \text{g Hg} \times \frac{1\ \text{mol Hg}}{200.6\ \text{g Hg}} = 3.000 \times 10^{-3}\ \text{mol Hg}$$

$$0.0480\ \text{g O} \times \frac{1\ \text{mol O}}{16.00\ \text{g O}} = 3.00 \times 10^{-3}\ \text{mol O}$$

The mol ratio between Hg and O is 1:1, so the empirical formula of compound I is HgO.

Compound II: mass Hg = 0.4172 g Hg_xO_y - 0.016 g O = 0.401 g Hg

$$0.401\ \text{g Hg} \times \frac{1\ \text{mol Hg}}{200.6\ \text{g Hg}} = 2.00 \times 10^{-3}\ \text{mol Hg};\quad 0.016\ \text{g O} \times \frac{1\ \text{mol O}}{16.00\ \text{g O}} = 1.0 \times 10^{-3}\ \text{mol O}$$

The mol ratio between Hg and O is 2:1, so the empirical formula is Hg_2O.

107. $2.00 \times 10^6\ \text{g CaCO}_3 \times \frac{1\ \text{mol CaCO}_3}{100.09\ \text{g CaCO}_3} \times \frac{1\ \text{mol CaO}}{\text{mol CaCO}_3} \times \frac{56.08\ \text{g CaO}}{\text{mol CaO}} = 1.12 \times 10^6\ \text{g CaO}$

109. $4\ Al + 3\ O_2 \rightarrow 2\ Al_2O_3$

a. $1.0\ \text{mol Al} \times \frac{3\ \text{mol O}_2}{4\ \text{mol Al}} = 0.75\ \text{mol O}_2$; Al is limiting since 1.0 mol O_2 is present.

b. $2.0\ \text{mol Al} \times \frac{3\ \text{mol O}_2}{4\ \text{mol Al}} = 1.5\ \text{mol O}_2$; Al is limiting since 4.0 mol O_2 is present.

c. $0.50\ \text{mol Al} \times \frac{3\ \text{mol O}_2}{4\ \text{mol Al}} = 0.38\ \text{mol O}_2$; Al is limiting since 0.75 mol O_2 is present.

d. $64.75 \text{ g Al} \times \frac{1 \text{ mol Al}}{26.98 \text{ g Al}} \times \frac{3 \text{ mol O}_2}{4 \text{ mol Al}} \times \frac{32.00 \text{ g O}_2}{\text{mol O}_2} = 57.60 \text{ g O}_2$; Al is limiting.

e. $75.89 \text{ g Al} \times \frac{1 \text{ mol Al}}{26.98 \text{ g Al}} \times \frac{3 \text{ mol O}_2}{4 \text{ mol Al}} \times \frac{32.00 \text{ g O}_2}{\text{mol O}_2} = 67.51 \text{ g O}_2$; Al is limiting.

f. $51.28 \text{ g Al} \times \frac{1 \text{ mol Al}}{26.98 \text{ g Al}} \times \frac{3 \text{ mol O}_2}{4 \text{ mol Al}} \times \frac{32.00 \text{ g O}_2}{\text{mol O}_2} = 45.62 \text{ g O}_2$; Al is limiting.

111. $C_6H_{10}O_4 + 2\ NH_3 + 4\ H_2 \rightarrow C_6H_{16}N_2 + 4\ H_2O$

Adip. (Adipic acid) HMD

a. $1.00 \times 10^3 \text{ g Adip.} \times \frac{1 \text{ mol Adip.}}{146.14 \text{ g Adip.}} \times \frac{1 \text{ mol HMD}}{\text{mol Adip.}} \times \frac{116.21 \text{ g HMD}}{\text{mol HMD}} = 795 \text{ g HMD}$

b. $\% \text{ yield} = \frac{765 \text{ g}}{795 \text{ g}} \times 100 = 96.2\%$

113. $2\ NaNO_3(s) \rightarrow 2\ NaNO_2(s) + O_2(g)$

The amount of $NaNO_3$ in the impure sample is:

$$0.2864 \text{ g NaNO}_2 \times \frac{1 \text{ mol NaNO}_2}{69.00 \text{ g NaNO}_2} \times \frac{2 \text{ mol NaNO}_3}{2 \text{ mol NaNO}_2} \times \frac{85.00 \text{ g NaNO}_3}{\text{mol NaNO}_3} = 0.3528 \text{ g NaNO}_3$$

$$\% \text{ NaNO}_3 = \frac{0.3528 \text{ g NaNO}_3}{0.4230 \text{ g sample}} \times 100 = 83.40\%$$

Challenge Problems

115. First, we will determine composition in mass percent. We assume all of the carbon in 0.213 g CO_2 came from 0.157 g of the compound and that all of the hydrogen in the 0.0310 g H_2O came from the 0.157 g of the compound.

$$0.213 \text{ g CO}_2 \times \frac{12.01 \text{ g C}}{44.01 \text{ g CO}_2} = 0.0581 \text{ g C}; \ \%\text{C} = \frac{0.0581 \text{ g C}}{0.157 \text{ g compound}} \times 100 = 37.0\% \text{ C}$$

$$0.0310 \text{ g H}_2\text{O} \times \frac{2.016 \text{ g H}}{18.02 \text{ g H}_2\text{O}} = 3.47 \times 10^{-3} \text{ g H}; \ \%\text{H} = \frac{3.47 \times 10^{-3} \text{ g}}{0.157 \text{ g}} = 2.21\% \text{ H}$$

We get %N from the second experiment:

$$0.0230 \text{ g NH}_3 \times \frac{14.01 \text{ g N}}{17.03 \text{ g NH}_3} = 1.89 \times 10^{-2} \text{ g N}$$

$$\%\text{N} = \frac{1.89 \times 10^{-2} \text{ g}}{0.103 \text{ g}} \times 100 = 18.3\% \text{ N}$$

The mass percent of oxygen is obtained by difference:

$$\%O = 100.00 - (37.0 + 2.21 + 18.3) = 42.5\%$$

So out of 100.00 g of compound, there are:

$$37.0 \text{ g C} \times \frac{1 \text{ mol C}}{12.01 \text{ g C}} = 3.08 \text{ mol C}; \quad 2.21 \text{ g H} \times \frac{1 \text{ mol H}}{1.008 \text{ g H}} = 2.19 \text{ mol H}$$

$$18.3 \text{ g N} \times \frac{1 \text{ mol N}}{14.01 \text{ g N}} = 1.31 \text{ mol N}; \quad 42.5 \text{ g O} \times \frac{1 \text{ mol O}}{16.00 \text{ g O}} = 2.66 \text{ mol O}$$

Lastly, and often the hardest part, we need to find simple whole number ratios. Divide all mole values by the smallest number:

$$\frac{3.08}{1.31} = 2.35; \quad \frac{2.19}{1.31} = 1.67; \quad \frac{1.31}{1.31} = 1.00; \quad \frac{2.66}{1.31} = 2.03$$

Multiplying all these ratios by 3 gives an empirical formula of $C_7H_5N_3O_6$.

117. $2\ C_3H_6 + 2\ NH_3 + 3\ O_2 \rightarrow 2\ C_3H_3N + 6\ H_2O$

a. An alternative method to determine the amount of product produced in a limiting reagent problem is to assume each reactant is limiting and calculate the amount of product produced from each reactant. The reactant that produces the smallest amount of product is the limiting reagent. Applying this method to the problem:

$$5.00 \times 10^2 \text{ g } C_3H_6 \times \frac{1 \text{ mol } C_3H_6}{42.08 \text{ g } C_3H_6} \times \frac{2 \text{ mol } C_3H_3N}{2 \text{ mol } C_3H_6} = 11.9 \text{ mol } C_3H_3N$$

$$5.00 \times 10^2 \text{ g } NH_3 \times \frac{1 \text{ mol } NH_3}{17.03 \text{ g } NH_3} \times \frac{2 \text{ mol } C_3H_3N}{2 \text{ mol } NH_3} = 29.4 \text{ mol } C_3H_3N$$

$$1.00 \times 10^3 \text{ g } O_2 \times \frac{1 \text{ mol } O_2}{32.00 \text{ g } O_2} \times \frac{2 \text{ mol } C_3H_3N}{3 \text{ mol } O_2} = 20.8 \text{ mol } C_3H_3N$$

Since C_3H_6 produces the smallest amount of product, then C_3H_6 is limiting and the mass of acrylonitrile produced is:

$$11.9 \text{ mol} \times \frac{53.06 \text{ g } C_3H_3N}{\text{mol}} = 631 \text{ g } C_3H_3N$$

b. $11.9 \text{ mol } C_3H_3N \times \frac{6 \text{ mol } H_2O}{2 \text{ mol } C_3H_3N} \times \frac{18.02 \text{ g } H_2O}{\text{mol } H_2O} = 643 \text{ g } H_2O$

Amount of NH_3 needed:

$$11.9 \text{ mol } C_3H_3N \times \frac{2 \text{ mol } NH_3}{2 \text{ mol } C_3H_3N} \times \frac{17.03 \text{ g } NH_3}{\text{mol } H_2O} = 203 \text{ g } NH_3$$

Amount NH_3 left = 500. g - 203 g = 297 g

Amount O_2 needed: $11.9 \text{ mol } C_3H_3N \times \frac{3 \text{ mol } O_2}{2 \text{ mol } C_3H_3N} \times \frac{32.00 \text{ g } O_2}{\text{mol } O_2} = 571 \text{ g } O_2$

Amount O_2 left = 1.00×10^3 g - 571 g = 430 g; 297 g NH_3 and 430 g O_2 left unreacted.

119. 10.00 g XCl_2 + excess $Cl_2 \rightarrow$ 12.55 g XCl_4; 2.55 g Cl reacted with XCl_2 to form XCl_4. XCl_4 contains 2.55 g Cl and 10.00 g XCl_2. From mol ratios, 10.00 g XCl_2 must also contain 2.55 g Cl with 10.00 - 2.45 = 7.45 g X.

$$2.55 \text{ g Cl} \times \frac{1 \text{ mol Cl}}{35.45 \text{ g Cl}} \times \frac{1 \text{ mol } XCl_2}{2 \text{ mol Cl}} \times \frac{1 \text{ mol X}}{\text{mol } XCl_2} = 3.60 \times 10^{-2} \text{ mol X}$$

So, 3.60×10^{-2} mol X must equal 7.45 g X. The molar mass of X is:

$$\frac{7.45 \text{ g X}}{3.60 \times 10^{-2} \text{ mol X}} = \frac{207 \text{ g}}{\text{mol X}}; \quad \text{X is Pb.}$$

121. Consider the case of aluminum plus oxygen. Aluminum forms Al^{3+} ions; oxygen forms O^{2-} anions. The simplest compound of the two elements is Al_2O_3. Similarly we would expect the formula of any group 6A element with Al to be Al_2X_3. Assuming this, out of 100.00 g of compound there are 18.56 g Al and 81.44 g of the unknown element, X. Let's use this information to determine the molar mass of X which will allow us to identify X from the periodic table.

$$18.56 \text{ g Al} \times \frac{1 \text{ mol Al}}{26.98 \text{ g Al}} \times \frac{3 \text{ mol X}}{2 \text{ mol Al}} = 1.032 \text{ mol X}$$

81.44 g of X must equal 1.032 mol of X.

$$\text{The molar mass of X} = \frac{81.44 \text{ g X}}{1.032 \text{ mol X}} = 78.91 \text{ g/mol X}$$

From the periodic table, the unknown element is selenium and the formula is Al_2Se_3.

123. $1.252 \text{ g Cu} \times \frac{1 \text{ mol Cu}}{63.55 \text{ g Cu}} = 1.970 \times 10^{-2} \text{ mol Cu}$

The molar mass of Cu_2O is 143.10 g/mol and the molar mass of CuO is 79.55 g/mol and note that Cu_2O contains twice the mol Cu as compared to CuO. Let x = g Cu_2O and y = g CuO, then x + y = 1.500 and:

$$2\left(\frac{x}{143.10}\right) + \frac{y}{79.55} = 1.970 \times 10^{-2} \text{ total mol Cu or } 1.112\,x + y = 1.567$$

Solving by the method of simultaneous equations:

$$\begin{array}{r} 1.112\,x + y = \;\; 1.567 \\ \underline{-x \;-\; y = -1.500} \\ 0.112\,x = \;\; 0.067 \end{array}$$

$x = 0.067/0.112 = 0.60\text{ g} = \text{mass } Cu_2O$

$$\%Cu_2O = \frac{0.60\text{ g}}{1.500\text{ g}} = 40.\%; \;\; \%CuO = 100. - 40. = 60.\%$$

CHAPTER FOUR

TYPES OF CHEMICAL REACTIONS AND SOLUTION STOICHIOMETRY

Questions

5. "Slightly soluble" refers to substances that dissolve only to a small extent. A slightly soluble salt may still dissociate completely to ions and, hence, be a strong electrolyte. An example of such a substance is $Mg(OH)_2$. It is a strong electrolyte, but not very soluble. A weak electrolyte is a substance that doesn't dissociate completely to produce ions. A weak electrolyte may be very soluble in water, or it may not be very soluble. Acetic acid is an example of a weak electrolyte that is very soluble in water.

Exercises

Aqueous solutions: Strong and Weak Electrolytes

7. a. $NaBr(s) \rightarrow Na^+(aq) + Br^-(aq)$

   b. $MgCl_2(s) \rightarrow Mg^{2+}(aq) + 2\ Cl^-(aq)$

   c. $Al(NO_3)_3(s) \rightarrow Al^{3+}(aq) + 3\ NO_3^-(aq)$

   d. $(NH_4)_2SO_4(s) \rightarrow 2NH_4^+(aq) + SO_4^{2-}(aq)$

   e. $HI(g) \rightarrow H^+(aq) + I^-(aq)$

   f. $FeSO_4(s) \rightarrow Fe^{2+}(aq) + SO_4^{2-}(aq)$

   g. $KMnO_4(s) \rightarrow K^+(aq) + MnO_4^-(aq)$

   h. $HClO_4(s) \rightarrow H^+(aq) + ClO_4^-(aq)$

   i. $NH_4C_2H_3O_2(s) \rightarrow NH_4^+(aq) + C_2H_3O_2^-(aq)$

9. $CaCl_2(s) \rightarrow Ca^{2+}(aq) + 2\ Cl^-(aq)$

Solution Concentration: Molarity

11. a. $5.623\text{ g NaHCO}_3 \times \dfrac{1\text{ mol NaHCO}_3}{84.01\text{ g NaHCO}_3} = 6.693 \times 10^{-2}\text{ mol NaHCO}_3$

$$M = \frac{6.693 \times 10^{-2}\text{ mol}}{250.0\text{ mL}} \times \frac{1000\text{ mL}}{\text{L}} = 0.2677\ M\text{ NaHCO}_3$$

b. $0.1846 \text{ g } K_2Cr_2O_7 \times \dfrac{1 \text{ mol } K_2Cr_2O_7}{294.20 \text{ g } K_2Cr_2O_7} = 6.275 \times 10^{-4} \text{ mol } K_2Cr_2O_7$

$M = \dfrac{6.275 \times 10^{-4} \text{ mol}}{500.0 \times 10^{-3} \text{ L}} = 1.255 \times 10^{-3}\, M\, K_2Cr_2O_7$

c. $0.1025 \text{ g Cu} \times \dfrac{1 \text{ mol Cu}}{63.55 \text{ g Cu}} = 1.613 \times 10^{-3} \text{ mol Cu} = 1.613 \times 10^{-3} \text{ mol } Cu^{2+}$

$M = \dfrac{1.613 \times 10^{-3} \text{ mol } Cu^{2+}}{200.0 \text{ mL}} \times \dfrac{1000 \text{ mL}}{\text{L}} = 8.065 \times 10^{-3}\, M\, Cu^{2+}$

13. a. $CaCl_2(s) \rightarrow Ca^{2+}(aq) + 2\, Cl^-(aq)$; $M_{Ca^{2+}} = 0.15\, M$; $M_{Cl^-} = 2(0.15) = 0.30\, M$

b. $Al(NO_3)_3(s) \rightarrow Al^{3+}(aq) + 3\, NO_3^-(aq)$; $M_{Al^{3+}} = 0.26\, M$; $M_{NO_3^-} = 3(0.26) = 0.78\, M$

c. $K_2Cr_2O_7(s) \rightarrow 2\, K^+(aq) + Cr_2O_7^{2-}(aq)$; $M_{K^+} = 2(0.25) = 0.50\, M$; $M_{Cr_2O_7^{2-}} = 0.25\, M$

d. $Al_2(SO_4)_3(s) \rightarrow 2\, Al^{3+}(aq) + 3\, SO_4^{2-}(aq)$

$M_{Al^{3+}} = \dfrac{2.0 \times 10^{-3} \text{ mol } Al_2(SO_4)_3}{\text{L}} \times \dfrac{2 \text{ mol } Al^{3+}}{\text{mol } Al_2(SO_4)_3} = 4.0 \times 10^{-3}\, M$

$M_{SO_4^{2-}} = \dfrac{2.0 \times 10^{-3} \text{ mol } Al_2(SO_4)_3}{\text{L}} \times \dfrac{3 \text{ mol } SO_4^{2-}}{\text{mol } Al_2(SO_4)_3} = 6.0 \times 10^{-3}\, M$

15. mol solute = volume (L) × molarity $\left(\dfrac{\text{mol}}{\text{L}}\right)$; $AlCl_3(s) \rightarrow Al^{3+}(aq) + 3\, Cl^-(aq)$

$\text{mol } Cl^- = 0.1000 \text{ L} \times \dfrac{0.30 \text{ mol } AlCl_3}{\text{L}} \times \dfrac{3 \text{ mol } Cl^-}{\text{mol } AlCl_3} = 9.0 \times 10^{-2} \text{ mol } Cl^-$

$MgCl_2(s) \rightarrow Mg^{2+}(aq) + 2\, Cl^-(aq)$

$\text{mol } Cl^- = 0.0500 \text{ L} \times \dfrac{0.60 \text{ mol } MgCl_2}{\text{L}} \times \dfrac{2 \text{ mol } Cl^-}{\text{mol } MgCl_2} = 6.0 \times 10^{-2} \text{ mol } Cl^-$

$NaCl(s) \rightarrow Na^+(aq) + Cl^-(aq)$

$\text{mol } Cl^- = 0.2000 \text{ L} \times \dfrac{0.40 \text{ mol NaCl}}{\text{L}} \times \dfrac{1 \text{ mol } Cl^-}{\text{mol NaCl}} = 8.0 \times 10^{-2} \text{ mol } Cl^-$

100.0 mL of 0.30 M $AlCl_3$ contains the largest moles of Cl^- ions.

17. Molar mass of $NaHCO_3$ = 22.99 + 1.008 + 12.01 + 3(16.00) = 84.01 g/mol

$\text{Volume} = 0.350 \text{ g } NaHCO_3 \times \dfrac{1 \text{ mol } NaHCO_3}{84.01 \text{ g } NaHCO_3} \times \dfrac{1 \text{ L}}{0.100 \text{ mol } NaHCO_3} = 0.0417 \text{ L} = 41.7 \text{ mL}$

41.7 mL of 0.100 M $NaHCO_3$ contains 0.350 g $NaHCO_3$.

19. a. $1.0 \text{ L} \times \dfrac{0.10 \text{ mol NaCl}}{\text{L}} \times \dfrac{58.44 \text{ g NaCl}}{\text{mol}} = 5.8 \text{ g NaCl}$

Place 5.8 g NaCl in a 1 L volumetric flask; add water to dissolve the NaCl and fill to the mark.

b. $1.0 \text{ L} \times \dfrac{0.10 \text{ mol NaCl}}{\text{L}} \times \dfrac{1 \text{ L stock}}{2.5 \text{ mol NaCl}} = 4.0 \times 10^{-2} \text{ L}$

Add 40. mL of 2.5 *M* solutiuon to a 1 L volumetric flask; fill to the mark with water.

c. $1.0 \text{ L} \times \dfrac{0.20 \text{ mol NaIO}_3}{\text{L}} \times \dfrac{197.9 \text{ g NaIO}_3}{\text{mol NaIO}_3} = 4.0 \times 10^1 \text{ g NaIO}_3$

As in a, instead using 40. g $NaIO_3$.

d. $1.0 \text{ L} \times \dfrac{0.010 \text{ mol NaIO}_3}{\text{L}} \times \dfrac{1 \text{ L stock}}{0.20 \text{ mol NaIO}_3} = 0.050 \text{ L}$

As in b, instead using 50. mL of the 0.20 *M* sodium iodate stock solution.

e. $1.0 \text{ L} \times \dfrac{0.050 \text{ mol KHP}}{\text{L}} \times \dfrac{204.22 \text{ g KHP}}{\text{mol KHP}} = 10.2 \text{ g KHP} = 10. \text{ g KHP}$

As in a, instead using 10. g KHP.

f. $1.0 \text{ L} \times \dfrac{0.040 \text{ mol KHP}}{\text{L}} \times \dfrac{1 \text{ L stock}}{0.50 \text{ mol KHP}} = 0.080 \text{ L}$

As in b, instead using 80. mL of the 0.50 *M* KHP stock solution.

21. $10.8 \text{ g (NH}_4)_2\text{SO}_4 \times \dfrac{1 \text{ mol}}{132.15 \text{ g}} = 8.17 \times 10^{-2} \text{ mol (NH}_4)_2\text{SO}_4$

$\text{Molarity} = \dfrac{8.17 \times 10^{-2} \text{ mol}}{100.0 \text{ mL}} \times \dfrac{1000 \text{ mL}}{\text{L}} = 0.817\, M \text{ (NH}_4)_2\text{SO}_4$

Moles of $(NH_4)_2SO_4$ in final solution:

$10.00 \times 10^{-3} \text{ L} \times \dfrac{0.817 \text{ mol}}{\text{L}} = 8.17 \times 10^{-3} \text{ mol}$

$\text{Molarity of final solution} = \dfrac{8.17 \times 10^{-3} \text{ mol}}{(10.00 + 50.00) \text{ mL}} \times \dfrac{1000 \text{ mL}}{\text{L}} = 0.136\, M \text{ (NH}_4)_2\text{SO}_4$

$(NH_4)_2SO_4(s) \rightarrow 2\, NH_4^+(aq) + SO_4^{2-}(aq)$; $M_{NH_4^+} = 2(0.136) = 0.272\, M$; $M_{SO_4^{2-}} = 0.136\, M$

23. $0.5842 \text{ g} \times \dfrac{1 \text{ mol}}{90.04 \text{ g}} = 6.488 \times 10^{-3} \text{ mol H}_2\text{C}_2\text{O}_4$

$$\frac{6.488 \times 10^{-3}\ \text{mol}}{100.0\ \text{mL}} \times \frac{1000\ \text{mL}}{\text{L}} = 6.488 \times 10^{-2}\ M$$

This is the concentration of the initial oxalic acid solution.

Consider, next, the dilution step:

$$10.00 \times 10^{-3}\ \text{L} \times \frac{6.488 \times 10^{-2}\ \text{mol}}{\text{L}} = 6.488 \times 10^{-4}\ \text{mol}\ H_2C_2O_4$$

The final solution contains 6.488×10^{-4} mol of oxalic acid in 250.0 mL of solution:

$$M = \frac{6.488 \times 10^{-4}\ \text{mol}}{0.2500\ \text{L}} = 2.595 \times 10^{-3}\ M\ H_2C_2O_4$$

Precipitation Reactions

25. In all these reactions, soluble ionic compounds are mixed together. To predict the precipitate, switch the anions and cations in the two reactant compounds to predict possible products, then use the solubility rules in Table 4.1 to predict if any of these possible products are insoluble (are the precipitate).

a. Possible products = $BaSO_4$ and NaCl; precipitate = $BaSO_4(s)$
b. Possible products = $PbCl_2$ and KNO_3; precipitate = $PbCl_2(s)$
c. Possible products = Ag_3PO_4 and $NaNO_3$; precipitate = $Ag_3PO_4(s)$
d. Possible products = $NaNO_3$ and $Fe(OH)_3$; precipitate = $Fe(OH)_3(s)$

27. For the following answers, the balanced molecular equation is first, followed by the complete ionic equation, then the net ionic equation.

a. $BaCl_2(aq) + Na_2SO_4(aq) \rightarrow BaSO_4(s) + 2\ NaCl(aq)$

$Ba^{2+}(aq) + 2\ Cl^-(aq) + 2\ Na^+(aq) + SO_4^{2-}(aq) \rightarrow BaSO_4(s) + 2\ Na^+(aq) + 2\ Cl^-(aq)$

$Ba^{2+}(aq) + SO_4^{2-}(aq) \rightarrow BaSO_4(s)$

b. $Pb(NO_3)_2(aq) + 2\ KCl(aq) \rightarrow PbCl_2(s) + 2\ KNO_3(aq)$

$Pb^{2+}(aq) + 2\ NO_3^-(aq) + 2\ K^+(aq) + 2\ Cl^-(aq) \rightarrow PbCl_2(s) + 2\ K^+(aq) + 2\ NO_3^-(aq)$

$Pb^{2+}(aq) + 2\ Cl^-(aq) \rightarrow PbCl_2(s)$

c. $3\ AgNO_3(aq) + Na_3PO_4(aq) \rightarrow Ag_3PO_4(s) + 3\ NaNO_3(aq)$

$3\ Ag^+(aq) + 3\ NO_3^-(aq) + 3\ Na^+(aq) + PO_4^{3-}(aq) \rightarrow Ag_3PO_4(s) + 3\ Na^+(aq) + 3\ NO_3^-(aq)$

$3\ Ag^+(aq) + PO_4^{3-}(aq) \rightarrow Ag_3PO_4(s)$

d. $3\ NaOH(aq) + Fe(NO_3)_3(aq) \rightarrow Fe(OH)_3(s) + 3\ NaNO_3(aq)$

$3\ Na^+(aq) + 3\ OH^-(aq) + Fe^{3+}(aq) + 3\ NO_3^-(aq) \rightarrow Fe(OH)_3(s) + 3\ Na^+(aq) + 3\ NO_3^-(aq)$

$Fe^{3+}(aq) + 3\ OH^-(aq) \rightarrow Fe(OH)_3(s)$

29. a. Silver iodide is insoluble. $AgNO_3(aq) + KI(aq) \rightarrow AgI(s) + KNO_3(aq)$

$Ag^+(aq) + I^-(aq) \rightarrow AgI(s)$

b. Copper(II) sulfide is insoluble. $CuSO_4(aq) + Na_2S(aq) \rightarrow CuS(s) + Na_2SO_4(aq)$

$Cu^{2+}(aq) + S^{2-}(aq) \rightarrow CuS(s)$

c. $CoCl_2(aq) + 2\ NaOH(aq) \rightarrow Co(OH)_2(s) + 2\ NaCl(aq)$

$Co^{2+}(aq) + 2\ OH^-(aq) \rightarrow Co(OH)_2(s)$

d. The potential products are $Ni(NO_3)_2$ and KCl. Both are soluble in water. Thus, no reaction occurs.

31. a. $(NH_4)_2SO_4(aq) + Ba(NO_3)_2(aq) \rightarrow 2\ NH_4NO_3(aq) + BaSO_4(s)$

$Ba^{2+}(aq) + SO_4^{2-}(aq) \rightarrow BaSO_4(s)$

b. $Pb(NO_3)_2(aq) + 2\ NaCl(aq) \rightarrow PbCl_2(s) + 2\ NaNO_3(aq)$

$Pb^{2+}(aq) + 2\ Cl^-(aq) \rightarrow PbCl_2(s)$

c. Potassium phosphate and sodium nitrate are both soluble in water. No reaction occurs.

d. No reaction occurs since all possible products are soluble.

e. $CuCl_2(aq) + 2\ NaOH(aq) \rightarrow Cu(OH)_2(s) + 2\ NaCl(aq)$

$Cu^{2+}(aq) + 2\ OH^-(aq) \rightarrow Cu(OH)_2(s)$

33. Three possibilities are:

Addition of K_2SO_4 solution to give a white ppt. of $PbSO_4$. Addition of NaCl solution to give a white ppt. of $PbCl_2$. Addition of K_2CrO_4 solution to give a bright yellow ppt. of $PbCrO_4$.

35. The reaction is: $AgNO_3(aq) + NaCl(aq) \rightarrow AgCl(s) + NaNO_3(aq)$

$$50.0 \times 10^{-3}\ L\ AgNO_3 \times \frac{0.0500\ mol\ AgNO_3}{L\ AgNO_3} \times \frac{1\ mol\ NaCl}{1\ mol\ AgNO_3} \times \frac{58.44\ g\ NaCl}{mol\ NaCl} = 0.146\ g\ NaCl$$

37. The reaction is: $AgNO_3(aq) + NaOH(aq) \rightarrow AgBr(s) + NaNO_3(aq)$

$$100.0\ mL\ AgNO_3 \times \frac{1\ L}{1000\ mL} \times \frac{0.150\ mol\ AgNO_3}{L\ AgNO_3} = 1.50 \times 10^{-2}\ mol\ AgNO_3$$

$$20.0\ mL\ NaBr \times \frac{1\ L}{1000\ mL} \times \frac{1.00\ mol\ NaBr}{L\ NaBr} = 2.00 \times 10^{-2}\ mol\ NaBr$$

From the balanced reaction, 1 mol of $AgNO_3$ is required to react with 1 mol of NaBr (1:1 mol ratio). The actual $AgNO_3$ to NaBr mol ratio is $1.50 \times 10^{-2}/2.00 \times 10^{-2} = 0.750$. Since the actual mol ratio is less than the required mol ratio, then $AgNO_3$ is the limiting reagent ($AgNO_3$ runs out first with NaBr in excess).

$$1.50 \times 10^{-2} \text{ mol AgNO}_3 \times \frac{1 \text{ mol AgBr}}{1 \text{ mol AgNO}_3} \times \frac{187.8 \text{ g AgBr}}{\text{mol AgBr}} = 2.82 \text{ g AgBr}$$

39. a. The balanced reaction is: $2\ KOH(aq) + Mg(NO_3)_2(aq) \rightarrow Mg(OH)_2(s) + 2\ KNO_3(aq)$

b. The precipitate is magnesium hydroxide.

c. $$0.1000 \text{ L KOH} \times \frac{0.200 \text{ mol KOH}}{\text{L KOH}} = 2.00 \times 10^{-2} \text{ mol KOH}$$

$$0.1000 \text{ L Mg(NO}_3)_2 \times \frac{0.200 \text{ mol Mg(NO}_3)_2}{\text{L Mg(NO}_3)_2} = 2.00 \times 10^{-2} \text{ mol Mg(NO}_3)_2$$

From the balanced equation, the required mol KOH to mol $Mg(NO_3)_2$ ratio is 2:1. The actual mol ratio present is 1:1. Not enough KOH is present to react with all of the $Mg(NO_3)_2$ present, so KOH is the limiting reagent.

$$0.0200 \text{ mol KOH} \times \frac{1 \text{ mol Mg(OH)}_2}{2 \text{ mol KOH}} \times \frac{58.33 \text{ g Mg(OH)}_2}{\text{mol Mg(OH)}_2} = 0.583 \text{ g Mg(OH)}_2$$

d. The net ionic equation for this reaction is: $Mg^{2+}(aq) + 2\ OH^-(aq) \rightarrow Mg(OH)_2(s)$

Since KOH was the limiting reagent, then all of the OH^- was used up in the reaction. So, $M_{OH^-} = 0\ M$. Note that K^+ is a spectator ion, so it is still present in solution after precipitation was complete. Also present will be the excess Mg^{2+} and NO_3^- (the other spectator ion).

$$\text{total Mg}^{2+} = 0.0200 \text{ mol Mg(NO}_3)_2 \times \frac{1 \text{ mol Mg}^{2+}}{\text{mol Mg(NO}_3)_2} = 0.0200 \text{ mol Mg}^{2+}$$

$$\text{mol Mg}^{2+} \text{ reacted} = 0.0200 \text{ mol KOH} \times \frac{1 \text{ mol Mg(NO}_3)_2}{2 \text{ mol KOH}} \times \frac{1 \text{ mol Mg}^{2+}}{\text{mol Mg(NO}_3)_2} = 0.0100 \text{ mol Mg}^{2+}$$

$$M_{Mg^{2+}} = \frac{\text{mol excess Mg}^{2+}}{\text{total volume}} = \frac{0.0200 - 0.0100 \text{ mol Mg}^{2+}}{0.1000 \text{ L} + 0.1000 \text{ L}} = 5.00 \times 10^{-2}\ M \text{ Mg}^{2+}$$

The spectator ions are K^+ and NO_3^-. The moles of each present are:

$$\text{mol K}^+ = 0.0200 \text{ mol KOH} \times \frac{1 \text{ mol K}^+}{\text{mol KOH}} = 0.0200 \text{ mol K}^+$$

$$\text{mol NO}_3^- = 0.0200 \text{ mol Mg(NO}_3)_2 \times \frac{2 \text{ mol NO}_3^-}{\text{mol Mg(NO}_3)_2} = 0.0400 \text{ mol NO}_3^-$$

The concentrations are:

$$M_{K^+} = \frac{0.0200 \text{ mol K}^+}{0.2000 \text{ L}} = 0.100\, M\, K^+; \quad M_{NO_3^-} = \frac{0.0400 \text{ mol NO}_3^-}{0.2000 \text{ L}} = 0.200\, M\, NO_3^-$$

Acid-Base Reactions

41. All the bases in this problem are soluble ionic compounds containing OH^-. The acids are either strong or weak electrolytes. The best way to determine if an acid is a strong or weak electrolyte is to memorize all the strong electrolytes (strong acids). Any other acid you encounter that is not a strong acid will be a weak electrolyte and should be kept together in a balanced equation. The strong acids to recognize are HCl, HBr, HI, HNO_3, $HClO_4$ and H_2SO_4. For the answers below, the order of the reactions are molecular, complete ionic and net ionic.

a. $2\ HClO_4(aq) + Mg(OH_2)(s) \rightarrow 2\ H_2O(l) + Mg(ClO_4)_2(aq)$

$2\ H^+(aq) + 2\ ClO_4^-(aq) + Mg(OH)_2(s) \rightarrow 2\ H_2O(l) + Mg^{2+}(aq) + 2\ ClO_4^-(aq)$

$2\ H^+(aq) + Mg(OH)_2(s) \rightarrow 2\ H_2O(l) + Mg^{2+}(aq)$

b. $HCN(aq) + NaOH(aq) \rightarrow H_2O(l) + NaCN(aq)$

$HCN(aq) + Na^+(aq) + OH^-(aq) \rightarrow H_2O(l) + Na^+(aq) + CN^-(aq)$

$HCN(aq) + OH^-(aq) \rightarrow H_2O(l) + CN^-(aq)$

c. $HCl(aq) + NaOH(aq) \rightarrow H_2O(l) + NaCl(aq)$

$H^+(aq) + Cl^-(aq) + Na^+(aq) + OH^-(aq) \rightarrow H_2O(l) + Na^+(aq) + Cl^-(aq)$

$H^+(aq) + OH^-(aq) \rightarrow H_2O(l)$

43. All the acids in this problem are strong electrolytes. The acids to recognize as strong electrolytes are HCl, HBr, HI, HNO_3, $HClO_4$ and H_2SO_4.

a. $KOH(aq) + HNO_3(aq) \rightarrow H_2O(l) + KNO_3(aq)$

$K^+(aq) + OH^-(aq) + H^+(aq) + NO_3^-(aq) \rightarrow H_2O(l) + K^+(aq) + NO_3^-(aq)$

$OH^-(aq) + H^+(aq) \rightarrow H_2O(l)$

b. $Ba(OH)_2(aq) + 2\ HCl(aq) \rightarrow 2\ H_2O(l) + BaCl_2(aq)$

$Ba^{2+}(aq) + 2\ OH^-(aq) + 2\ H^+(aq) + 2\ Cl^-(aq) \rightarrow 2\ H_2O(l) + Ba^{2+}(aq) + 2\ Cl^-(aq)$

$2\ OH^-(aq) + 2\ H^+(aq) \rightarrow 2\ H_2O(l)$ or $OH^-(aq) + H^+(aq) \rightarrow H_2O(l)$

c. $3\ HClO_4(aq) + Fe(OH)_3(s) \rightarrow 3\ H_2O(l) + Fe(ClO_4)_3(aq)$

$3\ H^+(aq) + 3\ ClO_4^-(aq) + Fe(OH)_3(s) \rightarrow 3\ H_2O(l) + Fe^{3+}(aq) + 3\ ClO_4^-(aq)$

$3\ H^+(aq) + Fe(OH)_3(s) \rightarrow 3\ H_2O(l) + Fe^{3+}(aq)$

45. If we begin with 50.00 mL of 0.200 *M* NaOH, then:

$$50.00 \times 10^{-3}\ L \times \frac{0.200\ mol}{L} = 1.00 \times 10^{-2}\ mol\ NaOH \text{ is to be neutralized.}$$

a. $NaOH(aq) + HCl(aq) \rightarrow NaCl(aq) + H_2O(l)$

$$1.00 \times 10^{-2}\ mol\ NaOH \times \frac{1\ mol\ HCl}{mol\ NaOH} \times \frac{1\ L\ soln}{0.100\ mol} = 0.100\ L \text{ or } 100.\ mL$$

b. $HNO_3(aq) + NaOH(aq) \rightarrow H_2O(l) + NaNO_3(aq)$

$$1.00 \times 10^{-2}\ mol\ NaOH \times \frac{1\ mol\ HNO_3}{mol\ NaOH} \times \frac{1\ L}{0.150\ mol\ HNO_3} = 6.67 \times 10^{-2}\ L \text{ or } 66.7\ mL$$

c. $HC_2H_3O_2(aq) + NaOH(aq) \rightarrow H_2O(l) + NaC_2H_3O_2(aq)$

$$1.00 \times 10^{-2}\ mol\ NaOH \times \frac{1\ mol\ HC_2H_3O_2}{mol\ NaOH} \times \frac{1\ L}{0.200\ mol\ HC_2H_3O_2} = 5.00 \times 10^{-2}\ L \text{ or } 50.0\ mL$$

47. $HNO_3(aq) + NaOH\ (aq) \rightarrow NaNO_3(aq) + H_2O(l)$

$$15.0\ g\ NaOH \times \frac{1\ mol\ NaOH}{40.00\ g} = 0.375\ mol\ NaOH$$

$$0.1500\ L \times \frac{0.250\ mol\ HNO_3}{L} = 0.0375\ mol\ HNO_3$$

We have added more moles of NaOH than mol of HNO_3 present. Since NaOH and HNO_3 react in a 1:1 mol ratio then NaOH is in excess and the solution will be basic. The ions present after reaction will be the excess OH^- ions and the spectator ions, Na^+ and NO_3^+. The moles of ions present initially are:

mol NaOH = mol Na^+ = mol OH^- = 0.375 mol

mol HNO_3 = mol H^+ = mol NO_3^- = 0.0375 mol

The net ionic reaction occurring is: $H^+(aq) + OH^-(aq) \rightarrow H_2O(l)$

The mol of excess OH^- remaining after reaction will be the initial mol of OH^- minus the amount of OH^- neutralized by reaction with H^+:

mol excess OH^- = 0.375 mol - 0.0375 mol = 0.338 mol OH^- excess

The concentration of ions present is:

$$M_{OH^-} = \frac{\text{mol OH}^- \text{ excess}}{\text{volume}} = \frac{0.338 \text{ mol OH}^-}{0.1500 \text{ L}} = 2.25\ M\ \text{OH}^-$$

$$M_{NO_3^-} = \frac{0.0375 \text{ mol NO}_3^-}{0.1500 \text{ L}} = 0.250\ M\ \text{NO}_3^-;\ M_{Na^+} = \frac{0.375 \text{ mol}}{0.1500 \text{ L}} = 2.50\ M\ \text{Na}^+$$

49. $HCl(aq) + NaOH(aq) \rightarrow H_2O(l) + NaCl(aq)$

$$24.16 \times 10^{-3} \text{ L NaOH} \times \frac{0.106 \text{ mol NaOH}}{\text{L NaOH}} \times \frac{1 \text{ mol HCl}}{\text{mol NaOH}} = 2.56 \times 10^{-3} \text{ mol HCl}$$

$$\text{Molarity of HCl} = \frac{2.56 \times 10^{-3} \text{ mol}}{25.00 \times 10^{-3} \text{ L}} = 0.102\ M\ \text{HCl}$$

51. Since KHP is a monoprotic acid, then the reaction is: $NaOH(aq) + KHP(aq) \rightarrow H_2O(l) + NaKP(aq)$

$$\text{mass KHP} = 0.02046 \text{ L NaOH} \times \frac{0.1000 \text{ mol NaOH}}{\text{L NaOH}} \times \frac{1 \text{ mol KHP}}{\text{mol NaOH}} \times \frac{204.22 \text{ g KHP}}{\text{mol KHP}}$$

$$= 0.4178 \text{ g KHP}$$

Oxidation-Reduction Reactions

53. Apply rules in Table 4.2.

a. $KMnO_4$ is composed of K^+ and MnO_4^- ions. Assign oxygen a value of -2 which gives manganese at +7 oxidation state since the sum of oxidation states for all atoms in MnO_4^- must equal the -1 charge on MnO_4^-. K, +1; O, -2; Mn, +7.

b. Assign O a -2 oxidation state, which gives nickle a +4 oxidation state. Ni, +4; O, -2.

c. $K_4Fe(CN)_6$ is composed of K^+ cations and $Fe(CN)_6^{4-}$ anions. $Fe(CN)_6^{4-}$ is composed of iron and CN^- anions. For an overall anion charge of -4, iron must have a +2 oxidation state.

d. $(NH_4)_2HPO_4$ is made of NH_4^+ cations and HPO_4^{2-} anions. Assign +1 as oxidation state of H and -2 as oxidation state of O. In NH_4^+, x + 4(+1) = +1, x = -3 = oxidation state of N. In HPO_4^{2-}, +1 + y + 4(-2) = -2, y = +5 = oxidation state of P.

e. O, -2; P, +3

f. O, -2; Fe, + 8/3

g. O, -2; F, -1; Xe, +6

h. F, -1; S, +4

i. O, -2; C, +2

j. Na, +1; O, -2; C, +3

55. OCl^-: Oxidation state of oxygen is (-2).

$-2 + x = -1,\ x = +1;$ The oxidation state of Cl in OCl^- is +1.

ClO_2^-: $2(-2) + x = -1$, $x = +3$; ClO_3^-: $3(-2) + x = -1$, $x = +5$; ClO_4^-: $4(-2) + x = -1$, $x = +7$

57. To determine if the reaction is an oxidation-reduction reaction, assign oxidation numbers. If the oxidation numbers change for some elements, then the reaction is a redox reaction. If the oxidation numbers do not change, then the reaction is not a redox reaction. In redox reactions the species oxidized (called the reducing agent) shows an increase in the oxidation numbers and the species reduced (called the oxidizing agent) shows a decrease in oxidation numbers.

	Redox?	Oxidizing Agent	Reducing Agent	Substance Oxidized	Substance Reduced
a.	Yes	O_2	CH_4	CH_4 (C)	O_2 (O)
b.	Yes	HCl	Zn	Zn	HCl (H)
c.	No	-	-	-	-
d.	Yes	O_3	NO	NO (N)	O_3 (O)
e.	Yes	H_2O_2	H_2O_2	H_2O_2 (O)	H_2O_2 (O)
f.	Yes	CuCl	CuCl	CuCl (Cu)	CuCl (Cu)

In c, no oxidation numbers change from reactants to products.

59. Use the method of half-reactions described in Section 4.10 of the text to balance these redox reactions. The first step always is to separate the reaction into the two half-reactions, then balance each half-reaction separately.

a. $Zn \rightarrow Zn^{2+} + 2\ e^-$ $\qquad$ $2e^- + 2\ HCl \rightarrow H_2 + 2Cl^-$

Adding the two balanced half-reactions, $Zn(s) + 2\ HCl(aq) \rightarrow H_2(g) + Zn^{2+}(aq) + 2Cl^-(aq)$

b. $3\ I^- \rightarrow I_3^- + 2e^-$ $\qquad$ $ClO^- \rightarrow Cl^-$

$2e^- + 2H^+ + ClO^- \rightarrow Cl^- + H_2O$

Adding the two balanced half-reactions so electrons cancel:

$$3\ I^-(aq) + 2\ H^+(aq) + ClO^-(aq) \rightarrow I_3^-(aq) + Cl^-(aq) + H_2O(l)$$

c. $As_2O_3 \rightarrow H_3AsO_4$ $\qquad$ $NO_3^- \rightarrow NO + 2\ H_2O$

$As_2O_3 \rightarrow 2\ H_3AsO_4$ $\qquad$ $4\ H^+ + NO_3^- \rightarrow NO + 2\ H_2O$

Left 3 - O; Right 8 - O $\qquad$ $(3\ e^- + 4\ H^+ + NO_3^- \rightarrow NO + 2\ H_2O) \times 4$

Right hand side has 5 extra O.

Balance the oxygen atoms first using H_2O, then balance H using H^+, and finally balance charge using electrons.

$$(5\ H_2O + As_2O_3 \rightarrow 2\ H_3AsO_4 + 4\ H^+ + 4\ e^-) \times 3$$

Common factor is a transfer of 12 e^-. Add half-reactions so electrons cancel.

$$12\ e^- + 16\ H^+ + 4\ NO_3^- \rightarrow 4\ NO + 8\ H_2O$$
$$15\ H_2O + 3\ As_2O_3 \rightarrow 6\ H_3AsO_4 + 12\ H^+ + 12\ e^-$$

$$7\ H_2O(l) + 4\ H^+(aq) + 3\ As_2O_3(s) + 4\ NO_3^-(aq) \rightarrow 4\ NO(g) + 6\ H_3AsO_4(aq)$$

d. $(2\ Br^- \rightarrow Br_2 + 2\ e^-) \times 5$

$MnO_4^- \rightarrow Mn^{2+} + 4\ H_2O$
$(5\ e^- + 8\ H^+ + MnO_4^- \rightarrow Mn^{2+} + 4\ H_2O) \times 2$

Common factor is a transfer of 10 e⁻.

$$10\ Br^- \rightarrow 5\ Br_2 + 10\ e^-$$
$$10\ e^- + 16\ H^+ + 2\ MnO_4^- \rightarrow 2\ Mn^{2+} + 8\ H_2O$$

$$16\ H^+(aq) + 2\ MnO_4^-(aq) + 10\ Br^-(aq) \rightarrow 5\ Br_2(l) + 2\ Mn^{2+}(aq) + 8\ H_2O(l)$$

e. $CH_3OH \rightarrow CH_2O$
$(CH_3OH \rightarrow CH_2O + 2\ H^+ + 2\ e^-) \times 3$

$Cr_2O_7^{2-} \rightarrow Cr^{3+}$
$14\ H^+ + Cr_2O_7^{2-} \rightarrow 2\ Cr^{3+} + 7\ H_2O$
$6\ e^- + 14\ H^+ + Cr_2O_7^{2-} \rightarrow 2\ Cr^{3+} + 7\ H_2O$

Common factor is a transfer of 6 e⁻.

$$3\ CH_3OH \rightarrow 3\ CH_2O + 6\ H^+ + 6\ e^-$$
$$6\ e^- + 14\ H^+ + Cr_2O_7^{2-} \rightarrow 2\ Cr^{3+} + 7\ H_2O$$

$$8\ H^+(aq) + 3\ CH_3OH(aq) + Cr_2O_7^{2-}(aq) \rightarrow 2\ Cr^{3+}(aq) + 3\ CH_2O(aq) + 7\ H_2O(l)$$

61. Use the same method as with acidic solutions. After the final balanced equation, then convert H^+ to OH^- as described in section 14.10 of the text. The extra step involves converting H^+ into H_2O by adding equal moles of OH^- to each side of the reaction. This converts the reaction to a basic solution while keeping it balanced.

a. $Al \rightarrow Al(OH)_4^-$
$4\ H_2O + Al \rightarrow Al(OH)_4^- + 4\ H^+$
$4\ H_2O + Al \rightarrow Al(OH)_4^- + 4\ H^+ + 3\ e^-$

$MnO_4^- \rightarrow MnO_2$
$3\ e^- + 4\ H^+ + MnO_4^- \rightarrow MnO_2 + 2\ H_2O$

$$4\ H_2O + Al \rightarrow Al(OH)_4^- + 4\ H^+ + 3\ e^-$$
$$3\ e^- + 4\ H^+ + MnO_4^- \rightarrow MnO_2 + 2\ H_2O$$

$$2\ H_2O(l) + Al(s) + MnO_4^-(aq) \rightarrow Al(OH)_4^-(aq) + MnO_2(s)$$

Since H^+ doesn't appear in the final balanced reaction, we are done.

b. $Cl_2 \rightarrow Cl^-$
$2\ e^- + Cl_2 \rightarrow 2\ Cl^-$

$Cl_2 \rightarrow ClO^-$
$2\ H_2O + Cl_2 \rightarrow 2\ ClO^- + 4\ H^+ + 2\ e^-$

$$2\ e^- + Cl_2 \rightarrow 2\ Cl^-$$
$$2\ H_2O + Cl_2 \rightarrow 2\ ClO^- + 4\ H^+ + 2\ e^-$$

$$2\ H_2O + 2\ Cl_2 \rightarrow 2\ Cl^- + 2\ ClO^- + 4\ H^+$$

Now convert to a basic solution. Add 4 OH^- to both sides of the equation. The 4 OH^- will react with the 4 H^+ on the product side to give 4 H_2O. After this step, cancel identical species on both sides (2 H_2O). Applying these steps gives: $4\ OH^- + 2\ Cl_2 \rightarrow 2\ Cl^- + 2\ ClO^- + 2\ H_2O$, which can be further simplified to:

$$2\ OH^-(aq) + Cl_2(g) \rightarrow Cl^-(aq) + ClO^-(aq) + H_2O(l)$$

c. $NO_2^- \rightarrow NH_3$ $\qquad$ $Al \rightarrow AlO_2^-$

$6\ e^- + 7\ H^+ + NO_2^- \rightarrow NH_3 + 2\ H_2O$ $\qquad$ $(2\ H_2O + Al \rightarrow AlO_2^- + 4\ H^+ + 3\ e^-) \times 2$

Common factor is a transfer of 6 e^-.

$$6e^- + 7\ H^+ + NO_2^- \rightarrow NH_3 + 2\ H_2O$$
$$4\ H_2O + 2\ Al \rightarrow 2\ AlO_2^- + 8\ H^+ + 6\ e^-$$

$$OH^- + 2\ H_2O + NO_2^- + 2\ Al \rightarrow NH_3 + 2\ AlO_2^- + H^+ + OH^-$$

Reducing gives: $OH^-(aq) + H_2O(l) + NO_2^-(aq) + 2\ Al(s) \rightarrow NH_3(g) + 2\ AlO_2^-(aq)$

63. $NaCl + H_2SO_4 + MnO_2 \rightarrow Na_2SO_4 + MnCl_2 + Cl_2 + H_2O$

We could balance this reaction by the half-reaction method or by inspection. Lets try inspection. To balance Cl^-, we need 4 NaCl:

$$4\ NaCl + H_2SO_4 + MnO_2 \rightarrow Na_2SO_4 + MnCl_2 + Cl_2 + H_2O$$

Balance the Na^+ and SO_4^{2-} ions next:

$$4\ NaCl + 2\ H_2SO_4 + MnO_2 \rightarrow 2\ Na_2SO_4 + MnCl_2 + Cl_2 + H_2O$$

On the left side: 4-H and 10-O; On the right side: 8-O not counting H_2O

We need 2 H_2O on the right side to balance H and O:

$$4\ NaCl(aq) + 2\ H_2SO_4(aq) + MnO_2(s) \rightarrow 2\ Na_2SO_4(aq) + MnCl_2(aq) + Cl_2(g) + 2\ H_2O(l)$$

Additional Exercises

65. $4.25\text{ g Ca} \times \frac{1\text{ mol Ca}}{40.08\text{ g Ca}} \times \frac{1\text{ mol Ca(OH)}_2}{\text{mol Ca}} \times \frac{2\text{ mol OH}^-}{\text{mol Ca(OH)}_2} = 0.212\text{ mol OH}^-$

$\text{Molarity} = \frac{0.212\text{ mol}}{225 \times 10^{-3}\text{ L}} = 0.942\ M\ OH^-$

67. For the following answers, the balanced molecular equation is first, followed by the complete ionic equation with the net ionic equation last.

a. $2\ AgNO_3(aq) + BaCl_2(aq) \rightarrow 2\ AgCl(s) + Ba(NO_3)_2(aq)$

$2\ Ag^+(aq) + 2\ NO_3^-(aq) + Ba^{2+}(aq) + 2\ Cl^-(aq) \rightarrow 2\ AgCl(s) + Ba^{2+}(aq) + 2\ NO_3^-(aq)$

$Ag^+(aq) + Cl^-(aq) \rightarrow AgCl(s)$

b. No reaction occurs since all of the possible products (NH_4Cl and KNO_3) are soluble.

c. $(NH_4)_2S(aq) + FeCl_2(aq) \rightarrow FeS(s) + 2\ NH_4Cl(aq)$

$2\ NH_4^+(aq) + S^{2-}(aq) + Fe^{2+}(aq) + 2\ Cl^-(aq) \rightarrow FeS(s) + 2\ NH_4^+(aq) + 2\ Cl^-(aq)$

$Fe^{2+}(aq) + S^{2-}(aq) \rightarrow FeS(s)$

d. $K_2CO_3(aq) + CuSO_4(aq) \rightarrow CuCO_3(s) + K_2SO_4(aq)$

$2\ K^+(aq) + CO_3^{2-}(aq) + Cu^{2+}(aq) + SO_4^{2-}(aq) \rightarrow CuCO_3(s) + 2\ K^+(aq) + SO_4^{2-}(aq)$

$Cu^{2+}(aq) + CO_3^{2-}(aq) \rightarrow CuCO_3(s)$

69. a. No; No element shows a change in oxidation number.

b. $1.0\ \text{L oxalic acid} \times \frac{0.14\ \text{mol oxalic acid}}{\text{L}} \times \frac{1\ \text{mol}\ Fe_2O_3}{6\ \text{mol oxalic acid}} \times \frac{159.70\ \text{g}\ Fe_2O_3}{\text{mol}\ Fe_2O_3} = 3.7\ \text{g}\ Fe_2O_3$

71. Use the silver nitrate data to calculate the mol Cl^- present, then use the formula of douglasite to convert from Cl^- to douglasite. The net ionic reaction is: $Ag^+ + Cl^- \rightarrow AgCl(s)$.

$$0.03720\ \text{L} \times \frac{0.1000\ \text{mol}\ Ag^+}{\text{L}} \times \frac{1\ \text{mol}\ Cl^-}{\text{mol}\ Ag^+} \times \frac{1\ \text{mol douglasite}}{4\ \text{mol}\ Cl^-}$$

$$\times \frac{311.88\ \text{g douglasite}}{\text{mol}} = 0.2900\ \text{g douglasite}$$

$$\text{Mass \% douglasite} = \frac{0.2900\ \text{g}}{0.4550\ \text{g}} \times 100 = 63.74\%$$

73. $0.104\ \text{g AgCl} \times \frac{35.45\ \text{g}\ Cl^-}{143.4\ \text{g AgCl}} = 2.57 \times 10^{-2}\ \text{g}\ Cl^- = Cl^-$ in chlorisondiamine

Molar mass of chlorisondiamine = 14(12.01) + 18(1.008) + 6(35.45) + 2(14.01) = 427.00 g/mol

There are 6(35.45) = 212.70 g chlorine for every mole (427.00 g) of chlorisondiamine.

$$2.57 \times 10^{-2}\ \text{g}\ Cl^- \times \frac{427.00\ \text{g drug}}{212.70\ \text{g}\ Cl^-} = 5.16 \times 10^{-2}\ \text{g drug};\ \ \%\ \text{drug} = \frac{5.16 \times 10^{-2}\ \text{g}}{1.28\ \text{g}} \times 100 = 4.03\%$$

75. $HC_2H_3O_2(aq) + NaOH(aq) \rightarrow H_2O(l) + NaC_2H_3O_2(aq)$

a. $16.58 \times 10^{-3}\ \text{L soln} \times \frac{0.5062\ \text{mol NaOH}}{\text{L soln}} \times \frac{1\ \text{mol acetic acid}}{\text{mol NaOH}} = 8.393 \times 10^{-3}\ \text{mol acetic acid}$

$$\text{Concentration of acetic acid} = \frac{8.393 \times 10^{-3}\ \text{mol}}{0.01000\ \text{L}} = 0.8393\ M$$

b. If we have 1.000 L of solution: total mass = $1000.\ \text{mL} \times \dfrac{1.006\ \text{g}}{\text{mL}} = 1006\ \text{g}$

$$\text{Mass of } HC_2H_3O_2 = 0.8393\ \text{mol} \times \frac{60.05\ \text{g}}{\text{mol}} = 50.40\ \text{g}$$

$$\text{Mass \% acetic acid} = \frac{50.40\ \text{g}}{1006\ \text{g}} \times 100 = 5.010\%$$

77. Using HA as an abbreviation for acetylsalicylic acid:

$$HA(aq) + NaOH(aq) \rightarrow H_2O(l) + NaA(aq)$$

$$\text{mol HA present} = 0.03517\ \text{L NaOH} \times \frac{0.5065\ \text{mol NaOH}}{\text{L NaOH}} \times \frac{1\ \text{mol HA}}{\text{mol NaOH}} = 1.781 \times 10^{-2}\ \text{mol HA}$$

From the problem, 3.210 g HA was reacted so:

$$3.210\ \text{g HA} = 1.781 \times 10^{-2}\ \text{mol HA, molar mass} = \frac{3.210\ \text{g HA}}{1.781 \times 10^{-2}\ \text{mol HA}} = 180.2\ \text{g/mol}$$

79. a. $4\ NH_3(g) + 5\ O_2(g) \rightarrow 4\ NO(g) + 6\ H_2O(g)$

-3 +1   0   +2 -2   +1 -2 oxidation numbers

$2\ NO(g) + O_2(g) \rightarrow 2\ NO_2(g)$

+2 -2   0   +4 -2

$3\ NO_2(g) + H_2O(l) \rightarrow 2\ HNO_3(aq) + NO(g)$

+4 -2   +1 -2   +1 +5 -2   +2 -2

All three reactions are oxidation-reduction reactions since there is a change in oxidation numbers of some of the elements in each reaction.

b. $4\ NH_3 + 5\ O_2 \rightarrow 4\ NO + 6\ H_2O$; O_2 is the oxidizing agent and NH_3 is the reducing agent.

$2\ NO + O_2 \rightarrow 2\ NO_2$; O_2 is the oxidizing agent and NO is the reducing agent.

$3\ NO_2 + H_2O \rightarrow 2\ HNO_3 + NO$; NO_2 is both the oxidizing and reducing agent.

81. Fe^{2+} will react with MnO_4^- (purple) producing Fe^{3+} and Mn^{2+}(almost colorless). There is no reaction between MnO_4^- and Fe^{3+}. Therefore, add a few drops of the potassium permanganate solution. If the purple color persists, the solution contains iron(III) sulfate. If the color disappears, iron(II) sulfate is present.

Challenge Problems

83. a. $0.308 \text{ g AgCl} \times \frac{35.45 \text{ g Cl}}{143.4 \text{ g AgCl}} = 0.0761 \text{ g Cl}$; $\%\text{Cl} = \frac{0.0761 \text{ g}}{0.256 \text{ g}} \times 100 = 29.7\% \text{ Cl}$

Cobalt(III) oxide, Co_2O_3: 2(58.93) + 3(16.00) = 165.86 g/mol

$0.145 \text{ g } Co_2O_3 \times \frac{117.86 \text{ g Co}}{165.86 \text{ g } Co_2O_3} = 0.103 \text{ g Co}$; $\%\text{Co} = \frac{0.103 \text{ g}}{0.416 \text{ g}} \times 100 = 24.8\% \text{ Co}$

The remainder, 100.0 - (29.7 + 24.8) = 45.5%, is water. Assuming 100.0 g of compound:

$$45.5 \text{ g } H_2O \times \frac{2.016 \text{ g H}}{18.02 \text{ g } H_2O} = 5.09 \text{ g H}; \ \%\text{H} = \frac{5.09 \text{ g H}}{100.0 \text{ g compound}} \times 100 = 5.09\% \text{ H}$$

$$45.5 \text{ g } H_2O \times \frac{16.00 \text{ g O}}{18.02 \text{ g } H_2O} = 40.4 \text{ g O}; \ \%\text{O} = \frac{40.4 \text{ g O}}{100.0 \text{ g compound}} \times 100 = 40.4\% \text{ O}$$

The mass percent composition is 24.8% Co, 29.7% Cl, 5.09% H and 40.4% O.

b. Out of 100.0 g of compound, there are:

$$24.8 \text{ g Co} \times \frac{1 \text{ mol}}{58.93 \text{ g Co}} = 0.421 \text{ mol Co}; \ 29.7 \text{ g Cl} \times \frac{1 \text{ mol}}{35.45 \text{ g Cl}} = 0.838 \text{ mol Cl}$$

$$5.09 \text{ g H} \times \frac{1 \text{ mol}}{1.008 \text{ g H}} = 5.05 \text{ mol H}; \ 40.4 \text{ g O} \times \frac{1 \text{ mol}}{16.00 \text{ g O}} = 2.53 \text{ mol O}$$

Dividing all results by 0.421, we get $CoCl_2 \bullet 6H_2O$.

c. $CoCl_2 \bullet 6H_2O(aq) + 2\ AgNO_3(aq) \rightarrow 2\ AgCl(s) + Co(NO_3)_2(aq) + 6\ H_2O(l)$

$CoCl_2 \bullet 6H_2O(aq) + 2\ NaOH(aq) \rightarrow Co(OH)_2(s) + 2\ NaCl(aq) + 6\ H_2O(l)$

$Co(OH)_2 \rightarrow Co_2O_3$ This is an oxidation-reduction reaction. Thus, we also need to include an oxidizing agent. The obvious choice is O_2.

$4\ Co(OH)_2(s) + O_2(g) \rightarrow 2\ Co_2O_3(s) + 4\ H_2O(l)$

85. $0.298 \text{ g } BaSO_4 \times \frac{96.07 \text{ g } SO_4^{2-}}{233.4 \text{ g } BaSO_4} = 0.123 \text{ g } SO_4^{2-}$; $\% \text{ sulfate} = \frac{0.123 \text{ g } SO_4^{2-}}{0.205 \text{ g}} = 60.0\%$

Assume we have 100.0 g of the mixture of Na_2SO_4 and K_2SO_4. There is:

$$60.0 \text{ g } SO_4^{2-} \times \frac{1 \text{ mol}}{96.07 \text{ g}} = 0.625 \text{ mol } SO_4^{2-}$$

There must be 2 × 0.625 = 1.25 mol of +1 cations to balance the 2- charge of SO_4^{2-}.

Let x = number of moles of K^+ and y = number of moles of Na^+, then x + y = 1.25.

The total mass of Na^+ and K^+ must be 40.0 g in the assumed 100.0 g of mixture. Setting up an equation:

$$x \text{ mol } K^+ \times \frac{39.10 \text{ g}}{\text{mol}} + y \text{ mol } Na^+ \times \frac{22.99 \text{ g}}{\text{mol}} = 40.0 \text{ g}$$

So, we have two equations with two unknowns: x + y = 1.25 and 39.10 x + 22.99 y = 40.0

Since x = 1.25 - y, then 39.10(1.25 - y) + 22.99 y = 40.0

48.9 - 39.10 y + 22.99 y = 40.0, -16.11 y = -8.9

y = 0.55 mol Na^+ and x = 1.25 - 0.55 = 0.70 mol K^+

Therefore:

$$0.70 \text{ mol } K^+ \times \frac{1 \text{ mol } K_2SO_4}{2 \text{ mol } K^+} = 0.35 \text{ mol } K_2SO_4;\ 0.35 \text{ mol } K_2SO_4 \times \frac{174.27 \text{ g}}{\text{mol}} = 61 \text{ g } K_2SO_4$$

Since we assumed 100.0 g, then the mixture is 61% K_2SO_4 and 39% Na_2SO_4.

87. $CaCO_3(s) + H_2SO_4(aq) \rightarrow CaSO_4(aq) + H_2O(l) + CO_2(g)$

89. a. $MgO(s) + 2\ HCl(aq) \rightarrow MgCl_2(aq) + H_2O(l)$

$Mg(OH)_2(s) + 2\ HCl(aq) \rightarrow MgCl_2(aq) + 2\ H_2O(l)$

$Al(OH)_3(s) + 3\ HCl(aq) \rightarrow AlCl_3(aq) + 3\ H_2O(l)$

b. Let's calculate the number of moles of HCl neutralized per gram of substance. We can get these directly from the balanced equations and the molar masses of the substances.

$$\frac{2 \text{ mol HCl}}{\text{mol MgO}} \times \frac{1 \text{ mol MgO}}{40.31 \text{ g MgO}} = \frac{4.962 \times 10^{-2} \text{ mol HCl}}{\text{g MgO}}$$

$$\frac{2 \text{ mol HCl}}{\text{mol Mg(OH)}_2} \times \frac{1 \text{ mol Mg(OH)}_2}{58.33 \text{ g Mg(OH)}_2} = \frac{3.429 \times 10^{-2} \text{ mol HCl}}{\text{g Mg(OH)}_2}$$

$$\frac{3 \text{ mol HCl}}{\text{mol Al(OH)}_3} \times \frac{1 \text{ mol Al(OH)}_3}{78.00 \text{ g Al(OH)}_3} = \frac{3.846 \times 10^{-2} \text{ mol HCl}}{\text{g Al(OH)}_3}$$

Therefore, one gram of magnesium oxide would neutralize the most 0.10 *M* HCl.

91. $H_2SO_4(aq) + 2\ NaOH(aq) \rightarrow Na_2SO_4(aq) + 2\ H_2O(l)$

$$0.02844 \text{ L} \times \frac{0.1000 \text{ mol NaOH}}{\text{L}} \times \frac{1 \text{ mol } H_2SO_4}{2 \text{ mol NaOH}} \times \frac{1 \text{ mol } SO_2}{\text{mol } H_2SO_4} \times \frac{32.07 \text{ g S}}{\text{mol } SO_2} = 4.560 \times 10^{-2} \text{ g S}$$

$$\%S = \frac{0.04560\text{ g}}{1.325\text{ g}} \times 100 = 3.442\% \text{ by mass}$$

93. First we will calculate the molarity of NaCl while ignoring the uncertainty.

$$0.150\text{ g} \times \frac{1\text{ mol}}{58.44\text{ g}} = 2.57 \times 10^{-3}\text{ moles; Molarity} = \frac{2.57 \times 10^{-3}\text{ mol}}{0.1000\text{ L}} = \frac{2.57 \times 10^{-2}\text{ mol}}{\text{L}}$$

The maximum value for the molarity is $= \dfrac{0.153\text{ g} \times \dfrac{1\text{ mol}}{58.44\text{ g}}}{0.0995\text{ L}} = \dfrac{2.63 \times 10^{-2}\text{ mol}}{\text{L}}$

The minimum value for the molarity is $= \dfrac{0.147\text{ g} \times \dfrac{1\text{ mol}}{58.44\text{ g}}}{0.1005\text{ L}} = \dfrac{2.50 \times 10^{-2}\text{ mol}}{\text{L}}$

The range of the NaCl molarity is 0.0250 *M* to 0.0263 *M* or we can express this range as 0.0257 ± 0.0007 *M*.

95. Desired uncertainty is 1% of 0.02 or ± 0.0002. So we want the solution to be 0.0200 ± 0.0002 *M* or the concentration should be between 0.0198 and 0.0202 *M*. We should use a 1-L volumetric flask to make the solution. They are good to ± 0.1%. We want to weigh out between 0.0198 mol and 0.0202 mol of KIO_3.

Molar mass of KIO_3 = 39.10 + 126.9 + 3(16.00) = 214.0 g/mol

$$0.0198\text{ mol} \times \frac{214.0\text{ g}}{\text{mol}} = 4.24\text{ g; } 0.0202\text{ mol} \times \frac{214.0\text{ g}}{\text{mol}} = 4.32\text{ g}$$

We should weigh out between 4.24 and 4.32 g of KIO_3. We should weigh it to the nearest mg or 0.1 mg. Dissolve the KIO_3 in water and dilute to the mark in a one liter volumetric flask. This will produce a solution whose concentration is within the limits and is known to at least the fourth decimal place.

CHAPTER FIVE

GASES

Questions

15. PV = nRT = constant at constant n and T. At two sets of conditions, P_1V_1 = constant = P_2V_2.

$P_1V_1 = P_2V_2$ (Boyle's law).

$\frac{V}{T} = \frac{nR}{P}$ = constant at constant n and P. At two sets of conditions, $\frac{V_1}{T_1}$ = constant = $\frac{V_2}{T_2}$.

$\frac{V_1}{T_1} = \frac{V_2}{T_2}$ (Charles's law)

17. For an ideal gas at constant n and T, the PV product should equal a constant value no matter what pressure or volume combination is used. The dashed line in Figure 5.6 is the PV vs P plot for an ideal gas. The real gas closest to the ideal plot is Ne, although O_2 is also fairly close to the ideal plot.

19. Method 1: molar mass = $\frac{dRT}{P}$

Determine the density of a gas at a measurable temperature and pressure then use the above equation to determine the molar mass.

Method 2: $\frac{\text{effusion rate for gas 1}}{\text{effusion rate for gas 2}} = \sqrt{\frac{(\text{molar mass})_1}{(\text{molar mass})_2}}$

Determine the relative effusion rate of the unknown gas to some known gas, then use Graham's law of effusion (the above equation) to determine the molar mass.

Exercises

Pressure

21. a. $4.8 \text{ atm} \times \frac{760 \text{ mm Hg}}{\text{atm}} = 3.6 \times 10^3$ mm Hg; b. $3.6 \times 10^3 \text{ mm Hg} \times \frac{1 \text{ torr}}{\text{mm Hg}} = 3.6 \times 10^3$ torr

c. $4.8 \text{ atm} \times \frac{1.013 \times 10^5 \text{ Pa}}{\text{atm}} = 4.9 \times 10^5$ Pa; d. $4.8 \text{ atm} \times \frac{14.7 \text{ psi}}{\text{atm}} = 71$ psi

23. $6.5 \text{ cm} \times \frac{10 \text{ mm}}{\text{cm}} = 65 \text{ mm Hg or } 65 \text{ torr}; \ 65 \text{ torr} \times \frac{1 \text{ atm}}{760 \text{ torr}} = 8.6 \times 10^{-2} \text{ atm}$

$$8.6 \times 10^{-2} \text{ atm} = \frac{1.013 \times 10^5 \text{ Pa}}{\text{atm}} = 8.7 \times 10^3 \text{ Pa}$$

25. If the levels of Hg in each arm of the manometer are equal, then the pressure in the flask is equal to atmospheric pressure. When they are unequal, the difference in height in mm will be equal to the difference in pressure in mm Hg between the flask and the atmosphere. Which level is higher will tell us whether the pressure in the flask is less than or greater than atmospheric.

a. $P_{flask} < P_{atm}; \ P_{flask} = 760. - 140. = 620. \text{ mm Hg} = 620. \text{ torr}; \ 620. \text{ torr} \times \frac{1 \text{ atm}}{760 \text{ torr}} = 0.816 \text{ atm}$

$$0.816 \text{ atm} \times \frac{1.013 \times 10^5 \text{ Pa}}{\text{atm}} = 8.27 \times 10^4 \text{ Pa}$$

b. $P_{flask} > P_{atm}; \ P_{flask} = 760. \text{ torr} + 175 \text{ torr} = 935 \text{ torr}; \ 935 \text{ torr} \times \frac{1 \text{ atm}}{760 \text{ torr}} = 1.23 \text{ atm}$

$$1.23 \text{ atm} \times \frac{1.013 \times 10^5 \text{ Pa}}{\text{atm}} = 1.25 \times 10^5 \text{ Pa}$$

c. $P_{flask} = 635 - 140. = 495 \text{ torr}; \ P_{flask} = 635 + 175 = 810. \text{ torr}$

Gas Laws

27. From Boyle's law, $P_1V_1 = P_2V_2$ at constant n and T.

$$P_2 = \frac{P_1V_1}{V_2} = \frac{5.20 \text{ atm} \times 0.400 \text{ L}}{2.14 \text{ L}} = 0.972 \text{ atm}$$

As expected, as the volume increased, the pressure decreased.

29. From Avogadro's law, $V_1/n_1 = V_2/n_2$ at constant T and P.

$$V_2 = \frac{V_1n_2}{n_1} = \frac{11.2 \text{ L} \times 2.00 \text{ mol}}{0.500 \text{ mol}} = 44.8 \text{ L}$$

As expected, as the mol of gas present increases, volume increases.

31. a. $PV = nRT, \ V = \frac{nRT}{P} = \frac{2.00 \text{ mol} \times \frac{0.08206 \text{ L atm}}{\text{mol K}} \times (155 + 273) \text{ K}}{5.00 \text{ atm}} = 14.0 \text{ L}$

b. $PV = nRT, \ n = \frac{PV}{RT} = \frac{0.300 \text{ atm} \times 2.00 \text{ L}}{\frac{0.08206 \text{ L atm}}{\text{mol K}} \times 155 \text{ K}} = 4.72 \times 10^{-2} \text{ mol}$

c. $PV = nRT,\ T = \frac{PV}{nR} = \frac{4.47 \text{ atm} \times 25.0 \text{ L}}{2.01 \text{ mol} \times \frac{0.08206 \text{ L atm}}{\text{mol K}}} = 678 \text{ K} = 405°\text{C}$

d. $PV = nRT,\ P = \frac{nRT}{V} = \frac{10.5 \text{ mol} \times \frac{0.08206 \text{ L atm}}{\text{mol K}} \times (273 + 75) \text{ K}}{2.25 \text{ L}} = 133 \text{ atm}$

33. $PV = nRT;\ n = \frac{PV}{RT} = \frac{145 \text{ atm} \times 75.0 \times 10^{-3} \text{ L}}{\frac{0.08206 \text{ L atm}}{\text{mol K}} \times 295 \text{ K}} = 0.449 \text{ mol } O_2$

35. a. $PV = nRT;\ 175 \text{ g Ar} \times \frac{1 \text{ mol Ar}}{39.95 \text{ g Ar}} = 4.38 \text{ mol Ar}$

$$T = \frac{PV}{nR} = \frac{10.0 \text{ atm} \times 2.50 \text{ L}}{4.38 \text{ mol} \times \frac{0.08206 \text{ L atm}}{\text{mol K}}} = 69.6 \text{ K}$$

b. $PV = nRT;\ P = \frac{nRT}{V} = \frac{4.38 \text{ mol} \times \frac{0.08206 \text{ L atm}}{\text{mol K}} \times 225 \text{ K}}{2.50 \text{ L}} = 32.3 \text{ atm}$

37. At constant n and T, PV = nRT = constant, $P_1V_1 = P_2V_2$; At sea level, P = 1.00 atm = 760. mm Hg.

$$V_2 = \frac{P_1V_1}{P_2} = \frac{760. \text{ mm Hg} \times 2.0 \text{ L}}{500. \text{ mm Hg}} = 3.0 \text{ L}$$

The balloon will burst at this pressure since the volume must expand beyond the 2.5 L limit of the balloon.

Note: To solve this problem, we did not have to convert the pressure units into atm; the units of mm Hg cancelled each other. In general, only convert units if you have to. Whenever the gas constant R is not used to solve a problem, pressure and volume units must only be consistent, and not necessarily in units of atm and L. The exception is temperature as T must <u>always</u> be converted to the Kelvin scale.

39. PV = nRT, V and n constant, so $\frac{P}{T} = \frac{nR}{V}$ = constant and $\frac{P_1}{T_1} = \frac{P_2}{T_2}$.

$$P_2 = \frac{P_1T_2}{T_1} = 13.7 \text{ MPa} \times \frac{(273 + 450.) \text{ K}}{(273 + 23) \text{ K}} = 33.5 \text{ MPa}$$

41. PV = nRT, n constant; $\frac{PV}{T}$ = nR = constant, $\frac{P_1V_1}{T_1} = \frac{P_2V_2}{T_2}$

$$P_2 = \frac{P_1V_1T_2}{V_2T_1} = 710. \text{ torr} \times \frac{5.0 \times 10^2 \text{ mL}}{25 \text{ mL}} \times \frac{(273 + 820.) \text{ K}}{(273 + 30.) \text{ K}} = 5.1 \times 10^4 \text{ torr}$$

43. PV = nRT, Assume n is constant. $\frac{PV}{T}$ = nR = constant, $\frac{P_1V_1}{T_1} = \frac{P_2V_2}{T_2}$

$$\frac{V_2}{V_1} = \frac{T_2P_1}{T_1P_2} = \frac{(273 + 15)\ K \times 720.\ torr}{(273 + 25)\ K \times 605\ torr} = 1.15;\ V_2 = 1.15 \times 855\ L = 983\ L$$

$V_2 = 1.15\ V_1$ or the volume has increased by 15% or $\Delta V = 983\ L - 855\ L = 128\ L$.

Gas Density, Molar Mass, and Reaction Stoichiometry

45. STP: T = 273 K and P = 1.00 atm; $n = \frac{PV}{RT} = \frac{1.00\ atm \times 1.5\ L}{\frac{0.08206\ L\ atm}{mol\ K} \times 273\ K} = 6.7 \times 10^{-2}$ mol He

Or we can use the fact that at STP, 1 mol of an ideal gas occupies 22.42 L.

$$1.5\ L \times \frac{1\ mol\ He}{22.42\ L} = 6.7 \times 10^{-2}\ mol\ He;\ 6.7 \times 10^{-2}\ mol\ He \times \frac{4.003\ g\ He}{mol\ He} = 0.27\ g\ He$$

47. The balanced equation is: $2\ C_8H_{18}(l) + 25\ O_2(g) \rightarrow 16\ CO_2(g) + 18\ H_2O(g)$

$$125\ g\ C_8H_{18} \times \frac{1\ mol\ C_8H_{18}}{114.22\ g\ C_8H_{18}} \times \frac{25\ mol\ O_2}{2\ mol\ C_8H_{18}} = 13.7\ mol\ O_2$$

$$V = \frac{nRT}{P} = \frac{13.7\ mol \times \frac{0.08206\ L\ atm}{mol\ K} \times 273\ K}{1.00\ atm} = 307\ L\ O_2$$

Or we can make use of the fact that at STP one mole of an ideal gas occupies a volume of 22.42 L. This can be calculated using the ideal gas law.

So: $13.7\ mol\ O_2 \times \frac{22.42\ L}{mol} = 307\ L\ O_2$

49. $$n_{H_2} = \frac{PV}{RT} = \frac{1.0\ atm \times \left[4800\ m^3 \times \left(\frac{100\ cm}{m}\right)^3 \times \frac{1\ L}{1000\ cm^3}\right]}{\frac{0.08206\ L\ atm}{mol\ K} \times 273\ K} = 2.1 \times 10^5\ mol$$

2.1×10^5 mol H_2 are in the balloon. This is 80.% of the total amount of H_2 that had to be generated:

0.80 (total mol H_2) = 2.1×10^5, total mol $H_2 = 2.6 \times 10^5$ mol H_2

$$2.6 \times 10^5\ mol\ H_2 \times \frac{1\ mol\ Fe}{mol\ H_2} \times \frac{55.85\ g\ Fe}{mol\ Fe} = 1.5 \times 10^7\ g\ Fe$$

$$2.6 \times 10^5 \text{ mol } H_2 \times \frac{1 \text{ mol } H_2SO_4}{\text{mol } H_2} \times \frac{98.09 \text{ g } H_2SO_4}{\text{mol } H_2SO_4} \times \frac{100 \text{ g reagent}}{98 \text{ g } H_2SO_4} = 2.6 \times 10^7 \text{ g of 98\% sulfuric acid}$$

51. $Xe(g) + 2\ F_2(g) \rightarrow XeF_4(s)$; $n_{Xe} = \frac{PV}{RT} = \frac{0.500 \text{ atm} \times 20.0 \text{ L}}{\frac{0.08206 \text{ L atm}}{\text{mol K}} \times 673 \text{ K}} = 0.181 \text{ mol Xe}$

We could do the same calculation for F_2. However, the only variable that changed is the pressure. Since the partial pressure of F_2 is triple that of Xe, then mol F_2 = 3(0.181) = 0.543 mol F_2. The balanced equation requires 2 mol of F_2 for every mol of Xe. The actual mol ratio is 3 mol F_2:1 mol Xe. Xe is the limiting reagent.

$$0.181 \text{ mol Xe} \times \frac{1 \text{ mol } XeF_4}{\text{mol Xe}} \times \frac{207.3 \text{ g } XeF_4}{\text{mol Xe}} = 37.5 \text{ g } XeF_4$$

53. a. $CH_4(g) + NH_3(g) + O_2(g) \rightarrow HCN(g) + H_2O(g)$; Balancing H first then O gives:

$CH_4 + NH_3 + \frac{3}{2}O_2 \rightarrow HCN + 3\ H_2O$ or $2\ CH_4(g) + 2\ NH_3(g) + 3\ O_2(g) \rightarrow 2\ HCN(g) + 6\ H_2O(g)$

b. PV = nRT, T and P constant; $\frac{V_1}{n_1} = \frac{V_2}{n_2}$, $\frac{V_1}{V_2} = \frac{n_1}{n_2}$

Since the volumes are all measured at constant T and P, then the volumes of gas present are directly proportional to the mol of gas present (Avogadro's law). Because Avogadro's law applies, the balanced reaction gives mol relationships as well as volume relationships. Therefore, 2 L of CH_4, 2 L of NH_3 and 3 L of O_2 are required by the balanced equation for the production of 2 L of HCN. The actual volume ratio is 20 L CH_4:20 L NH_3:20 L O_2 (or 1:1:1). The volume of O_2 required to react with all of the CH_4 and NH_3 present is 20 L ×(3/2) = 30 L. Since only 20.0 L of O_2 are present, then O_2 is the limiting reagent. The volume of HCN produced is:

$$20.0 \text{ L } O_2 \times \frac{2 \text{ L HCN}}{3 \text{ L } O_2} = 13.3 \text{ L HCN}$$

55. One of these equations developed in the text to determine molar mass is:

molar mass = $\frac{dRT}{P}$ where d = density in units of g/L

$$\text{molar mass} = \frac{1.65 \text{ g/L} \times \frac{0.08206 \text{ L atm}}{\text{mol K}} \times (273 + 27) \text{ K}}{734 \text{ torr} \times \frac{1 \text{ atm}}{760 \text{ torr}}} = 42.1 \text{ g/mol}$$

The empirical formula mass of CH_2 = 12.01 + 2(1.008) = 14.03 g/mol.

$\frac{42.1}{14.03} = 3.00$; Molecular formula = C_3H_6

57. $P \times (\text{molar mass}) = dRT$, d = density $= \dfrac{P \times (\text{molar mass})}{RT}$

For $SiCl_4$, molar mass = M = 28.09 + 4(35.45) = 169.89 g/mol

$$d = \frac{(758 \text{ torr} \times \frac{1 \text{ atm}}{760 \text{ torr}}) \times \frac{169.89 \text{ g}}{\text{mol}}}{\frac{0.08206 \text{ L atm}}{\text{mol K}} \times 358 \text{ K}} = 5.77 \text{ g/L for } SiCl_4$$

For $SiHCl_3$, molar mass = M = 28.09 + 1.008 + 3(35.45) = 135.45 g/mol

$$d = \frac{PM}{RT} = \frac{(758 \text{ torr} \times \frac{1 \text{ atm}}{760 \text{ torr}}) \times \frac{135.45 \text{ g}}{\text{mol}}}{\frac{0.08206 \text{ L atm}}{\text{mol K}} \times 358 \text{ K}} = 4.60 \text{ g/L for } SiHCl_3$$

Partial Pressure

59. $$P_{CO_2} = \frac{nRT}{V} = \frac{\left(7.8 \text{ g} \times \frac{1 \text{ mol}}{44.01 \text{ g}}\right) \times \frac{0.08206 \text{ L atm}}{\text{mol K}} \times 300. \text{ K}}{4.0 \text{ L}} = 1.1 \text{ atm}$$

With air present, the partial pressure of CO_2 will still be 1.1 atm. The total pressure will be the sum of the partial pressures, $P_{total} = P_{CO_2} + P_{air}$.

$$P_{total} = 1.1 \text{ atm} + \left(740 \text{ torr} \times \frac{1 \text{ atm}}{760 \text{ torr}}\right) = 1.1 + 0.97 = 2.1 \text{ atm}$$

61. Use the relationship $P_1V_1 = P_2V_2$ for each gas, since T and n for each gas is constant.

For H_2: $P_2 = \dfrac{P_1V_1}{V_2} = 475 \text{ torr} \times \dfrac{2.00 \text{ L}}{3.00 \text{ L}} = 317 \text{ torr}$

For N_2: $P_2 = 0.200 \text{ atm} \times \dfrac{1.00 \text{ L}}{3.00 \text{ L}} = 0.0667 \text{ atm}$; $0.0667 \text{ atm} \times \dfrac{760 \text{ torr}}{\text{atm}} = 50.7 \text{ torr}$

$P_{total} = P_{H_2} + P_{N_2} = 317 + 50.7 = 368 \text{ torr}$

63. a. mol fraction $CH_4 = \chi_{CH_4} = \dfrac{P_{CH_4}}{P_{total}} = \dfrac{0.175 \text{ atm}}{0.175 \text{ atm} + 0.250 \text{ atm}} = 0.412$; $\chi_{O_2} = 1.000 - 0.412 = 0.588$

b. $PV = nRT$, $n_{total} = \dfrac{P_{total} \times V}{RT} = \dfrac{0.425 \text{ atm} \times 10.5 \text{ L}}{\frac{0.08206 \text{ L atm}}{\text{mol K}} \times 338 \text{ K}} = 0.161 \text{ mol}$

c. $\chi_{CH_4} = \frac{n_{CH_4}}{n_{total}}$, $n_{CH_4} = \chi_{CH_4} \times n_{total} = 0.412 \times 0.161 \text{ mol} = 6.63 \times 10^{-2} \text{ mol } CH_4$

$$6.63 \times 10^{-2} \text{ mol } CH_4 \times \frac{16.04 \text{ g } CH_4}{\text{mol } CH_4} = 1.06 \text{ g } CH_4$$

$$n_{O_2} = 0.588 \times 0.161 \text{ mol} = 9.47 \times 10^{-2} \text{ mol } O_2;\ 9.47 \times 10^{-2} \text{ mol } O_2 \times \frac{32.00 \text{ g } O_2}{\text{mol } O_2} = 3.03 \text{ g } O_2$$

65. $P_{total} = 1.00 \text{ atm} = 760. \text{ torr} = P_{N_2} + P_{H_2O} = P_{N_2} + 17.5 \text{ torr},\ P_{N_2} = 743 \text{ torr}$

$$PV = nRT;\ n_{N_2} = \frac{P_{N_2} \times V}{RT} = \frac{(743 \text{ torr} \times \frac{1 \text{ atm}}{760 \text{ torr}}) \times (2.50 \times 10^2 \text{ mL} \times \frac{1 \text{ L}}{1000 \text{ mL}})}{\frac{0.08206 \text{ L atm}}{\text{mol K}} \times 293 \text{ K}} = 1.02 \times 10^{-2} \text{ mol } N_2$$

$$1.02 \times 10^{-2} \text{ mol } N_2 \times \frac{28.02 \text{ g } N_2}{\text{mol } N_2} = 0.286 \text{ g } N_2$$

67. $3.70 \text{ g } KClO_3 \times \frac{1 \text{ mol } KClO_3}{122.55 \text{ g } KClO_3} \times \frac{3 \text{ mol } O_2}{2 \text{ mol } KClO_3} = 4.53 \times 10^{-2} \text{ mol } O_2$

$$P_{total} = P_{O_2} + P_{H_2O},\ P_{O_2} = P_{total} - P_{H_2O} = 735 - 26.7 = 708 \text{ torr} \times \frac{1 \text{ atm}}{760 \text{ torr}} = 0.932 \text{ atm}$$

$$V = \frac{n_{O_2} \times RT}{P_{O_2}} = \frac{4.53 \times 10^{-2} \text{ mol} \times \frac{0.08206 \text{ L atm}}{\text{mol K}} \times 300. \text{ K}}{0.932 \text{ atm}} = 1.20 \text{ L}$$

Kinetic Molecular Theory and Real Gases

69. $(KE)_{avg} = (3/2)\,RT$; At 273 K: $(KE)_{avg} = \frac{3}{2} \times \frac{8.3145 \text{ J}}{\text{mol K}} \times 273 \text{ K} = 3.40 \times 10^3 \text{ J/mol}$

At 546 K: $(KE)_{avg} = \frac{3}{2} \times \frac{8.3145 \text{ J}}{\text{mol K}} \times 546 \text{ K} = 6.81 \times 10^3 \text{ J/mol}$

71. $u_{rms} = \left(\frac{3RT}{M}\right)^{1/2}$, where $R = \frac{8.3145 \text{ J}}{\text{mol K}}$ and M = molar mass in kg = 1.604×10^{-2} kg/mol for CH_4

For CH_4 at 273 K: $u_{rms} = \left(\frac{\frac{3 \times 8.3145 \text{ J}}{\text{mol K}} \times 273 \text{ K}}{1.604 \times 10^{-2} \text{ kg/mol}}\right)^{1/2} = 652 \text{ m/s}$

Similarly u_{rms} for CH_4 at 546 K is 921 m/s.

73. No, the number calculated in 5.69 is the average kinetic energy. There is a distribution of energies.

75. $KE_{ave} = (3/2)\ RT$ and $KE = (1/2)\ mv^2$; As the temperature increases, the average kinetic energy of the gas sample will increase. The average kinetic energy increases because the increased temperature results in an increase in the average velocity of the gas molecules.

77. a. They will all have the same average kinetic energy since they are all at the same temperature.

b. Flask C; H_2 has the smallest molar mass. At constant T, the lightest molecules are the fastest (on the average). This must be true in order for the average kinetic energies to be constant.

79. Graham's law of effusion:

$$\frac{\text{Rate}_1}{\text{Rate}_2} = \left(\frac{M_2}{M_1}\right)^{1/2} \text{ where M = molar mass;} \quad \frac{31.50}{30.50} = \left(\frac{32.00}{M}\right)^{1/2} = 1.033$$

$\frac{32.00}{M} = 1.067$, so M = 29.99 g/mol; Of the choices, the gas would be NO, nitrogen monoxide.

81. $$\frac{\text{Rate}_1}{\text{Rate}_2} = \left(\frac{M_2}{M_1}\right)^{1/2}, \quad \frac{{}^{12}C\,{}^{17}O}{{}^{12}C\,{}^{18}O} = \left(\frac{30.0}{29.0}\right)^{1/2} = 1.02; \quad \frac{{}^{12}C\,{}^{16}O}{{}^{12}C\,{}^{18}O} = \left(\frac{30.0}{28.0}\right)^{1/2} = 1.04$$

The relative rates of effusion of ${}^{12}C^{16}O$: ${}^{12}C^{17}O$: ${}^{12}C^{18}O$ are 1.04: 1.02: 1.00.

Advantage: CO_2 isn't as toxic as CO.

Major disadvantages of using CO_2 instead of CO:

1. Can get a mixture of oxygen isotopes in CO_2.

2. Some species, e.g., ${}^{12}C^{16}O^{18}O$ and ${}^{12}C^{17}O_2$, would effuse at about the same rate since the masses are about equal. Thus, some species cannot be separated from each other.

83. a. $$P = \frac{nRT}{V} = \frac{0.5000\text{ mol} \times \frac{0.08206\text{ L atm}}{\text{mol K}} \times (25.0 + 273.2)\text{ K}}{1.0000\text{ L}} = 12.24\text{ atm}$$

b. $$\left[P + a\left(\frac{n}{V}\right)^2\right] \times (V - nb) = nRT;$$ For N_2: $a = 1.39$ atm L^2/mol^2 and $b = 0.0391$ L/mol

$$\left[P + 1.39\left(\frac{0.5000}{1.0000}\right)^2 \text{atm}\right] \times (1.0000\text{ L} - 0.5000 \times 0.0391\text{ L}) = 12.24\text{ L atm}$$

$$(P + 0.348\text{ atm}) \times (0.9805\text{ L}) = 12.24\text{ L atm}$$

$$P = \frac{12.24\text{ L atm}}{0.9805\text{ L}} - 0.348\text{ atm} = 12.48 - 0.348 = 12.13\text{ atm}$$

c. The ideal gas law is high by 0.11 atm or $\frac{0.11}{12.13} \times 100 = 0.91\%$.

Atmospheric Chemistry

85. $\chi_{NO} = 5 \times 10^{-7}$ from Table 5.4. $P_{NO} = \chi_{NO} \times P_{total} = 5 \times 10^{-7} \times 1.0\text{ atm} = 5 \times 10^{-7}\text{ atm}$

$$PV = nRT,\ \frac{n}{V} = \frac{P}{RT} = \frac{5 \times 10^{-7}\text{ atm}}{\frac{0.08206\text{ L atm}}{\text{mol K}} \times 273\text{ K}} = 2 \times 10^{-8}\text{ mol NO/L}$$

$$\frac{2 \times 10^{-8}\text{ mol}}{\text{L}} \times \frac{1\text{ L}}{1000\text{ cm}^3} \times \frac{6.022 \times 10^{23}\text{ molecules}}{\text{mol}} = 1 \times 10^{13}\text{ molecules NO/cm}^3$$

87. At 100. km, $T \approx -75°C$ and $P \approx 10^{-4.5} \approx 3 \times 10^{-5}$ atm.

$$PV = nRT,\ \frac{PV}{T} = nR = \text{Constant},\ \frac{P_1V_1}{T_1} = \frac{P_2V_2}{T_2}$$

$$V_2 = \frac{V_1P_1T_2}{P_2T_1} = \frac{10.0\text{ L} \times (3 \times 10^{-5}\text{ atm}) \times 273\text{ K}}{1.0\text{ atm} \times 198\text{ K}} = 4 \times 10^{-4}\text{ L} = 0.4\text{ mL}$$

89. $N_2(g) + O_2(g) \rightarrow 2\ NO(g)$, automobile combustion or formed by lightning

$2\ NO(g) + O_2(g) \rightarrow 2\ NO_2(g)$, reaction with atmospheric O_2

$2\ NO_2(g) + H_2O(l) \rightarrow HNO_3(aq) + HNO_2(aq)$, reaction with atmospheric H_2O

$S(s) + O_2(g) \rightarrow SO_2(g)$, combustion of coal

$2\ SO_2(g) + O_2(g) \rightarrow 2SO_3(g)$, reaction with atmospheric O_2

$H_2O(l) + SO_3(g) \rightarrow H_2SO_4(aq)$, reaction with atmospheric H_2O

Additional Exercises

91. a. $PV = nRT$

$PV = \text{Constant}$

b. $PV = nRT$

$P = \left(\frac{nR}{V}\right) \times T = \text{Const} \times T$

c. $PV = nRT$

$T = \left(\frac{P}{nR}\right) \times V = \text{Const} \times V$

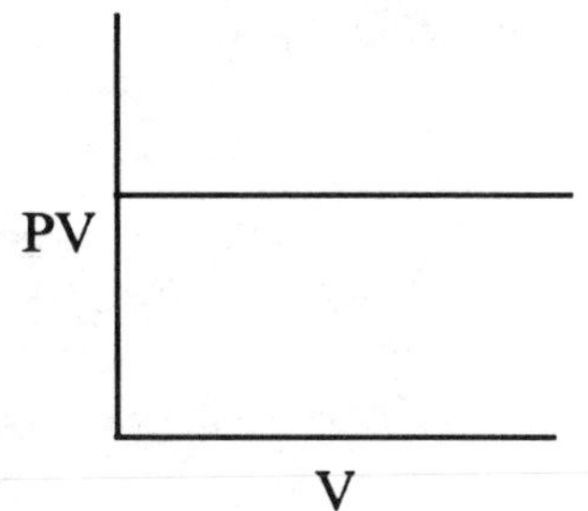

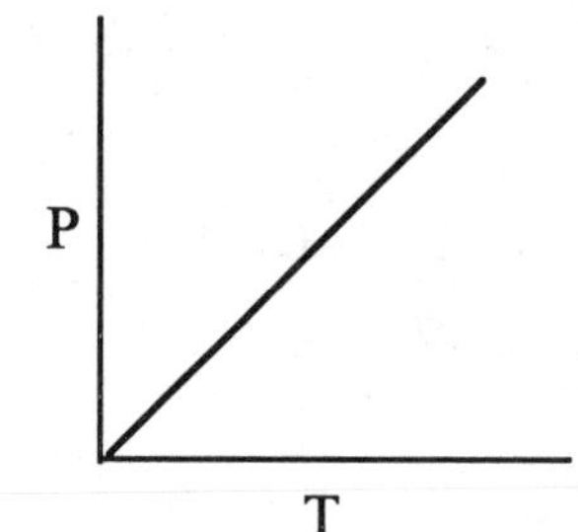

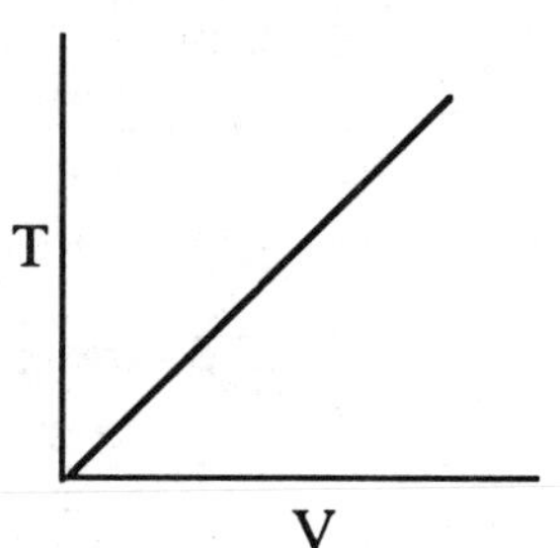

d. $PV = nRT$

$PV = \text{Constant}$

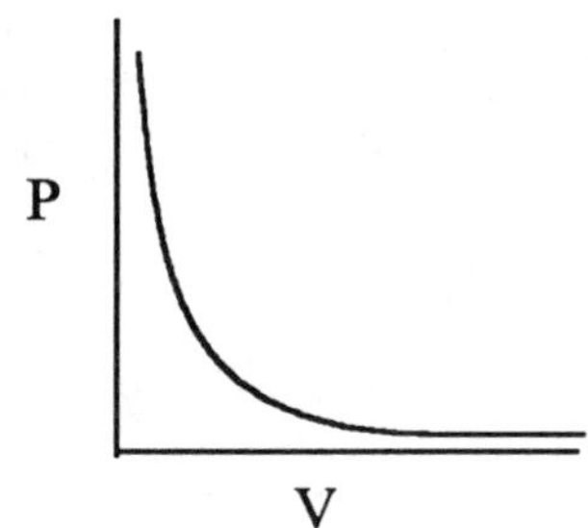

e. $P = \frac{nRT}{V} = \frac{\text{Constant}}{V}$

$P = \text{Constant} \times \frac{1}{V}$

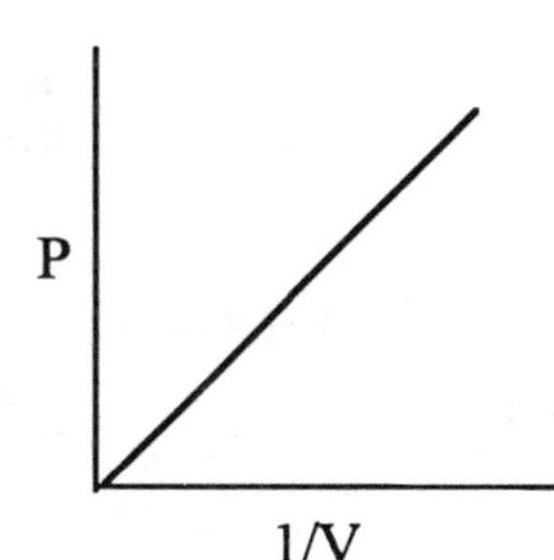

f. $PV = nRT$

$\frac{PV}{T} = nR = \text{Constant}$

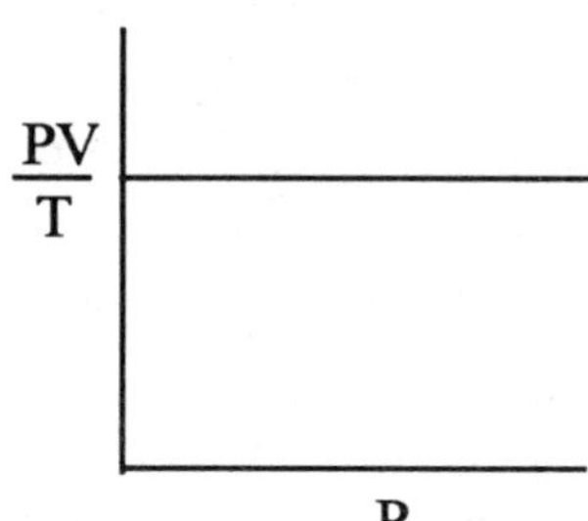

93. Processes a, c and e will all result in a doubling of the pressure. Process a has the effect of halving the volume, which would double the pressure (Boyle's law). Process c doubles the pressure because the absolute temperature is doubled (from 200. K to 400. K). Process e doubles the pressure because the moles of gas are doubled (28 g N_2 is 1 mol of N_2 and 32 g O_2 is 1 mol of O_2). Process b won't double the pressure since the absolute temperature is not doubled (303 K to 333 K). Process d won't double the pressure because 28 g O_2 is less than one mol of gas.

95. We will apply Boyle's law to solve. PV = nRT = constant, $P_1V_1 = P_2V_2$

Let condition (1) correspond to He from the tank that can be used to fill balloons. We must leave 1.0 atm of He in the tank, so P_1 = 200. atm - 1.00 = 199 atm and V_1 = 15.0 L. Condition (2) will correspond to the filled balloons with P_2= 1.00 atm and V_2 = N(2.00 L) where N is the number of filled balloons, each at a volume of 2.00 L.

199 atm × 15.0 L = 1.00 atm × N(2.00 L), N = 1492.5; We can't fill 0.5 of a balloon. So N = 1492 balloons or to 3 significant figures, 1490 balloons.

97. $P_1V_1 = P_2V_2$; The total volume is 1.00 L + 1.00 L + 2.00 L = 4.00 L.

For He: $P_2 = \frac{P_1V_1}{V_2} = 180.\ \text{torr} \times \frac{1.00\ \text{L}}{4.00\ \text{L}} = 45.0\ \text{torr He}$

For Ne: $P_2 = 0.450\ \text{atm} \times \frac{1.00\ \text{L}}{4.00\ \text{L}} = 0.113\ \text{atm}$; $0.113\ \text{atm} \times \frac{760\ \text{torr}}{\text{atm}} = 85.9\ \text{torr Ne}$

For Ar: $P_2 = 25.0\ \text{kPa} \times \frac{2.0\ \text{L}}{4.0\ \text{L}} = 12.5\ \text{kPa}$; $12.5\ \text{kPa} \times \frac{1\ \text{atm}}{101.3\ \text{kPa}} \times \frac{760\ \text{torr}}{\text{atm}} = 93.8\ \text{torr Ar}$

$P_{total} = P_{He} + P_{Ne} + P_{Ar} = 45.0 + 85.9 + 93.8 = 224.7\ \text{torr}$

99. For O_2, n and T are constant, so $P_1V_1 = P_2V_2$.

$P_1 = \frac{P_2V_2}{V_1} = 785\ \text{torr} \times \frac{1.94\ \text{L}}{2.00\ \text{L}} = 761\ \text{torr} = P_{O_2}$

$P_{total} = P_{O_2} + P_{H_2O}$, $P_{H_2O} = 785 - 761 = 24$ torr

101. $750.\ \text{mL juice} \times \dfrac{12\ \text{mL}\ C_2H_5OH}{100\ \text{mL juice}} = 90.\ \text{mL}\ C_2H_5OH$ present

$$90.\ \text{mL}\ C_2H_5OH \times \frac{0.79\ \text{g}\ C_2H_5OH}{\text{mL}\ C_2H_5OH} \times \frac{1\ \text{mol}\ C_2H_5OH}{46.07\ \text{g}\ C_2H_5OH} \times \frac{2\ \text{mol}\ CO_2}{2\ \text{mol}\ C_2H_5OH} = 1.5\ \text{mol}\ CO_2$$

The CO_2 will occupy (825 - 750. =) 75 mL not occupied by the liquid (headspace).

$$P_{CO_2} = \frac{n_{CO_2} \times RT}{V} = \frac{1.5\ \text{mol} \times \frac{0.08206\ \text{L atm}}{\text{mol K}} \times 298\ \text{K}}{75 \times 10^{-3}\ \text{L}} = 490\ \text{atm}$$

Actually, enough CO_2 will dissolve in the wine to lower the pressure of CO_2 to a much more reasonable value.

103. $$d_{UF_6} = \frac{P \times (\text{molar mass})}{RT} = \frac{\left(745\ \text{torr} \times \frac{1\ \text{atm}}{760\ \text{torr}}\right) \times 352.0\ \text{g/mol}}{\frac{0.08206\ \text{L atm}}{\text{mol K}} \times 333\ \text{K}} = 12.6\ \text{g/L}$$

105. Out of 100.0 g of compound, there are:

$$87.4\ \text{g N} \times \frac{1\ \text{mol N}}{14.01\ \text{g N}} = 6.24\ \text{mol N};\ \frac{6.24}{6.24} = 1.00$$

$$12.6\ \text{g H} \times \frac{1\ \text{mol H}}{1.008\ \text{g H}} = 12.5\ \text{mol H};\ \frac{12.5}{6.24} = 2.00$$

Empirical formula is NH_2.

$$\text{molar mass} = M = \frac{dRT}{P} = \frac{\frac{0.977\ \text{g}}{\text{L}} \times \frac{0.08206\ \text{L atm}}{\text{mol K}} \times 373\ \text{K}}{710.\ \text{torr} \times \frac{1\ \text{atm}}{760\ \text{torr}}} = 32.0\ \text{g/mol}$$

Empirical formula mass of NH_2 = 16.0 g. Therefore, molecular formula is N_2H_4.

107. $P_{total} = P_{N_2} + P_{H_2O}$, $P_{N_2} = 726\ \text{torr} - 23.8\ \text{torr} = 702\ \text{torr} \times \dfrac{1\ \text{atm}}{760\ \text{torr}} = 0.924\ \text{atm}$

$$PV = nRT,\ n_{N_2} = \frac{P_{N_2} \times V}{RT} = \frac{0.924\ \text{atm} \times 31.8 \times 10^{-3}\ \text{L}}{\frac{0.08206\ \text{L atm}}{\text{mol K}} \times 298\ \text{K}} = 1.20 \times 10^{-3}\ \text{mol}\ N_2$$

$$\text{Mass of N in compound} = 1.20 \times 10^{-3}\ \text{mol} \times \frac{28.02\ \text{g}\ N_2}{\text{mol}} = 3.36 \times 10^{-2}\ \text{g}$$

$$\% \text{N} = \frac{3.36 \times 10^{-2} \text{ g}}{0.253 \text{ g}} \times 100 = 13.3\% \text{ N}$$

109. The pressure will increase because the lighter H_2 molecules will effuse into container A faster than air will escape.

111. The values of a are: H_2, $\frac{0.244 \text{ L}^2 \text{ atm}}{\text{mol}^2}$; CO_2, 3.59; N_2, 1.39; CH_4, 2.25

Since a is a measure interparticle attractions, the attractions are greatest for CO_2.

Challenge Problems

113. $BaO(s) + CO_2(g) \rightarrow BaCO_3(s)$; $CaO(s) + CO_2(g) \rightarrow CaCO_3(s)$

$$n_i = \frac{P_iV}{RT} = \text{initial moles of } CO_2 = \frac{\frac{750.}{760} \text{ atm} \times 1.50 \text{ L}}{\frac{0.08206 \text{ L atm}}{\text{mol K}} \times 303.2 \text{ K}} = 0.0595 \text{ mol } CO_2$$

$$n_f = \frac{P_fV}{RT} = \text{final moles of } CO_2 = \frac{\frac{230.}{760} \text{ atm} \times 1.50 \text{ L}}{\frac{0.08206 \text{ L atm}}{\text{mol K}} \times 303.2 \text{ K}} = 0.0182 \text{ mol } CO_2$$

$0.0595 - 0.0182 = 0.0413$ mol CO_2 reacted.

Since each metal reacts 1:1 with CO_2, then the mixture contains 0.0413 mol of BaO and CaO. The molar masses of BaO and CaO are 153.3 g/mol and 56.08 g/mol, respectively.

Let x = g BaO and y = g CaO, so:

$$x + y = 5.14 \text{ g and } \frac{x}{153.3} + \frac{y}{56.08} = 0.0413 \text{ mol}$$

Solving by simultaneous equations:

$$\begin{array}{rcl} x + 2.734\,y &=& 6.33 \\ -x \qquad\;\; -y &=& -5.14 \\ \hline 1.734\,y &=& 1.19 \end{array}$$

$y = 0.686$ g CaO and $5.14 - y = x = 4.45$ g BaO

$$\% \text{ BaO} = \frac{4.45 \text{ g BaO}}{5.14 \text{ g}} \times 100 = 86.6\% \text{ BaO}; \ \% \text{ CaO} = 100.0 - 86.6 = 13.4\% \text{ CaO}$$

115. a. The reaction is: $CH_4(g) + 2\ O_2(g) \rightarrow CO_2(g) + 2\ H_2O(g)$

$$PV = nRT;\ \frac{PV}{n} = RT = \text{constant};\ \frac{P_{CH_4}V_{CH_4}}{n_{CH_4}} = \frac{P_{O_2}V_{O_2}}{n_{O_2}}$$

For three fold excess of O_2: $n_{O_2} = 6\ n_{CH_4}$, $\frac{n_{O_2}}{n_{CH_4}} = 6$; $P_{O_2} = 0.21\ P_{air} = 0.21$ atm

In one minute:

$$V_{O_2} = V_{CH_4} \times \frac{n_{O_2}}{n_{CH_4}} \times \frac{P_{CH_4}}{P_{O_2}} = 200.\ \text{L} \times 6 \times \frac{1.50\ \text{atm}}{0.21\ \text{atm}} = 8.6 \times 10^3\ \text{L}\ O_2$$

$$\text{We need: } \frac{8.6 \times 10^3\ \text{L}\ O_2}{\text{min}} \times \frac{100\ \text{L air}}{21\ \text{L}\ O_2} = 4.1 \times 10^4\ \text{L air/min}$$

b. If n moles of CH_4 were reacted, then 6 n mol O_2 were added, producing 0.950 n mol CO_2 and 0.050 n mol of CO. In addition, 2 n mol H_2O must be produced to balance the hydrogens.

$CH_4 + 2\ O_2 \rightarrow CO_2 + 2\ H_2O$; $CH_4 + 3/2\ O_2 \rightarrow CO + 2\ H_2O$

Amount O_2 reacted:

$$0.950\ \text{n mol}\ CO_2 \times \frac{2\ \text{mol}\ O_2}{\text{mol}\ CO_2} = 1.90\ \text{n mol}\ O_2$$

$$0.050\ \text{n mol CO} \times \frac{1.5\ \text{mol}\ O_2}{\text{mol CO}} = 0.075\ \text{n mol}\ O_2$$

Amount of O_2 left in reaction mixture = 6.00 n - 1.90 n - 0.075 n = 4.03 n mol O_2

$$\text{Amount of } N_2 \text{ remaining} = 6.00\ \text{n mol}\ O_2 \times \frac{79\ \text{mol}\ N_2}{21\ \text{mol}\ O_2} = 22.6\ \text{n} \approx 23\ \text{n mol}\ N_2$$

The reaction mixture contains:

0.950 n mol CO_2 + 0.050 n mol CO + 4.03 n mol O_2 + 2.00 n mol H_2O

+ 23 n mol N_2 = 30. n total mol of gas

$$\chi_{CO} = \frac{0.050\ \text{n}}{30.\ \text{n}} = 0.0017;\ \chi_{CO_2} = \frac{0.950\ \text{n}}{30.\ \text{n}} = 0.032;\ \chi_{O_2} = \frac{4.03\ \text{n}}{30.\ \text{n}} = 0.13;$$

$$\chi_{H_2O} = \frac{2.00\ \text{n}}{30.\ \text{n}} = 0.067;\quad \chi_{N_2} = \frac{23\ \text{n}}{30.\ \text{n}} = 0.77$$

117. a. Volume of hot air: $V = \frac{4}{3}\pi^3 = \frac{4}{3}(2.50\ \text{m})\pi^3 = 65.4\ \text{m}^3$

(Note: radius = diameter/2 = 5.00/2 = 2.50 m)

$$65.4\ \text{m}^3 \times \left(\frac{10\ \text{dm}}{\text{m}}\right)^3 \times \frac{1\ \text{L}}{\text{dm}^3} = 6.54 \times 10^4\ \text{L}$$

$$n = \frac{PV}{RT} = \frac{\left(745\ \text{torr} \times \frac{1\ \text{atm}}{760\ \text{torr}}\right) \times 6.54 \times 10^4\ \text{L}}{\frac{0.08206\ \text{L atm}}{\text{mol K}} \times (273 + 65)\ \text{K}} = 2.31 \times 10^3\ \text{mol air}$$

$$\text{Mass of hot air} = 2.31 \times 10^3\ \text{mol} \times \frac{29.0\ \text{g}}{\text{mol}} = 6.70 \times 10^4\ \text{g}$$

Mass of air displaced:

$$n = \frac{PV}{RT} = \frac{\frac{745}{760}\ \text{atm} \times 6.54 \times 10^4\ \text{L}}{\frac{0.08206\ \text{L atm}}{\text{mol K}} \times (273 + 21)\ \text{K}} = 2.66 \times 10^3\ \text{mol air}$$

$$\text{Mass} = 2.66 \times 10^3\ \text{mol} \times \frac{29.0\ \text{g}}{\text{mol}} = 7.71 \times 10^4\ \text{g of air displaced}$$

Lift = 7.71×10^4 g - 6.70×10^4 g = 1.01×10^4 g

b. Mass of air displaced is the same, 7.71×10^4 g. Moles of He in balloon will be the same as moles of air displaced, 2.66×10^3 mol, since P, V and T are the same.

$$\text{Mass of He} = 2.66 \times 10^3\ \text{mol} \times \frac{4.003\ \text{g}}{\text{mol}} = 1.06 \times 10^4\ \text{g}$$

Lift = 7.71×10^4g - 1.06×10^4 g = 6.65×10^4 g

c. Mass of hot air:

$$n = \frac{PV}{RT} = \frac{\frac{630.}{760}\ \text{atm} \times 6.54 \times 10^4\ \text{L}}{\frac{0.08206\ \text{L atm}}{\text{mol K}} \times 338\ \text{K}} = 1.95 \times 10^3\ \text{mol air}$$

$$1.95 \times 10^3\ \text{mol} \times \frac{29.0\ \text{g}}{\text{mol}} = 5.66 \times 10^4\ \text{g of hot air}$$

Mass of air displaced:

$$n = \frac{PV}{RT} = \frac{\frac{630.}{760}\ \text{atm} \times 6.54 \times 10^4\ \text{L}}{\frac{0.08206\ \text{L atm}}{\text{mol K}} \times 294\ \text{K}} = 2.25 \times 10^3\ \text{mol air}$$

$$2.25 \times 10^3\ \text{mol} \times \frac{29.0\ \text{g}}{\text{mol}} = 6.53 \times 10^4\ \text{g of air displaced}$$

Lift = 6.53×10^4 g - 5.66×10^4 g = 8.7×10^3 g

119. a. If we have 1.0×10^6 L of air, then there are 3.0×10^2 L of CO.

$$P_{CO} = \chi_{CO} \times P_{total}; \ \chi_{CO} = \frac{V_{CO}}{V_{total}} \text{ since } V \propto n; \ P_{CO} = \frac{3.0 \times 10^2 \text{ L}}{1.0 \times 10^6 \text{ L}} \times 628 \text{ torr} = 0.19 \text{ torr}$$

b. $n_{CO} = \dfrac{P_{CO} \times V}{RT}$; Assuming 1.0 cm^3 of air = 1.0 mL = 1.0×10^{-3} L:

$$n_{CO} = \frac{\frac{0.19}{760} \text{ atm} \times 1.0 \times 10^{-3} \text{ L}}{\frac{0.08206 \text{ L atm}}{\text{mol K}} \times 273 \text{ K}} = 1.1 \times 10^{-8} \text{ mol CO}$$

$$1.1 \times 10^{-8} \text{ mol} \times \frac{6.022 \times 10^{23} \text{ molecules}}{\text{mol}} = 6.6 \times 10^{15} \text{ molecules CO in the 1.0 cm}^3 \text{ of air}$$

CHAPTER SIX

THERMOCHEMISTRY

Questions

9. A coffee-cup calorimeter is at constant (atmospheric) pressure. The heat released or gained at constant pressure is ΔH. A bomb calorimeter is at constant volume. The heat released or gained at constant volume is ΔE.

11. The specific heat capacities are: 0.89 J/g•°C (Al) and 0.45 J/g•°C (Fe)
Al would be the better choice. It has a higher heat capacity and a lower density than Fe. Using Al, the same amount of heat could be dissipated by a smaller mass, keeping the mass of the amplifier down.

13. A state function is a function whose change depends only on the initial and final states and not on how one got from the initial to the final state. If H and E were not state functions, the law of conservation of energy (first law) would not be true.

15. In order to compare values of ΔH to each other, a common reference (or zero) point must be chosen. The definition of ΔH_f° establishes the pure elements in their standard states as that common reference point.

Exercises

Potential and Kinetic Energy

17. $KE = \frac{1}{2}mv^2$; Convert mass and velocity to SI units. $1\ J = \frac{1\ kg\ m^2}{s^2}$

$$\text{Mass} = 5.25\ oz \times \frac{1\ lb}{16\ oz} \times \frac{1\ kg}{2.205\ lb} = 0.149\ kg$$

$$\text{Velocity} = \frac{1.0 \times 10^2\ mi}{hr} \times \frac{1\ hr}{60\ min} \times \frac{1\ min}{60\ s} \times \frac{1760\ yd}{mi} \times \frac{1\ m}{1.094\ yd} = \frac{45\ m}{s}$$

$$KE = \frac{1}{2}mv^2 = \frac{1}{2} \times 0.149\ kg \times \left(\frac{45\ m}{s}\right)^2 = 150\ J$$

19. $KE = \frac{1}{2}mv^2 = \frac{1}{2} \times 2.0\text{ kg} \times \left(\frac{1.0\text{ m}}{\text{s}}\right)^2 = 1.0\text{ J}$; $KE = \frac{1}{2}mv^2 = \frac{1}{2} \times 1.0\text{ kg} \times \left(\frac{2.0\text{ m}}{\text{s}}\right)^2 = 2.0\text{ J}$

The 1.0 kg object with a velocity of 2.0 m/s has the greater kinetic energy.

Heat and Work

21. a. $\Delta E = q + w = 51\text{ kJ} + (-15\text{ kJ}) = 36\text{ kJ}$

b. $\Delta E = 100.\text{ kJ} + (-65\text{ kJ}) = 35\text{ kJ}$ c. $\Delta E = -65 + (-20.) = -85\text{ kJ}$

d. When the system delivers work to the surroundings, $w < 0$. This is the case in all these examples, a, b and c.

23. $\Delta E = q + w = 45\text{ kJ} + (-29\text{ kJ}) = 16\text{ kJ}$

25. $w = -P\Delta V = -P \times (V_f - V_i) = -2.0\text{ atm} \times (5.0 \times 10^{-3}\text{ L} - 5.0\text{ L}) = -2.0\text{ atm} \times (-5.0\text{ L}) = 10.\text{ L atm}$

We can also calculate the work in Joules.

$1\text{ atm} = 1.013 \times 10^5\text{ Pa} = 1.013 \times 10^5 \frac{\text{kg}}{\text{m s}^2}$; $1\text{ L} = 1000\text{ cm}^3 = 1 \times 10^{-3}\text{ m}^3$

$1\text{ L atm} = 1 \times 10^{-3}\text{ m}^3 \times 1.013 \times 10^5 \frac{\text{kg}}{\text{m s}^2} = 101.3 \frac{\text{kg m}^2}{\text{s}^2} = 101.3\text{ J}$

$w = 10.\text{ L atm} \times \frac{101.3\text{ J}}{\text{L atm}} = 1013\text{ J} = 1.0 \times 10^3\text{ J}$

27. $q = \text{molar heat capacity} \times \text{mol} \times \Delta T = \frac{20.8\text{ J}}{°\text{C mol}} \times 39.1\text{ mol} \times (38.0 - 0.0)\text{ °C} = 30{,}900\text{ J} = 30.9\text{ kJ}$

$w = -P\Delta V = -1.00\text{ atm} \times (998\text{ L} - 876\text{ L}) = -122\text{ L atm} \times \frac{101.3\text{ J}}{\text{L atm}} = -12{,}400\text{ J} = -12.4\text{ kJ}$

$\Delta E = q + w = 30.9\text{ kJ} + (-12.4\text{ kJ}) = 18.5\text{ kJ}$

Properties of Enthalpy

29. Since the sign of ΔH is negative, the reaction is exothermic. Heat is evolved by the system to the surroundings.

31. Heat is absorbed in endothermic processes and heat is released in exothermic processes.

a. endothermic b. exothermic c. exothermic d. endothermic

33. $S(s) + O_2(g) \rightarrow SO_2(g) \quad \Delta H = \frac{-296 \text{ kJ}}{\text{mol}}$; Molar mass of SO_2 = 64.07 g/mol

a. $275 \text{ g S} \times \frac{1 \text{ mol S}}{32.07 \text{ g S}} \times \frac{-296 \text{ kJ}}{\text{mol S}} = -2.54 \times 10^3$ kJ heat released

b. $25 \text{ mol S} \times \frac{-296 \text{ kJ}}{\text{mol S}} = -7.4 \times 10^3 \text{ kJ}$; c. $150. \text{ g } SO_2 \times \frac{1 \text{ mol } SO_2}{64.07 \text{ g } SO_2} \times \frac{-296 \text{ kJ}}{\text{mol } SO_2} = -693 \text{ kJ}$

35. From Sample Exercise 6.3, $q = 1.3 \times 10^8$ J. Molar mass of C_3H_8 = 44.09 g/mol

$$\text{mass } C_3H_8 = 1.3 \times 10^8 \text{ J} \times \frac{1 \text{ mol } C_3H_8}{2221 \times 10^3 \text{ J}} \times \frac{44.09 \text{ g } C_3H_8}{\text{mol } C_3H_8} = 2.6 \times 10^3 \text{ g } C_3H_8$$

Calorimetry and Heat Capacity

37. a. $\text{energy} = s \times m \times \Delta T = \frac{0.900 \text{ J}}{\text{g °C}} \times 850. \text{ g} \times (94.6 - 22.8)\text{°C} = 5.49 \times 10^4 \text{ J}$ or 54.9 kJ

b. $\frac{0.900 \text{ J}}{\text{g °C}} \times \frac{26.98 \text{ g}}{\text{mol Al}} = \frac{24.3 \text{ J}}{\text{mol °C}}$

39. The units for specific heat capacity (s) are J/g•°C. $s = \frac{78.2 \text{ J}}{45.6 \text{ g} \times 13.3\text{°C}} = \frac{0.129 \text{ J}}{\text{g °C}}$

$$\text{Molar heat capacity} = \frac{0.129 \text{ J}}{\text{g °C}} \times \frac{207.2 \text{ g}}{\text{mol Pb}} = \frac{26.7 \text{ J}}{\text{mol °C}}$$

41. | Heat loss by hot water | = | Heat gain by cooler water |

The magnitude of heat loss and heat gain are equal in calorimetry problem. The only difference is the sign (positive and negative). To avoid sign errors, keep all quantities positive and, if necessary, deduce the correct signs at the end of the problem. Water has a specific heat capacity = s = 4.18 J/°C•g = 4.18 J/K•g (ΔT in °C = ΔT in K).

$$\text{Heat loss by hot water} = s \times m \times \Delta T = \frac{4.18 \text{ J}}{\text{g K}} \times 50.0 \text{ g} \times (330. \text{ K} - T_f)$$

Heat gain by cooler water = $\frac{4.18 \text{ J}}{\text{g K}} \times 30.0 \text{ g} \times (T_f - 280. \text{ K})$; Heat loss = Heat gain, so:

$$\frac{209 \text{ J}}{\text{K}} \times (330. \text{ K} - T_f) = \frac{125 \text{ J}}{\text{K}} \times (T_f - 280. \text{ K}),\ 6.90 \times 10^4 - 209\ T_f = 125\ T_f - 3.50 \times 10^4$$

$$334\ T_f = 1.040 \times 10^5,\ T_f = 311 \text{ K}$$

Note that the final temperature is closer to the temperature of the more massive hot water, which is as it should be.

43. Heat gained by water = Heat lost by nickel = s × m × ΔT where s = specific heat capacity

$$\text{Heat gain} = \frac{4.18\text{ J}}{\text{g °C}} \times 150.0\text{ g} \times (25.0\text{°C} - 23.5\text{°C}) = 940\text{ J}$$

Note: A temperature <u>change</u> of one Kelvin is the same as a temperature change of one degree Celsius.

A common error in calorimetery problems are sign errors. Keeping all quantities positive helps eliminate sign errors. Therefore:

$$\text{Heat loss} = 940\text{ J} = s \times 28.2\text{ g} \times (99.8 - 25.0)\text{°C},\ \ s = \frac{940\text{ J}}{28.2\text{ g} \times 74.8\text{°C}} = \frac{0.45\text{ J}}{\text{g °C}}$$

45. Heat lost by solution = Heat gained by KBr; Mass of solution = 125 g + 10.5 g = 136 g

Note: Sign errors are common with calorimetry problems. However, the correct sign for ΔH can easily be obtained from the ΔT data. When working calorimetry problems, keep all quantities positive (ignore signs). When finished, deduce the correct sign for ΔH. For this problem, T decreases as KBr dissolves so ΔH is positive; the dissolution of KBr is endothermic (absorbs heat).

$$\text{Heat lost by solution} = \frac{4.18\text{ J}}{\text{g °C}} \times 136\text{ g} \times (24.2\text{°C} - 21.1\text{°C}) = 1800\text{ J} = \text{Heat gained by KBr}$$

$$\Delta H \text{ in units of J/g} = \frac{1800\text{ J}}{10.5\text{ g KBr}} = 170\text{ J/g}$$

$$\Delta H \text{ in units of kJ/mol} = \frac{170\text{J}}{\text{g KBr}} \times \frac{119.0\text{ g KBr}}{\text{mol KBr}} \times \frac{1\text{ kJ}}{1000\text{ J}} = 20.\text{ kJ/mol}$$

47. 50.0×10^{-3} L × 0.100 mol/L = 5.00×10^{-3} mol of both $AgNO_3$ and HCl are reacted. Thus, 5.00×10^{-3} mol of AgCl will be produced since there is a 1:1 mol ratio between reactants.

Heat lost by chemicals = Heat gained by solution

$$\text{Heat gain} = \frac{4.18\text{ J}}{\text{g °C}} \times 100.0\text{ g} \times (23.40 - 22.60)\text{°C} = 330\text{ J}$$

Heat loss = 330 J; This is the heat evolved (exothermic reaction) when 5.00×10^{-3} mol of AgCl is produced. So q = -330 J and ΔH (heat per mol AgCl formed) is negative with a value of:

$$\Delta H = \frac{-330\text{ J}}{5.00 \times 10^{-3}\text{ mol}} \times \frac{1\text{ kJ}}{1000\text{ J}} = -66\text{ kJ/mol}$$

49. Heat lost by camphor = Heat gained by calorimeter

$$\text{Heat lost by combustion of camphor} = 0.1204\text{ g} \times \frac{1\text{ mol}}{152.23\text{ g}} \times \frac{5903.6\text{ kJ}}{\text{mol}} = 4.669\text{ kJ}$$

Let C_{cal} = heat capacity of the calorimeter in units of kJ/°C, then:

Heat gained by calorimter = $C_{cal} \times \Delta T$, 4.669 kJ = $C_{cal} \times 2.28°C$, C_{cal} = 2.05 kJ/°C

Hess's Law

51. Information given:

$$C(s) + O_2(g) \rightarrow CO_2(g) \qquad \Delta H = -393.7 \text{ kJ}$$
$$CO(g) + 1/2\ O_2(g) \rightarrow CO_2(g) \qquad \Delta H = -283.3 \text{ kJ}$$

Using Hess's Law:

$$2\ C(s) + 2\ O_2(g) \rightarrow 2\ CO_2(g) \qquad \Delta H_1 = 2(-393.7 \text{ kJ})$$
$$2\ CO_2(g) \rightarrow 2\ CO(g) + O_2(g) \qquad \Delta H_2 = -2(-283.3 \text{ kJ})$$

$$2\ C(s) + O_2(g) \rightarrow 2\ CO(g) \qquad \Delta H = \Delta H_1 + \Delta H_2 = -220.8 \text{ kJ}$$

Note: The enthalpy change for a reaction that is reversed is the negative quantity of the enthalpy change for the original reaction. If the coefficients in a balanced reaction are multiplied by an integer, then the value of ΔH is multiplied by the same integer.

53.

$$S + 3/2\ O_2 \rightarrow SO_3 \qquad \Delta H = -395.2 \text{ kJ}$$
$$SO_3 \rightarrow SO_2 + 1/2\ O_2 \qquad \Delta H = -1/2(-198.2 \text{ kJ}) = 99.1 \text{ kJ}$$

$$S(s) + O_2(g) \rightarrow SO_2(g) \qquad \Delta H = -296.1 \text{ kJ}$$

55.

$$NO + O_3 \rightarrow NO_2 + O_2 \qquad \Delta H = -199 \text{ kJ}$$
$$3/2\ O_2 \rightarrow O_3 \qquad \Delta H = -1/2(-427 \text{ kJ})$$
$$O \rightarrow 1/2\ O_2 \qquad \Delta H = -1/2(495 \text{ kJ})$$

$$NO(g) + O(g) \rightarrow NO_2(g) \qquad \Delta H = -233 \text{ kJ}$$

57.

$$4\ HNO_3 \rightarrow 2\ N_2O_5 + 2\ H_2O \qquad \Delta H = -2(-76.6 \text{ kJ})$$
$$2\ N_2 + 6\ O_2 + 2\ H_2 \rightarrow 4\ HNO_3 \qquad \Delta H = 4(-174.1 \text{ kJ})$$
$$2\ H_2O \rightarrow 2\ H_2 + O_2 \qquad \Delta H = -2(-285.8 \text{ kJ})$$

$$2\ N_2(g) + 5\ O_2(g) \rightarrow 2\ N_2O_5(g) \qquad \Delta H = 28.4 \text{ kJ}$$

Standard Enthalpies of Formation

59. The change in enthalpy that accompanies the formation of one mole of a compound from its elements, with all substances in their standard states is the standard enthalpy of formation for a compound. The reactions that refer to ΔH_f° are:

$$Na(s) + 1/2\ Cl_2(g) \rightarrow NaCl(s); \quad H_2(g) + 1/2\ O_2(g) \rightarrow H_2O(l)$$

$6\ C(graphite, s) + 6\ H_2(g) + 3\ O_2(g) \rightarrow C_6H_{12}O_6(s)$; $Pb(s) + S(s) + 2\ O_2(g) \rightarrow PbSO_4(s)$

61. In general: $\Delta H° = \Sigma n_p \Delta H°_{f,\ products} - \Sigma n_r \Delta H°_{f,\ reactants}$ and all elements in their standard state have $\Delta H°_f = 0$ by definition.

a. $2\ NH_3(g) + 3\ O_2(g) + 2\ CH_4(g) \rightarrow 2\ HCN(g) + 6\ H_2O(g)$

$$\Delta H° = [2 \text{ mol HCN} \times \Delta H°_{f,\ HCN} + 6 \text{ mol } H_2O(g) \times \Delta H°_{f,\ H_2O}] - [2 \text{ mol } NH_3 \times \Delta H°_{f,\ NH_3} + 2 \text{ mol } CH_4 \times \Delta H°_{f,\ CH_4}]$$

$$\Delta H° = [2(135.1) + 6(-242)] - [2(-46) + 2(-75)] = -940.\ kJ$$

b. $Ca_3(PO_4)_2(s) + 3\ H_2SO_4(l) \rightarrow 3\ CaSO_4(s) + 2\ H_3PO_4(l)$

$$\Delta H° = \left[3 \text{ mol } CaSO_4\left(\frac{-1433 \text{ kJ}}{\text{mol}}\right) + 2 \text{ mol } H_3PO_4(l)\left(\frac{-1267 \text{ kJ}}{\text{mol}}\right)\right] - \left[1 \text{ mol } Ca_3(PO_4)_2\left(\frac{-4126 \text{ kJ}}{\text{mol}}\right) + 3 \text{ mol } H_2SO_4(l)\left(\frac{-814 \text{ kJ}}{\text{mol}}\right)\right]$$

$$\Delta H° = -6833 \text{ kJ} - (-6568 \text{ kJ}) = -265 \text{ kJ}$$

c. $NH_3(g) + HCl(g) \rightarrow NH_4Cl(s)$

$$\Delta H° = [1 \text{ mol } NH_4Cl \times \Delta H°_{f,\ NH_4Cl}] - [1 \text{ mol } NH_3 \times \Delta H°_{f,\ NH_3} + 1 \text{ mol HCl} \times \Delta H°_{f,\ HCl}]$$

$$\Delta H° = \left[1 \text{ mol}\left(\frac{-314 \text{ kJ}}{\text{mol}}\right)\right] - \left[1 \text{ mol}\left(\frac{-46 \text{ kJ}}{\text{mol}}\right) + 1 \text{ mol}\left(\frac{-92 \text{ kJ}}{\text{mol}}\right)\right]$$

$$\Delta H° = -314 \text{ kJ} + 138 \text{ kJ} = -176 \text{ kJ}$$

63. a. $4\ NH_3(g) + 5\ O_2(g) \rightarrow 4\ NO(g) + 6\ H_2O(g)$; $\Delta H° = \Sigma n_p \Delta H°_{f,\ products} - \Sigma n_r \Delta H°_{f,\ reactants}$

$$\Delta H° = \left[4 \text{ mol}\left(\frac{90.\ \text{kJ}}{\text{mol}}\right) + 6 \text{ mol}\left(\frac{-242 \text{ kJ}}{\text{mol}}\right)\right] - \left[4 \text{ mol}\left(\frac{-46 \text{ kJ}}{\text{mol}}\right)\right] = -908 \text{ kJ}$$

$2\ NO(g) + O_2(g) \rightarrow 2\ NO_2(g)$

$$\Delta H° = \left[2 \text{ mol}\left(\frac{34 \text{ kJ}}{\text{mol}}\right)\right] - \left[2 \text{ mol}\left(\frac{90.\ \text{kJ}}{\text{mol}}\right)\right] = -112 \text{ kJ}$$

$3\ NO_2(g) + H_2O(l) \rightarrow 2\ HNO_3(aq) + NO(g)$

$$\Delta H° = \left[2\ mol\left(\frac{-207\ kJ}{mol}\right) + 1\ mol\left(\frac{90.\ kJ}{mol}\right)\right] - \left[3\ mol\left(\frac{34\ kJ}{mol}\right) + 1\ mol\left(\frac{-286\ kJ}{mol}\right)\right] = -140.\ kJ$$

Note: All ΔH_f° values are assumed ± 1 kJ.

b. $12\ NH_3(g) + 15\ O_2(g) \rightarrow 12\ NO(g) + 18\ H_2O(g)$
$12\ NO(g) + 6\ O_2(g) \rightarrow 12\ NO_2(g)$
$12\ NO_2(g) + 4\ H_2O(l) \rightarrow 8\ HNO_3(aq) + 4\ NO(g)$
$4\ H_2O(g) \rightarrow 4\ H_2O(l)$

$12\ NH_3(g) + 21\ O_2(g) \rightarrow 8\ HNO_3(aq) + 4\ NO(g) + 14\ H_2O(g)$

The overall reaction is exothermic since each step is exothermic.

65. $3\ Al(s) + 3\ NH_4ClO_4(s) \rightarrow Al_2O_3(s) + AlCl_3(s) + 3\ NO(g) + 6\ H_2O(g)$

$$\Delta H° = \left[6\ mol\left(\frac{-242\ kJ}{mol}\right) + 3\ mol\left(\frac{90.\ kJ}{mol}\right) + 1\ mol\left(\frac{-704\ kJ}{mol}\right) + 1\ mol\left(\frac{-1676\ kJ}{mol}\right)\right] - \left[3\ mol\left(\frac{-295\ kJ}{mol}\right)\right] = -2677\ kJ$$

67. $2\ ClF_3(g) + 2\ NH_3(g) \rightarrow N_2(g) + 6\ HF(g) + Cl_2(g) \qquad \Delta H° = -1196\ kJ$

$$\Delta H° = [6\ \Delta H^\circ_{f,\ HF}] - [2\ \Delta H^\circ_{f,\ ClF_3} + 2\ \Delta H^\circ_{f,\ NH_3}]$$

$$-1196\ kJ = 6\ mol\left(\frac{-271\ kJ}{mol}\right) - 2\ \Delta H^\circ_{f,\ ClF_3} - 2\ mol\left(\frac{-46\ kJ}{mol}\right)$$

$$-1196\ kJ = -1626\ kJ - 2\ \Delta H^\circ_{f,\ ClF_3} + 92\ kJ,\ \Delta H^\circ_{f,\ ClF_3} = \frac{(-1626 + 92 + 1196)\ kJ}{2\ mol} = \frac{-169\ kJ}{mol}$$

Energy Consumption and Sources

69. $C(s) + H_2O(g) \rightarrow H_2(g) + CO(g),\ \Delta H° = -110.5\ kJ - (-242\ kJ) = 132\ kJ$

71. $C_3H_8(g) + 5\ O_2(g) \rightarrow 3\ CO_2(g) + 4\ H_2O(l)$

$$\Delta H° = [3(-393.5\ kJ) + 4(-286\ kJ)] - [-104\ kJ] = -2221\ kJ/mol\ C_3H_8$$

$$\frac{-2221\ kJ}{mol} \times \frac{1\ mol}{44.09\ g} = \frac{-50.37\ kJ}{g}$$ vs. -47.7 kJ/g for octane (Sample Exercise 6.11)

The fuel values are very close. An advantage of propane is that it burns more cleanly. The boiling point of propane is -42°C. Thus, it is more difficult to store propane and there are extra safety hazards associated with using high pressure compressed gas tanks.

73. The molar volume of a gas at STP is 22.42 L.

$$4.19 \times 10^6 \text{ kJ} \times \frac{1 \text{ mol } CH_4}{891 \text{ kJ}} \times \frac{22.42 \text{ L } CH_4}{\text{mol } CH_4} = 1.05 \times 10^5 \text{ L } CH_4$$

Additional Exercises

75. $w = -P\Delta V$; Δn = mol gaseous products - mol gaseous reactants. Only gases can do PV work. When a balanced reaction has more mol of product gases than mol of reactant gases (Δn positive), then the reaction will expand in volume (ΔV positive) and the system does work on the surroundings. For example, in reaction e, $\Delta n = 6 - 0 = 6$ mol and this reaction would do expansion work against the surroundings. When a balanced reaction has a decrease in mol gas from reactants to products (Δn negative), then the reaction will contract in volume (ΔV negative) and the surroundings does compression work on the system. When there is no change in mol of gas from reactants to products, then $\Delta V = 0$ and $w = 0$.

When $\Delta V > 0$ ($\Delta n > 0$), then $w < 0$ and system does work on the surroundings (e and f).

When $\Delta V < 0$ ($\Delta n < 0$), then $w > 0$ and the surroundings does work on the system (a, c and d).

When $\Delta V = 0$ ($\Delta n = 0$), then $w = 0$ (b).

77. $2 \text{ K(s)} + 2 \text{ } H_2O(l) \rightarrow 2 \text{ KOH(aq)} + H_2(g)$, $\Delta H° = 2(-481 \text{ kJ}) - 2(-286 \text{ kJ}) = -390. \text{ kJ}$

$$5.00 \text{ g K} \times \frac{1 \text{ mol K}}{39.10 \text{ g K}} \times \frac{-390. \text{ kJ}}{2 \text{ mol K}} = -24.9 \text{ kJ}$$ of heat released upon reaction of 5.00 g of potassium.

$$24{,}900 \text{ J} = \frac{4.18 \text{ J}}{\text{g °C}} \times (1.00 \times 10^3 \text{ g}) \times \Delta T, \quad \Delta T = \frac{24{,}900}{4.18 \times 1.00 \times 10^3} = 5.96°\text{C}$$

Final temperature = 24.0 + 5.96 = 30.0°C

79. The specific heat of water is 4.18 J/g•°C, which is equal to 4.18 kJ/kg•°C.

We have 1.00 kg of H_2O, so: $1.00 \text{ kg} \times \frac{4.18 \text{ kJ}}{\text{kg °C}} = 4.18 \text{ kJ/°C}$

This is the portion of the heat capacity that can be attributed to H_2O.

Total heat capacity = $C_{cal} + C_{H_2O}$, $C_{cal} = 10.84 - 4.18 = 6.66$ kJ/°C

81. First, we need to get the heat capacity of the calorimeter from the combustion of benzoic acid.

Heat lost by combustion = Heat gained by calorimeter

$$\text{Heat loss} = 0.1584\ \text{g} \times \frac{26.42\ \text{kJ}}{\text{g}} = 4.185\ \text{kJ}$$

$$\text{Heat gain} = 4.185\ \text{kJ} = C_{cal} \times \Delta T,\ C_{cal} = \frac{4.185\ \text{kJ}}{2.54°\text{C}} = 1.65\ \text{kJ/°C}$$

Now we can calculate the heat of combustion of vanillin. Heat loss = Heat gain

$$\text{Heat gain by calorimeter} = \frac{1.65\ \text{kJ}}{°\text{C}} \times 3.25°\text{C} = 5.36\ \text{kJ}$$

Heat loss = 5.36 kJ which is the heat evolved by the combustion of the vanillin.

$$\Delta E_{comb} = \frac{-5.36\ \text{kJ}}{0.2130\ \text{g}} = -25.2\ \text{kJ/g};\ \Delta E_{comb} = \frac{-25.2\ \text{kJ}}{\text{g}} \times \frac{152.14\ \text{g}}{\text{mol}} = -3830\ \text{kJ/mol}$$

83. The combustion of phosphorus is exothermic. Thus the product, P_4O_{10}, is lower in energy than either red or white phosphorus. Since the conversion of white phosphorus to red phosphorus is exothermic, red phosphorus is lower in energy than white phosphorus. Thus, white phosphorus will release more heat when burned in air since there is a larger energy difference between white phosphorus and products.

85. a. $C_2H_4(g) + O_3(g) \rightarrow CH_3CHO(g) + O_2(g)$, $\Delta H° = -166\ \text{kJ} - [143\ \text{kJ} + 52\ \text{kJ}] = -361\ \text{kJ}$

b. $O_3(g) + NO(g) \rightarrow NO_2(g) + O_2(g)$, $\Delta H° = 34\ \text{kJ} - [90.\ \text{kJ} + 143\ \text{kJ}] = -199\ \text{kJ}$

c. $SO_3(g) + H_2O(l) \rightarrow H_2SO_4(aq)$, $\Delta H° = -909\ \text{kJ} - [-396\ \text{kJ} + (-286\ \text{kJ})] = -227\ \text{kJ}$

d. $2\ NO(g) + O_2(g) \rightarrow 2\ NO_2(g)$, $\Delta H° = 2(34)\ \text{kJ} - 2(90.)\ \text{kJ} = -112\ \text{kJ}$

Challenge Problems

87. a. $C_{12}H_{22}O_{11}(s) + 12\ O_2(g) \rightarrow 12\ CO_2(g) + 11\ H_2O(l)$

b. A bomb calorimeter is at constant volume, so heat released = $q_v = \Delta E$:

$$\Delta E = \frac{-24.00\ \text{kJ}}{1.46\ \text{g}} \times \frac{342.30\ \text{g}}{\text{mol}} = -5630\ \text{kJ/mol}\ C_{12}H_{22}O_{11}$$

c. Since $PV = nRT$, then $P\Delta V = RT\Delta n$ where Δn = mol gaseous products - mol gaseous reactants.

$$\Delta H = \Delta E + P\Delta V = \Delta E + RT\Delta n$$

For this reaction $\Delta n = 12 - 12 = 0$, so $\Delta H = \Delta E = -5630$ kJ/mol.

89. $$\text{Energy used in 8.0 hours} = 40.\ \text{kWh} = \frac{40.\ \text{kJ h}}{\text{s}} \times \frac{3600\ \text{s}}{\text{h}} = 1.4 \times 10^5\ \text{kJ}$$

$$\text{Energy from the sun in 8.0 hours} = \frac{1.0\ \text{kJ}}{\text{s m}^2} \times \frac{60\ \text{s}}{\text{min}} \times \frac{60\ \text{min}}{\text{h}} \times 8.0\ \text{h} = 2.9 \times 10^4\ \text{kJ/m}^2$$

Only 13% of the sunlight is converted into electricity:

$$0.13 \times (2.9 \times 10^4 \text{ kJ/m}^2) \times \text{Area} = 1.4 \times 10^5 \text{ kJ}, \quad \text{Area} = 37 \text{ m}^2$$

91. $400 \text{ kcal} \times \dfrac{4.18 \text{ kJ}}{\text{kcal}} = 1.67 \times 10^3 \text{ kJ} \approx 2 \times 10^3 \text{ kJ}$

$$\text{PE} = \text{mgz} = \left(180 \text{ lb} \times \frac{1 \text{ kg}}{2.205 \text{ lb}}\right) \times \frac{9.8 \text{ m}}{\text{s}^2} \times \left(8 \text{ in} \times \frac{2.54 \text{ cm}}{\text{in}} \times \frac{1 \text{ m}}{100 \text{ cm}}\right) = 160 \text{ J} \approx 200 \text{ J}$$

200 J of energy are needed to climb one step. The total numer of steps to climb are:

$$2 \times 10^6 \text{ J} \times \frac{1 \text{ step}}{200 \text{ J}} = 1 \times 10^4 \text{ steps}$$

CHAPTER SEVEN

ATOMIC STRUCTURE AND PERIODICITY

Questions

13. Planck found that heated bodies only give off certain frequencies of light and Einstein's study of the photoelectric effect.

15. Only very small particles with a minuscule mass exhibit quantum effects, e.g., an electron. These tiny particles generally have large velocities.

17. a. A discrete bundle of light energy.

    b. A number describing a discrete energy state of an electron.

    c. The lowest energy state of the electron(s) in an atom or ion.

    d. An allowed energy state that is higher in energy than the ground state.

19. The 2p orbitals differ from each other in the direction in which they point in space.

21. A nodal surface in an atomic orbital is a surface in which the probability of finding an electron is zero.

23. No, the spin is a convenient model. Since we cannot locate or "see" the electron, we cannot see if it is spinning.

25. The outermost electrons are the valence electrons. When atoms interact with each other, it will be the outermost electrons that are involved in these interactions. In addition, how tightly the nucleus holds these outermost electrons determines atomic size, ionization energy and other properties of atoms.

27. If one more electron is added to a half-filled subshell, electron-electron repulsions will increase since two electrons must now occupy the same atomic orbital. This may slightly decrease the stability of the atom. Hence, half-filled subshells minimize electron-electron repulsions.

29. As electrons are removed, the nuclear charge exerted on the remaining electrons increases. Since the remaining electrons are 'held' more strongly by the nucleus, the energy required to remove these electrons increases.

31. For hydrogen all atomic orbitals with the same n value have the same energy. For polyatomic atoms/ions, the energy of the atomic orbitals also depends on ℓ. Since there are more nondegenerate energy levels for polyatomic atoms/ions as compared to hydrogen, then there are many more possible electronic transitions resulting in more complicated line spectra.

33. Yes, the maximum number of unpaired electrons in any configuration corresponds to a minimum in electron-electron repulsions.

Exercises

Light and Matter

35. $\nu = \frac{c}{\lambda} = \frac{2.998 \times 10^8 \text{ m/s}}{780. \times 10^{-9} \text{ m}} = 3.84 \times 10^{14} \text{ s}^{-1}$

37. $\nu = \frac{c}{\lambda} = \frac{3.00 \times 10^8 \text{ m/s}}{1.0 \times 10^{-2} \text{ m}} = 3.0 \times 10^{10} \text{ s}^{-1}$

$E = h\nu = 6.63 \times 10^{-34} \text{ J s} \times 3.0 \times 10^{10} \text{ s}^{-1} = 2.0 \times 10^{-23}$ J/photon

$$\frac{2.0 \times 10^{-23} \text{ J}}{\text{photon}} \times \frac{6.02 \times 10^{23} \text{ photons}}{\text{mol}} = 12 \text{ J/mol}$$

39. The energy needed to remove a single electron is:

$$\frac{279.7 \text{ kJ}}{\text{mol}} \times \frac{1 \text{ mol}}{6.0221 \times 10^{23}} = 4.645 \times 10^{-22} \text{ kJ} = 4.645 \times 10^{-19} \text{ J}$$

$$E = \frac{hc}{\lambda}, \ \lambda = \frac{hc}{E} = \frac{6.6261 \times 10^{-34} \text{ J s} \times 2.9979 \times 10^8 \text{ m/s}}{4.645 \times 10^{-19} \text{ J}} = 4.277 \times 10^{-7} \text{ m} = 427.7 \text{ nm}$$

41. Ionization energy = energy to remove an electron = $7.21 \times 10^{-19} = E_{photon}$

$E_{photon} = h\nu$ and $\lambda\nu = c$. So, $\nu = \frac{c}{\lambda}$ and $E = \frac{hc}{\lambda}$

$$\lambda = \frac{hc}{E_{photon}} = \frac{6.626 \times 10^{-34} \text{ J s} \times 2.998 \times 10^8 \text{ m/s}}{7.21 \times 10^{-19} \text{ J}} = 2.76 \times 10^{-7} \text{ m} = 276 \text{ nm}$$

43. a. 5.0% of speed of light = $0.050 \times 3.00 \times 10^8 \text{ m/s} = 1.5 \times 10^7$ m/s

$$\lambda = \frac{h}{mv}, \ \lambda = \frac{6.63 \times 10^{-34} \text{ J s}}{1.67 \times 10^{-27} \text{ kg} \times 1.5 \times 10^7 \text{ m/s}} = 2.6 \times 10^{-14} \text{ m} = 2.6 \times 10^{-5} \text{ nm}$$

Note: For units to come out, the mass must be in kg since $1 \text{ J} = \frac{1 \text{ kg m}^2}{\text{s}^2}$.

b. $\text{mass} = 5.2 \text{ oz} \times \dfrac{1 \text{ lb}}{16 \text{ oz}} \times \dfrac{1 \text{ kg}}{2.205 \text{ lb}} = 0.15 \text{ kg}$

$$\text{velocity} = \frac{100.8 \text{ mi}}{\text{hr}} \times \frac{1 \text{ hr}}{3600 \text{ s}} \times \frac{1760 \text{ yd}}{\text{mi}} \times \frac{1 \text{ m}}{1.0936 \text{ yd}} = 45.06 \text{ m/s}$$

$$\lambda = \frac{h}{mv} = \frac{6.63 \times 10^{-34} \text{ J s}}{0.15 \text{ kg} \times 45.06 \text{ m/s}} = 9.8 \times 10^{-35} \text{ m} = 9.8 \times 10^{-26} \text{ nm}$$

This number is so small that it is essentially zero. We cannot detect a wavelength this small. The meaning of this number is that we do not have to consider the wave properties of large objects.

45. $\lambda = \dfrac{h}{mv} = \dfrac{6.63 \times 10^{-34} \text{ J s}}{9.11 \times 10^{-31} \text{ kg} \times (1.0 \times 10^{-3} \times 3.00 \times 10^{8} \text{ m/s})} = 2.4 \times 10^{-9} \text{ m} = 2.4 \text{ nm}$

Hydrogen Atom: The Bohr Model

47. For the H-atom (Z = 1): $E_n = -2.178 \times 10^{-18} \text{ J}/n^2$; For a spectral transition, $\Delta E = E_f - E_i$:

$$\Delta E = -2.178 \times 10^{-18} \text{ J} \left(\frac{1}{n_f^2} - \frac{1}{n_i^2} \right)$$

where n_i and n_f are the levels of the initial and final states, respectively. A positive value of ΔE always corresponds to an absorption of light and a negative value of ΔE corresponds to an emission of light.

a. $\Delta E = -2.178 \times 10^{-18} \text{ J} \left(\dfrac{1}{2^2} - \dfrac{1}{3^2} \right) = -2.178 \times 10^{-18} \text{ J} \left(\dfrac{1}{4} - \dfrac{1}{9} \right)$

$$\Delta E = -2.178 \times 10^{-18} \text{ J} \times (0.2500 - 0.1111) = -3.025 \times 10^{-19} \text{ J}$$

The photon of light must have precisely this energy (3.025×10^{-19} J).

$$|\Delta E| = E_{photon} = h\nu = \frac{hc}{\lambda} \text{ or } \lambda = \frac{hc}{|\Delta E|} = \frac{6.6261 \times 10^{-34} \text{ J s} \times 2.9979 \times 10^{8} \text{ m/s}}{3.025 \times 10^{-19} \text{ J}}$$

$$= 6.567 \times 10^{-7} \text{ m} = 656.7 \text{ nm}$$

b. $\Delta E = -2.178 \times 10^{-18} \text{ J} \left(\dfrac{1}{2^2} - \dfrac{1}{4^2} \right) = -4.084 \times 10^{-19} \text{ J}$

$$\lambda = \frac{hc}{|\Delta E|} = \frac{6.6261 \times 10^{-34} \text{ J s} \times 2.9979 \times 10^{8} \text{ m/s}}{4.084 \times 10^{-19} \text{ J}} = 4.864 \times 10^{-7} \text{ m} = 486.4 \text{ nm}$$

c. $\Delta E = -2.178 \times 10^{-18}\ J\left(\frac{1}{1^2} - \frac{1}{2^2}\right) = -1.634 \times 10^{-18}\ J$

$$\lambda = \frac{6.6261 \times 10^{-34}\ J\ s \times 2.9979 \times 10^{8}\ m/s}{1.634 \times 10^{-18}\ J} = 1.216 \times 10^{-7}\ m = 121.6\ nm$$

49.

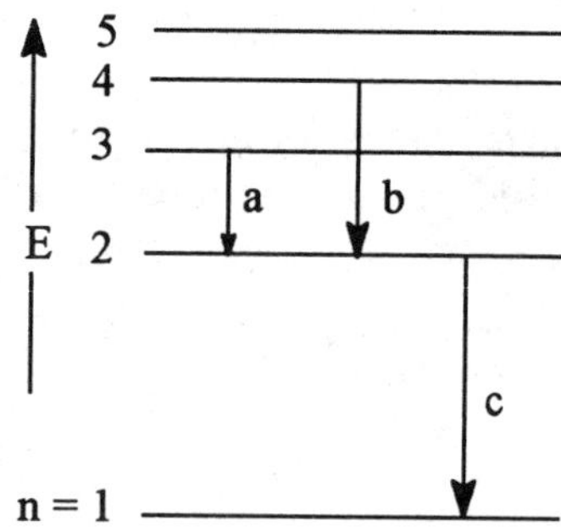

a. $3 \rightarrow 2$

b. $4 \rightarrow 2$

c. $2 \rightarrow 1$

Energy levels are not to scale.

51. The longest wavelength light emitted will correspond to the transition with the lowest energy change. This is the transition from n = 6 to n = 5.

$$\Delta E = -2.178 \times 10^{-18}\ J\left(\frac{1}{5^2} - \frac{1}{6^2}\right) = -2.662 \times 10^{-20}\ J$$

$$\lambda = \frac{hc}{|\Delta E|} = \frac{6.6261 \times 10^{-34}\ J\ s \times 2.9979 \times 10^{8}\ m/s}{2.662 \times 10^{-20}\ J} = 7.462 \times 10^{-6}\ m = 7462\ nm$$

The shortest wavelength emitted will correspond to the largest ΔE; this is $n = 6 \rightarrow n = 1$.

$$\Delta E = -2.178 \times 10^{-18}\ J\left(\frac{1}{1^2} - \frac{1}{6^2}\right) = -2.118 \times 10^{-18}\ J$$

$$\lambda = \frac{hc}{|\Delta E|} = \frac{6.6261 \times 10^{-34}\ J\ s \times 2.9979 \times 10^{8}\ m/s}{2.118 \times 10^{-18}\ J} = 9.379 \times 10^{-8}\ m = 93.79\ nm$$

53. Ionization from n = 1 corresponds to the transition $n_i = 1 \rightarrow n_f = \infty$ where $E_\infty = 0$.

$$\Delta E = E_\infty - E_1 = -E_1 = 2.178 \times 10^{-18}\left(\frac{1}{1^2}\right) = 2.178 \times 10^{-18}\ J = E_{photon}$$

$$\lambda = \frac{hc}{E} = \frac{6.6261 \times 10^{-34}\ J\ s \times 2.9979 \times 10^{8}\ m/s}{2.178 \times 10^{-18}\ J} = 9.120 \times 10^{-8}\ m = 91.20\ nm$$

To ionize from n = 2, $\Delta E = E_\infty - E_2 = -E_2 = 2.178 \times 10^{-18}\left(\frac{1}{2^2}\right) = 5.445 \times 10^{-19}\ J$

$$\lambda = \frac{6.6261 \times 10^{-34}\ J\ s \times 2.9979 \times 10^{8}\ m/s}{5.445 \times 10^{-19}\ J} = 3.648 \times 10^{-7}\ m = 364.8\ nm$$

Quantum Mechanics, Quantum Numbers, and Orbitals

55. $\Delta mv = m\Delta v = 9.11 \times 10^{-31}$ kg × 0.100 m/s = $\dfrac{9.11 \times 10^{-32}\ \text{kg m}}{\text{s}}$

$$\Delta x \Delta mv \geq \frac{h}{4\pi},\ \Delta x = \frac{h}{4\pi\Delta mv} = \frac{6.626 \times 10^{-34}\ \text{J s}}{4 \times 3.142 \times 9.11 \times 10^{-32}\ \text{kg m/s}} = 5.79 \times 10^{-4}\ \text{m}$$

$\Delta x = 5.79 \times 10^{-4}$ m $= 5.79 \times 10^{-2}$ cm $= 5.79 \times 10^{5}$ nm

Diameter of H atom is roughly 1.0×10^{-8} cm. The uncertainty in position is much larger than the size of the atom.

57. n = 1, 2, 3, ... ; ℓ = 0, 1, 2, ... (n - 1); $m_\ell = -\ell$ ... -2, -1, 0, 1, 2, ...$+\ell$

59. b. ℓ must be smaller than n. d. For $\ell = 0$, $m_\ell = 0$ is the only allowed value.

61. ψ^2 gives the probability of finding the electron at that point.

Polyelectronic Atoms

63. 5p: three orbitals; $3d_{z^2}$: one orbital; 4d: five orbitals

n = 5: ℓ = 0 (1 orbital), ℓ = 1 (3 orbitals), ℓ = 2 (5 orbitals), ℓ = 3 (7 orbitals), ℓ = 4 (9 orbitals)

Total for n = 5 is 25 orbitals.

n = 4: ℓ = 0 (1), ℓ = 1 (3), ℓ = 2 (5), ℓ = 3 (7); Total for n = 4 is 16 orbitals.

65. a. n = 4: ℓ can be 0, 1, 2, or 3. Thus we have s (2 e^-), p (6 e^-), d (10 e^-) and f (14 e^-) orbitals present. Total number of electrons to fill these orbitals is 32.

b. n = 5, m_ℓ = +1: For n = 5, ℓ = 0, 1, 2, 3, 4. For ℓ = 1, 2, 3, 4, all can have m_ℓ = +1. Four distinct orbitals, thus 8 electrons.

c. n = 5, m_s = +1/2: For n = 5, ℓ = 0, 1, 2, 3, 4. Number of orbitals = 1, 3, 5, 7, 9 for each value of ℓ, respectively. There are 25 orbitals with n = 5. They can hold 50 electrons and 25 of these electrons can have m_s = +1/2.

d. n = 3, ℓ = 2: These quantum numbers define a set of 3d orbitals. There are 5 degenerate 3d orbitals which can hold a total of 10 electrons.

e. n = 2, ℓ = 1: These define a set of 2p orbitals. There are 3 degenerate 2p orbitals which can hold a total of 6 electrons.

67. Si: $1s^22s^22p^63s^23p^2$ or $[Ne]3s^23p^2$; Ga: $1s^22s^22p^63s^23p^64s^23d^{10}4p^1$ or $[Ar]4s^23d^{10}4p^1$

As: $[Ar]4s^23d^{10}4p^3$; Ge: $[Ar]4s^23d^{10}4p^2$; Al: $[Ne]3s^23p^1$; Cd: $[Kr]5s^24d^{10}$

S: $[Ne]3s^23p^4$; Se: $[Ar]4s^23d^{10}4p^4$

69. The following are complete electron configurations. Noble gas shorthand notation could also be used.

Sc: $1s^22s^22p^63s^2 3p^64s^23d^1$; Fe: $1s^22s^22p^63s^2 3p^64s^23d^6$

P: $1s^22s^22p^63s^2 3p^3$; Cs: $1s^22s^22p^63s^2 3p^64s^23d^{10}4p^65s^24d^{10}5p^66s^1$

Eu: $1s^22s^22p^63s^2 3p^64s^23d^{10}4p^65s^24d^{10}5p^66s^24f^65d^1$*

Pt: $1s^22s^22p^63s^2 3p^64s^23d^{10}4p^65s^24d^{10}5p^66s^24f^{14}5d^8$*

Xe: $1s^22s^22p^63s^2 3p^64s^23d^{10}4p^65s^24d^{10}5p^6$; Br: $1s^22s^22p^63s^2 3p^64s^23d^{10}4p^5$

*Note: These electron configurations were written down using only the periodic table.

Actual electron configurations are: Eu: $[Xe]6s^24f^7$ and Pt: $[Xe]6s^14f^{14}5d^9$

71. Exceptions: Cr, Cu, Nb, Mo, Tc, Ru, Rh, Pd, Ag, Pt, Au; Tc, Ru, Rh, Pd and Pt do not correspond to the supposed extra stability of half-filled and filled subshells.

73. a. The lightest halogen is fluorine: $1s^22s^22p^5$ b. K: $1s^22s^22p^63s^23p^64s^1$

c. Be: $1s^22s^2$; Mg: $1s^22s^22p^63s^2$; Ca: $1s^22s^22p^63s^23p^64s^2$

d. In: $[Kr]5s^24d^{10}5p^1$ e. C: $1s^22s^22p^2$; Si: $1s^22s^22p^63s^23p^2$

f. This will be element #118: $[Rn]7s^25f^{14}6d^{10}7p^6$

75. B : $1s^22s^22p^1$

	n	ℓ	m_ℓ	m_s
1s	1	0	0	+1/2
1s	1	0	0	-1/2
2s	2	0	0	+1/2
2s	2	0	0	-1/2
2p*	2	1	-1	+1/2

*This is only one of several possibilities for the 2p electron. The 2p electron in B could have m_ℓ = -1, 0 or +1, and m_s = +1/2 or -1/2, a total of six possibilities.

N : $1s^2 2s^2 2p^3$

	n	ℓ	m_ℓ	m_s	
1s	1	0	0	+1/2	
1s	1	0	0	-1/2	
2s	2	0	0	+1/2	
2s	2	0	0	-1/2	
2p	2	1	-1	+1/2	
2p	2	1	0	+1/2	(Or all 2p electrons could have m_s = -1/2.)
2p	2	1	+1	+1/2	

77. O: $1s^2 2s^2 2p_x^2 2p_y^2$ (↑↓ ↑↓ __); There are no unpaired electrons in this oxygen atom. This configuration would be an excited state and in going to the more stable ground state (↑↓ ↑ ↑), energy would be released.

79. We get the number of unpaired electrons by looking at the incompletely filled subshells.

Sc: $[Ar]4s^2 3d^1$	$3d^1$:	↑ __ __ __ __	one unpaired e^-
Ti: $[Ar]4s^2 3d^2$	$3d^2$:	↑ ↑ __ __ __	two unpaired e^-
Al: $[Ne]3s^2 3p^1$	$3p^1$:	↑ __ __	one unpaired e^-
Sn: $[Kr]5s^2 4d^{10} 5p^2$	$5p^2$:	↑ ↑ __	two unpaired e^-
Te: $[Ar]5s^2 4d^{10} 5p^4$	$5p^4$:	↑↓ ↑ ↑	two unpaired e^-
Br: $[Ar]4s^2 3d^{10} 4p^5$	$4p^5$:	↑↓ ↑↓ ↑	one unpaired e^-

The Periodic Table and Periodic Properties

81. Size decreases left to right across the periodic table and size increases from top to bottom of the periodic table.

a. Be < Mg < Ca b. Xe < I < Te c. Ge < Ga < In

83. The ionization energy trend is the opposite of the radii trend (see Exercise 7.81).

a. Ca < Mg < Be b. Te < I < Xe c. In < Ga < Ge

85. Ge: $[Ar]4s^2 3d^{10} 4p^2$; As: $[Ar]4s^2 3d^{10} 4p^3$; Se: $[Ar]4s^2 3d^{10} 4p^4$; There are extra electron-electron repulsions in Se because two electrons are in the same 4p orbital, resulting in a lower ionization energy. Se is an exception to the general ionization energy trend.

87. a. Li b. P

c. O^+. This ion has the fewest electrons as compared to the other oxygen species present. O^+ has the smallest amount of electron-electron repulsions which makes it the smallest ion with the largest ionization energy.

d. From the radii trend, Ar < Cl < S and Kr > Ar. Since variation in size down a family is greater than the variation across a period, we would predict Cl to be the smallest of the three.

e. Cu

89. a. 106: $[Rn]7s^2 5f^{14} 6d^4$ b. W c. SgO_3 and SgO_4^{2-} (similar to Cr; Sg = 106)

91. $P(g) \rightarrow P^+(g) + e^-$; IE refers to atoms in the gas phase.

93. a. More favorable EA: C and Br; The electron affinity trend is very erratic. Both N and Ar have positive EA values (unfavorable) due to their electron configuration (see text for detailed explanation).

b. Higher IE: N and Ar (follows the IE trend)

c. Larger size: C and Br (follows the radii trend)

95. The electron affinity trend is very erratic. In general, EA decreases down the periodic table and the trend across the table is too erratic to be of much use.

a. Se < S; S is most exothermic. b. I < Br < F < Cl; Cl is most exothermic (F is an exception).

97. Electron-electron repulsions are much greater in O^- than in S^- because the electron goes into a smaller 2p orbital vs. the larger 3p orbital in sulfur. This results in a more favorable (more exothermic) EA for sulfur.

99. a. The electron affinity of Mg^{2+} is ΔH for: $Mg^{2+}(g) + e^- \rightarrow Mg^+(g)$; This is just the reverse of the second ionization energy, or: $EA(Mg^{2+}) = -IE_2(Mg) = -1445$ kJ/mol (Table 7.5)

b. EA of Al^+ is ΔH for: $Al^+(g) + e^- \rightarrow Al(g)$; $EA(Al^+) = -IE_1(Al) = -580$ kJ/mol (Table 7.5)

Alkali and Alkaline Earth Elements

101. It should be potassium peroxide, K_2O_2, since K^+ ions are stable. K^{2+} ions are not stable; the second ionization energy of K is very large as compared to the first.

103. $\nu = \dfrac{c}{\lambda} = \dfrac{2.9979 \times 10^8 \text{ m/s}}{455.5 \times 10^{-9} \text{ m}} = 6.582 \times 10^{14} \text{ s}^{-1}$

$E = h\nu = 6.6261 \times 10^{-34} \text{ J s} \times 6.582 \times 10^{14} \text{ s}^{-1} = 4.361 \times 10^{-19} \text{ J}$

105. a. Li_3N; lithium nitride b. NaBr; sodium bromide c. K_2S; potassium sulfide

107. a. $4\ Li(s) + O_2(g) \rightarrow 2\ Li_2O(s)$ b. $2\ K(s) + S(s) \rightarrow K_2S(s)$

Additional Exercises

109. $$E = \frac{310.\ kJ}{mol} \times \frac{1\ mol}{6.022 \times 10^{23}} = 5.15 \times 10^{-22}\ kJ = 5.15 \times 10^{-19}\ J$$

$$E = \frac{hc}{\lambda},\ \lambda = \frac{hc}{E} = \frac{6.626 \times 10^{-34}\ J\ s \times 2.998 \times 10^{8}\ m/s}{5.15 \times 10^{-19}} = 3.86 \times 10^{-7}\ m = 386\ nm$$

111. There are 4 possible transitions for an electron in the n = 5 level ($5 \rightarrow 4$, $5 \rightarrow 3$, $5 \rightarrow 2$ and $5 \rightarrow 1$). If an electron initially drops to the n = 4 level, three additional transitions can occur ($4 \rightarrow 3$, $4 \rightarrow 2$ and $4 \rightarrow 1$). Similarily, there are two more transitions from the n = 3 level ($3 \rightarrow 2$, $3 \rightarrow 1$) and one more transition for the n = 2 level ($2 \rightarrow 1$). There are a total of 10 possible transitions for an electron in the n = 5 level for a possible total of 10 different wavelength emissions.

113. a. n = 3; We can have 3s, 3p, and 3d orbitals. Nine orbitals can hold 18 electrons.

b. n = 2, $\ell = 0$; This is a 2s orbital. 2 electrons

c. n = 2, $\ell = 2$; Not possible. No electrons can have this combination of quantum numbers.

d. These four quantum numbers completely specify a single electron.

115. b and f are the only possible sets of quantum numbers.

a. For $\ell = 0$, m_ℓ can only be 0.

c. m_s can only be + 1/2 or -1/2.

d. For n = 1, ℓ can only be 0.

e. For $\ell = 2$, m_ℓ cannot be -3. The lowest allowed m_ℓ value is -2.

117. Sb: $1s^2 2s^2 2p^6 3s^2 3p^6 4s^2 3d^{10} 4p^6 5s^2 4d^{10} 5p^3$

a. $\ell = 1$: Designates p orbitals. There are 21 electrons in p orbitals.

b. $m_\ell = 0$: All s electrons, 2 out of each set of 2p, 3p, 4p electrons, 2 out of each set of 3d and 4d electrons, and one of the 5p electrons have $m_\ell = 0$. $10 + 6 + 4 + 1 = 21\ e^-$ with $m_\ell = 0$.

c. $m_\ell = 1$: 2 out of each set of 2p, 3p, and 4p electrons, 2 out of each set of 3d and 4d electrons, and one of the 5p electrons have $m_\ell = 1$. $6 + 4 + 1 = 11\ e^-$ with $m_\ell = 1$.

119. a. As: $1s^2 2s^2 2p^6 3s^2 3p^6 4s^2 3d^{10} 4p^3$

b. Element 116 will be below Po in the periodic table: $[Rn]7s^2 5f^{14} 6d^{10} 7p^4$

c. Ta: $[Xe]6s^24f^{14}5d^3$ or Ir: $[Xe]6s^24f^{14}5d^7$

d. Ti: $[Ar]4s^23d^2$; Ni: $[Ar]4s^23d^8$; Os: $[Xe]6s^24f^{14}5d^6$

121. a. The 4+ ion contains 20 electrons. Thus, the electrically neutral atom will contain 24 electrons. The atomic number is 24.

b. The ground state electron configuration of the ion must be: $1s^22s^22p^63s^23p^64s^03d^2$; There are 6 electrons in s orbitals.

c. 12 d. 2

e. Because of the mass, this is an isotope of $^{50}_{24}Cr$. There are 26 neutrons in the nucleus.

f. $1s^22s^22p^63s^23p^64s^13d^5$ is the ground state electron configuration for Cr. Cr is an exception to the normal filling order.

123. Electron-electron repulsions become more important when we try to add electrons to an atom. From the standpoint of electron-electron repulsions, larger atoms would have more favorable (more exothermic) electron affinities. Considering only electron-nucleus attractions, smaller atoms would be expected to have the more favorable (more exothermic) EA's. These trends are exactly the opposite of each other. Thus, the overall variation in EA is not as great as ionization energy in which attractions to the nucleus dominate.

125. a.

$$Na(g) \rightarrow Na^+(g) + e^- \qquad I_1 = 495 \text{ kJ}$$
$$Cl(g) + e^- \rightarrow Cl^-(g) \qquad EA = -348.7 \text{ kJ}$$
$$\overline{Na(g) + Cl(g) \rightarrow Na^+(g) + Cl^-(g) \qquad \Delta H = 146 \text{ kJ}}$$

b.

$$Mg(g) \rightarrow Mg^+(g) + e^- \qquad I_1 = 735 \text{ kJ}$$
$$F(g) + e^- \rightarrow F^-(g) \qquad EA = -327.8 \text{ kJ}$$
$$\overline{Mg(g) + F(g) \rightarrow Mg^+(g) + F^-(g) \qquad \Delta H = 407 \text{ kJ}}$$

c.

$$Mg^+(g) \rightarrow Mg^{2+}(g) + e^- \qquad I_2 = 1445 \text{ kJ}$$
$$F(g) + e^- \rightarrow F^-(g) \qquad EA = -327.8 \text{ kJ}$$
$$\overline{Mg^+(g) + F(g) \rightarrow Mg^{2+}(g) + F^-(g) \qquad \Delta H = 1117 \text{ kJ}}$$

d. From parts b and c we get:

$$Mg(g) + F(g) \rightarrow Mg^+(g) + F^-(g) \qquad \Delta H = 407 \text{ kJ}$$
$$Mg^+(g) + F(g) \rightarrow Mg^{2+}(g) + F^-(g) \qquad \Delta H = 1117 \text{ kJ}$$
$$\overline{Mg(g) + 2\ F(g) \rightarrow Mg^{2+}(g) + 2\ F^-(g) \qquad \Delta H = 1524 \text{ kJ}}$$

127. The IE is for removal of the electron from the atom in the gas phase. The work function is for the removal of an electron from the solid.

$M(g) \rightarrow M^+(g) + e^-$ IE; $M(s) \rightarrow M^+(s) + e^-$ work function

Challenge Problems

129. $E_{photon} = \frac{hc}{\lambda} = \frac{6.6261 \times 10^{-34}\text{ J s} \times 2.9979 \times 10^{8}\text{ m/s}}{253.4 \times 10^{-9}\text{ m}} = 7.839 \times 10^{-19}$ J; $\Delta E = -7.839 \times 10^{-19}$ J

$$\Delta E = -2.178 \times 10^{-18}\text{ J }(Z)^2 \left(\frac{1}{n_f^2} - \frac{1}{n_i^2}\right),\ Z = 4 \text{ for } Be^{3+}$$

$$-7.839 \times 10^{-19}\text{ J} = -2.178 \times 10^{-18}\,(4)^2 \left(\frac{1}{n_f^2} - \frac{1}{5^2}\right)$$

$$\frac{7.839 \times 10^{-19}}{2.178 \times 10^{-18} \times 16} + \frac{1}{25} = \frac{1}{n_f^2},\ \frac{1}{n_f^2} = 0.06249,\ n_f = 4$$

This emission line corresponds to the $n = 5 \rightarrow n = 4$ electronic transition.

131. a. 1st period: p = 1, q = 1, r = 0, s = ± 1/2 (2 elements)

2nd period: p = 2, q = 1, r = 0, s = ± 1/2 (2 elements)

3rd period: p = 3, q = 1, r = 0, s = ± 1/2 (2 elements)

p = 3, q = 3, r = -2, s = ± 1/2 (2 elements)

p = 3, q = 3, r = 0, s = ± 1/2 (2 elements)

p = 3, q = 3, r = +2, s = ± 1/2 (2 elements)

4th period: p = 4; q and r values are the same as with p = 3 (8 total elements)

1							2
3							4
5	6	7	8	9	10	11	12
13	14	15	16	17	18	19	20

b. Elements 2, 4, 12 and 20 all have filled shells and will be least reactive.

c. Draw similarities to the modern periodic table.

XY could be X^+Y^-, $X^{2+}Y^{2-}$ or $X^{3+}Y^{3-}$. Possible ions for each are:

X^+ could be elements 1, 3, 5 or 13; Y^- could be 11 or 19.

X^{2+} could be 6 or 14; Y^{2-} could be 10 or 18.

X^{3+} could be 7 or 15; Y^{3-} could be 9 or 17.

Note: X^{4+} and Y^{4-} ions probably won't form.

XY_2 will be $X^{2+}(Y^-)_2$; See above for possible ions.

X_2Y will be $(X^+)_2Y^{2-}$; See above for possible ions.

XY_3 will be $X^{3+}(Y^-)_3$; See above for possible ions.

X_2Y_3 will be $(X^{3+})_2(Y^{2-})_3$; See above for possible ions.

d. $p = 4,\ q = 3,\ r = -2,\ s = \pm 1/2$ (2)

$p = 4,\ q = 3,\ r = 0,\ s = \pm 1/2$ (2)

$p = 4,\ q = 3,\ r = +2,\ s = \pm 1/2$ (2)

A total of 6 electrons can have $p = 4$ and $q = 3$.

e. $p = 3,\ q = 0,\ r = 0$: This is not allowed; q must be odd. Zero electrons can have these quantum numbers.

f. $p = 6,\ q = 1,\ r = 0,\ s = \pm 1/2$ (2)

$p = 6,\ q = 3,\ r = -2, 0, +2;\ s = \pm 1/2$ (6)

$p = 6,\ q = 5,\ r = -4, -2, 0, +2, 4;\ s = \pm 1/2$ (10)

Eighteen electrons can have $p = 6$.

133. a. As we remove succeeding electrons, the electron being removed is closer to the nucleus and there are fewer electrons left repelling it. The remaining electrons are more strongly attracted to the nucleus and it takes more energy to remove these electrons.

b. Al : $1s^22s^22p^63s^23p^1$; For I_4, we begin removing an electron with $n = 2$. For I_3, we removed an electron with $n = 3$. In going from $n = 3$ to $n = 2$ there is a big jump in ionization energy because the $n = 2$ electrons are much closer to the nucleus on the average than $n = 3$ electrons. Since the $n = 2$ electrons are closer to the nucleus, then they are held more tightly and require a much larger amount of energy to remove them as compared to the $n = 3$ electrons.

c. Al^{4+}; The electron affinity for Al^{4+} is ΔH for the reaction:

$$Al^{4+}(g) + e^- \rightarrow Al^{3+}(g) \quad \Delta H = -I_4 = -11{,}600 \text{ kJ/mol}$$

d. The greater the number of electrons, the greater the size.

Size trend: $Al^{4+} < Al^{3+} < Al^{2+} < Al^+ < Al$

CHAPTER EIGHT

BONDING: GENERAL CONCEPTS

Questions

11. Isoelectronic: same number of electrons. There are two variables, number of protons and number of electrons, that will determine the size of an ion. Keeping the number of electrons constant we only have to consider the number of protons to predict trends in size.

13. The two general requirements for a polar molecular are:

    1. polar bonds

    2. a structure such that the bond dipoles of the polar bonds do not cancel.

Exercises

Chemical Bonds and Electronegativity

15. The general trend for electronegativity is:
    1) increase as we go from left to right across a period and
    2) decrease as we go down a group

    Using these trends, the expected orders are:

    a. C < N < O b. Se < S < Cl c. Sn < Ge < Si d. Tl < Ge < S

17. The most polar bond will have the greatest difference in electronegativity between the two atoms. From positions in the periodic table, we would predict:

    a. Ge–F b. P–Cl c. S–F d. Ti–Cl

19. The general trends in electronegativity used on Exercises 8.15 and 8.17 are only rules of thumb. In this exercise we use experimental values of electronegativities and can begin to see several exceptions. The order of EN from Figure 8.3 is:

    a. C (2.5) < N (3.0) < O (3.5) same as predicted

    b. Se (2.4) < S (2.5) < Cl (3.0) same

c. Si = Ge = Sn (1.8) different

d. Tl (1.8) = Ge (1.8) < S (2.5) different

Most polar bonds using actual EN values:

a. Si–F and Ge–F equal polarity (Ge–F predicted)

b. P–Cl (same as predicted)

c. S–F (same as predicted) d. Ti–Cl (same as predicted)

21. Electronegativity values increase from left to right across the periodic table. The order of electronegativities for the atoms from smallest to largest electronegativity will be H = P < C < N < O < F. The most polar bond will be F–H since it will have the largest difference in electronegativities and the least polar bond will be P–H since it will have the smallest difference in electronegativities (ΔEN = 0). The order of the bonds in decreasing polarity will be F–H > O–H > N–H > C–H > P–H.

Ions and Ionic Compounds

23. Rb^+: $[Ar]4s^23d^{10}4p^6$; Ba^{2+}: $[Kr]5s^24d^{10}5p^6$; Se^{2-}: $[Ar]4s^23d^{10}4p^6$

I^-: $[Kr]5s^24d^{10}5p^6$

25. a. Mg^{2+}: $1s^22s^22p^6$; K^+: $1s^22s^22p^63s^23p^6$; Al^{3+}: $1s^22s^22p^6$

b. N^{3-}, O^{2-} and F^-: $1s^22s^22p^6$; Te^{2-}: $[Kr]5s^24d^{10}5p^6$

27. a. Sc^{3+} b. Te^{2-} c. Ce^{4+} and Ti^{4+} d. Ba^{2+}

All of these have the number of electrons of a noble gas.

29. There are many possible ions with 10 electrons. Some are: N^{3-}, O^{2-}, F^-, Na^+, Mg^{2+} and Al^{3+}. In terms of size, the ion with the most protons will hold the electrons the tightest and will be the smallest. The largest ion will be the ion with the fewest protons. The size trend is:

$Al^{3+} < Mg^{2+} < Na^+ < F^- < O^{2-} < N^{3-}$
smallest largest

31. a. $Cu > Cu^+ > Cu^{2+}$ b. $Pt^{2+} > Pd^{2+} > Ni^{2+}$ c. $O^{2-} > O^- > O$

d. $La^{3+} > Eu^{3+} > Gd^{3+} > Yb^{3+}$ e. $Te^{2-} > I^- > Cs^+ > Ba^{2+} > La^{3+}$

For answer a, as electrons are removed from an atom, size decreases. Answers b and d follow the radii trend. For answer c, as electrons are added to an atom, size increases. Answer e follows the trend for an isoelectronic series, i.e., the smallest ion has the most protons.

33. a. Li^+ and N^{3-} are the expected ions. The formula of the compound would be Li_3N (lithium nitride).

b. Ga^{3+} and O^{2-}; Ga_2O_3, gallium(III) oxide or gallium oxide

c. Rb^+ and Cl^-; RbCl, rubidium chloride

d. Ba^{2+} and S^{2-}; BaS, barium sulfide

35. Lattice energy is proportional to Q_1Q_2/r where Q is the charge of the ions and r is the distance between the ions. In general, charge effects on lattice energy are much greater than size effects.

a. NaCl; Na^+ is smaller than K^+.

b. LiF; F^- is smaller than Cl^-.

c. MgO; O^{2-} has a greater charge than OH^-.

d. $Fe(OH)_3$; Fe^{3+} has a greater charge than Fe^{2+}.

e. Na_2O; O^{2-} has a greater charge than Cl^-.

f. MgO; The ions are smaller in MgO.

37.

$Na(s) \rightarrow Na(g)$	$\Delta H = 109$ kJ (sublimation)
$Na(g) \rightarrow Na^+(g) + e^-$	$\Delta H = 495$ kJ (ionization energy)
$1/2\ Cl_2(g) \rightarrow Cl(g)$	$\Delta H = 239/2$ kJ (bond energy)
$Cl(g) + e^- \rightarrow Cl^-(g)$	$\Delta H = -349$ kJ (electron affinity)
$Na^+(g) + Cl^-(g) \rightarrow NaCl(s)$	$\Delta H = -786$ kJ (lattice energy)
$Na(s) + 1/2\ Cl_2(g) \rightarrow NaCl(s)$	$\Delta H_f^\circ = -412$ kJ/mol

39. From the data given, it costs less energy to produce $Mg^+(g) + O^-(g)$ than to produce $Mg^{2+}(g) + O^{2-}(g)$. However, the lattice energy for $Mg^{2+}O^{2-}$ will be much more exothermic than for Mg^+O^- (due to the greater charges in $Mg^{2+}O^{2-}$). The favorable lattice energy term will dominate and $Mg^{2+}O^{2-}$ forms.

41. Ca^{2+} has greater charge than Na^+, and Se^{2-} is smaller than Te^{2-}. The effect of charge on the lattice energy is greater than the effect of size. We expect the trend from most exothermic to least exothermic to be:

CaSe	>	CaTe	>	Na_2Se	>	Na_2Te	
(-2862)		(-2721)		(-2130)		(-2095)	This is what we observe.

Bond Energies

43. a. $H-H + Cl-Cl \rightarrow 2\ H-Cl$

Bonds broken:	Bonds formed:
1 H – H (432 kJ/mol)	2 H – Cl (427 kJ/mol)
1 Cl – Cl (239 kJ/mol)	

$\Delta H = \Sigma D_{broken} - \Sigma D_{formed}$, $\Delta H = 432\ kJ + 239\ kJ - 2(427)\ kJ = -183\ kJ$

b.

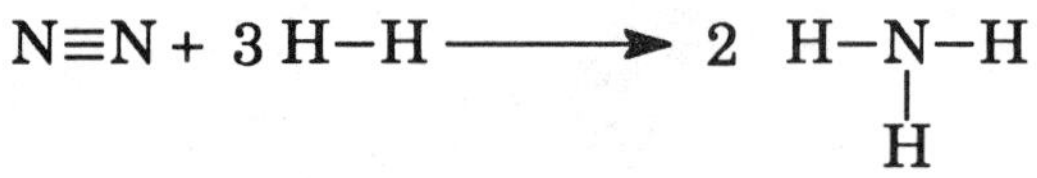

$$N{\equiv}N + 3\,H{-}H \longrightarrow 2\ H{-}\underset{\substack{|\\H}}{N}{-}H$$

Bonds broken:

1 N ≡ N (941 kJ/mol)
3 H – H (432 kJ/mol)

Bonds formed:

6 N – H (391 kJ/mol)

$\Delta H = 941\text{ kJ} + 3(432)\text{ kJ} - 6(391)\text{ kJ} = -109\text{ kJ}$

45.

$$H{-}\underset{\substack{|\\H}}{\overset{\substack{H\\|}}{C}}{-}N{\equiv}C \longrightarrow H{-}\underset{\substack{|\\H}}{\overset{\substack{H\\|}}{C}}{-}C{\equiv}N$$

Bonds broken: 1 C – N (305 kJ/mol)

Bonds formed: 1 C – C (347 kJ/mol)

$\Delta H = \Sigma D_{broken} - \Sigma D_{formed}$, $\Delta H = 305 - 347 = -42\text{ kJ}$

Note: Sometimes some of the bonds remain the same between reactants and products. To save time, only break and form bonds that are involved in the reaction.

47. $H{-}C{\equiv}C{-}H + 5/2\ O{=}O \rightarrow 2\ O{=}C{=}O + H{-}O{-}H$

Bonds broken:

1 C ≡ C (839 kJ/mol)
2 C – H (413 kJ/mol)
5/2 O = O (495 kJ/mol)

Bonds formed:

2 × 2 C = O (799 kJ/mol)
2 O – H (467 kJ/mol)

$\Delta H = 839 + 2(413) + 5/2\,(495) - [4(799) + 2(467)] = -1228\text{ kJ}$

49.

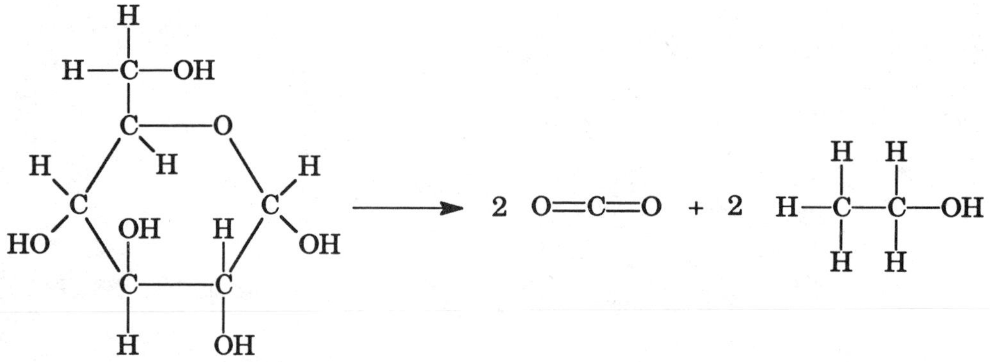

(Glucose ring structure: H–C(H)(OH)– attached to ring C—O; ring carbons bearing H, HO; OH, H; H, OH; H; OH) $\longrightarrow$ 2 O=C=O + 2 $H{-}\underset{\substack{|\\H}}{\overset{\substack{H\\|}}{C}}{-}\underset{\substack{|\\H}}{\overset{\substack{H\\|}}{C}}{-}OH$

The molecules are complicated enough that it will be easier to break all bonds in glucose and make all the bonds in CO_2 and CH_3CH_2OH.

Bonds broken:	Bonds formed:
5 C − C (347 kJ/mol)	2 × 2 C = O (799 kJ/mol)
7 C − O (358 kJ/mol)	2 × 5 C − H (413 kJ/mol)
5 O − H (467 kJ/mol)	2 C − O (358 kJ/mol)
7 C − H (413 kJ/mol)	2 O − H (467 kJ/mol)
	2 C − C (347 kJ/mol)

$$\Delta H = 5(347) + 7(358) + 5(467) + 7(413)$$

$$- [4(799) + 10(413) + 2(358) + 2(467) + 2(347)] = -203 \text{ kJ}$$

51. a. $\Delta H° = 2\ \Delta H^\circ_{f,\,HCl} = 2$ mol (-92 kJ/mol) = -184 kJ (= -183 kJ from bond energies)

b. $\Delta H° = 2\ \Delta H^\circ_{f,\,NH_3} = 2$ mol (-46 kJ/mol) = -92 kJ (= -109 kJ from bond energies)

Comparing the values for each reaction, bond energies seem to give a reasonably good estimate of the enthalpy change for a reaction. The estimate is especially good for gas phase reactions.

53. a. Using SF_4 data: $SF_4(g) \rightarrow S(g) + 4\ F(g)$

$$\Delta H° = 4\ D_{SF} = 278.8 + 4\ (79.0) - (-775) = 1370.\ \text{kJ}$$

$$D_{SF} = \frac{1370.\ \text{kJ}}{4 \text{ mol SF bonds}} = 342.5 \text{ kJ/mol}$$

Using SF_6 data: $SF_6(g) \rightarrow S(g) + 6\ F(g)$

$$\Delta H° = 6\ D_{SF} = 278.8 + 6\ (79.0) - (-1209) = 1962 \text{ kJ}$$

$$D_{SF} = \frac{1962 \text{ kJ}}{6} = 327.0 \text{ kJ/mol}$$

b. The S – F bond energy in the table is 327 kJ/mol. The value in the table was based on the S –F bond in SF_6.

c. S(g) and F(g) are not the most stable form of the element at 25 °C. The most stable forms are $S_8(s)$ and $F_2(g)$; $\Delta H^\circ_f = 0$ for these two species.

55. $N_2 + 3\ H_2 \rightarrow 2\ NH_3$; $\Delta H = D_{N_2} + 3\ D_{HN_2} - 6\ D_{NH}$; $\Delta H° = 2(-46 \text{ kJ}) = -92 \text{ kJ}$

$-92 \text{ kJ} = 941 \text{ kJ} + 3(432 \text{ kJ}) - (6\ D_{N\text{-}H})$, $6\ D_{N\text{-}H} = 2329$ kJ, $D_{N\text{-}H} = 388.2$ kJ/mol

Table in text: 391 kJ/mol; There is good agreement.

Lewis Structures and Resonance

57. Drawing Lewis structures is mostly trial and error. However, the first two steps are always the same. These steps are 1) count the valence electrons available in the molecule and 2) attach all atoms to each other with single bonds (called the skeletal structure). Unless noted otherwise, the atom listed first is assumed to be the atom in the middle (called the central atom) and all other atoms in the formulas are attached to this atom. The most notable exceptions to the rule are formulas which begin with H, e.g., H_2O, H_2CO, etc. Hydrogen can never be a central atom since this would require H to have more than two electrons.

After counting valence electrons and drawing the skeletal structure, the rest is trial and error. We place the remaining electrons around the various atoms in an attempt to satisfy the octet rule (or duet rule for H). Keep in mind that practice makes perfect. After practicing you can (and will) become very adept at drawing Lewis structures.

a. HCN has 1 + 4 + 5 = 10 valence electrons.

H—C—N H—C≡N:

Skeletal structure Lewis structure

Skeletal structures uses 4 e^- ; 6 e^- remain

b. PH_3 has 5 + 3(1) = 8 valence electrons.

H—P—H (with H below P) H—P̈—H (with H below P)

Skeletal structure Lewis structure

Skeletal structure uses 6 e^-; 2 e^- remain

c. $CHCl_3$ has 4 + 1 + 3(7) = 26 valence electrons.

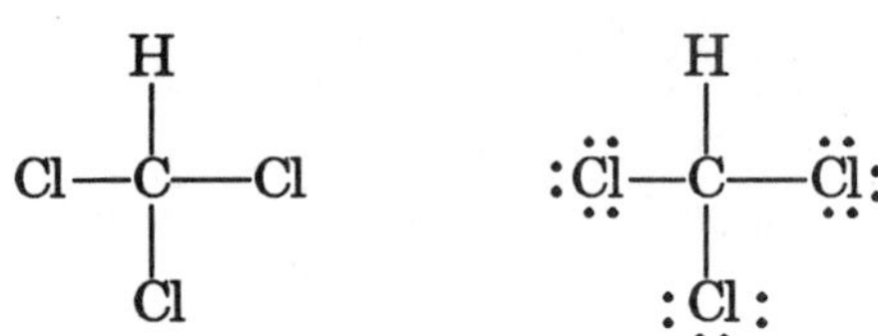

Skeletal structure Lewis structure

Skeletal structure uses 8 e^-; 18 e^- remain

d. NH_4^+ has 5 + 4(1) - 1 = 8 valence electrons.

[H—N—H (with H above and below N)]$^+$

Lewis structure

Note: Subtract valence electrons for positive charged ions.

e. H_2CO has 2(1) + 4 + 6 = 12 valence electrons.

:O: (double bonded to C, with C bonded to H and H)

f. SeF_2 has 6 + 2(7) = 20 valence electrons.

:F̤̈—S̤̈e—F̤̈:

g. CO_2 has 4 + 2(6) = 16 valence electrons.

O=C=O

h. O_2 has 2(6) = 12 valence electrons.

O=O

i. HBr has 1 + 7 = 8 valence electrons.

H—Br:

59. In each case in this problem, the octet rule cannot be satisfied for the central atom. BeH_2 and BH_3 all have too few electrons around the central atom and all the others have to many electrons around the central atom. Always try to satisfy the octet rule for every atom, but when it is impossible, the central atom is the species which will disobey the octet rule.

PF_5, 5 + 5(7) = 40 e^-

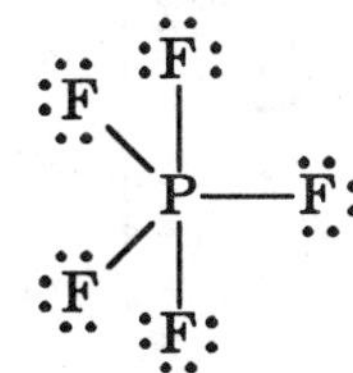

BeH_2, 2 + 2(1) = 4 e^-

H – Be – H

BH_3, 3 + 3(1) = 6 e^-

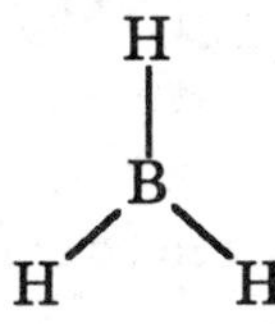

Br_3^-, 3(7) + 1 = 22 e^-

[:Br—Br—Br:]⁻

SF_4, 6 + 4(7) = 34 e^-

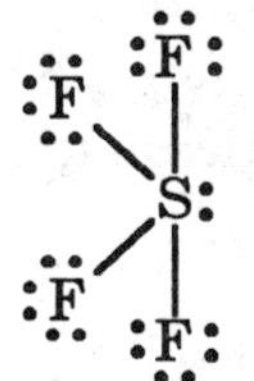

XeF_4, 8 + 4(7) = 36 e^-

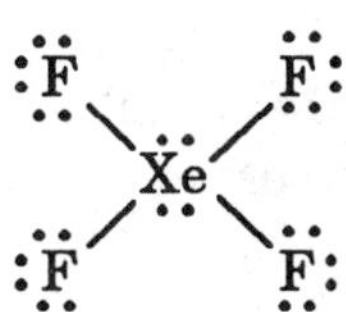

ClF_5, 7 + 5(7) = 42 e^-

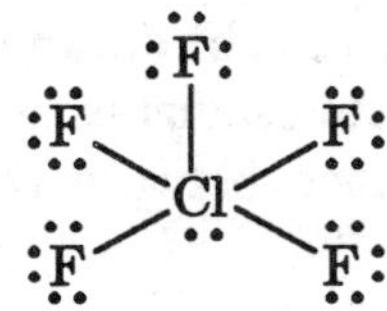

SF_6, 6 + 6(7) = 48 e^-

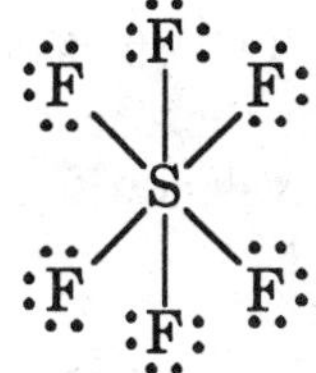

61. a. NO_2^- has 5 + 2(6) + 1 = 18 valence electrons. The skeletal structure is: O – N – O

To get an octet about the nitrogen and only use 18 e^-, we must form a double bond to one of the oxygen atoms.

$$[\,O=N-O:\,]^- \longleftrightarrow [\,:O-N=O\,]^-$$

Since there is no reason to have the double bond to a particular oxygen atom, we can draw two resonance structures. Each Lewis structure uses the correct number of electrons and satisfies the octet rules so each is a valid Lewis structure. Resonance structures occur when you have multiple bonds that can be in various positions. We say the actual structure is an average of these two resonance structures.

NO_3^- has 5 + 3(6) + 1 = 24 valence electrons. We can draw three resonance structures for NO_3^-, with the double bond rotating between the three oxygen atoms.

$$[\,O{=}N(-O)(-O)\,]^- \longleftrightarrow [\,O{-}N(=O)(-O)\,]^- \longleftrightarrow [\,O{-}N(-O)(=O)\,]^-$$

b. OCN^- has 6 + 4 + 5 + 1 = 16 valence electrons. We can draw three resonance structures for OCN^-.

$$[\,:O-C\equiv N:\,]^- \longleftrightarrow [\,O=C=N\,]^- \longleftrightarrow [\,:O\equiv C-N:\,]^-$$

SCN^- has 6 + 4 + 5 + 1 = 16 valence electrons. Three resonance structures can be drawn.

$$[\,:S-C\equiv N:\,]^- \longleftrightarrow [\,S=C=N\,]^- \longleftrightarrow [\,:S\equiv C-N:\,]^-$$

N_3^- has 3(5) + 1 = 16 valence electrons. As with OCN^- and SCN^-, three different resonance structures can be drawn.

$$[\,:N-N\equiv N:\,]^- \longleftrightarrow [\,N=N=N\,]^- \longleftrightarrow [\,:N\equiv N-N:\,]^-$$

63. Benzene has 6(4) + 6(1) = 30 valence electrons. Two resonance structures can be drawn for benzene. The actual structure of benzene is an average of these two resonance structures, that is, all carbon-carbon bonds are equivalent with a bond length and bond strength somewhere between a single and a double bond.

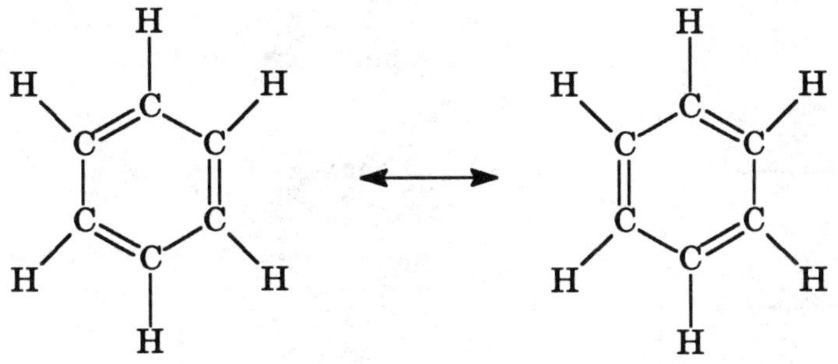

65. We will use a hexagon to represent the six membered carbon ring and we will omit the 4 hydrogen atoms and the three lone pairs of electrons on each chlorine. If no resonance exists, we could draw 4 different molecules:

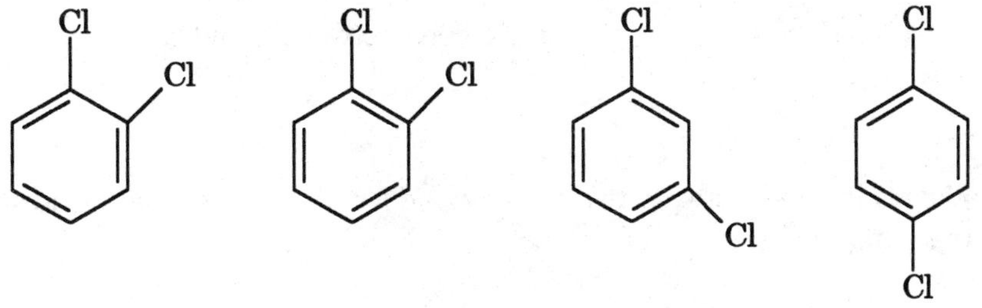

If the double bonds in the benzene ring exhibit resonance, then we can draw only three different dichlorobenzenes. The circle in the hexagon represents the delocalization of the three double bonds in the benzene ring (see Exercise 8.63).

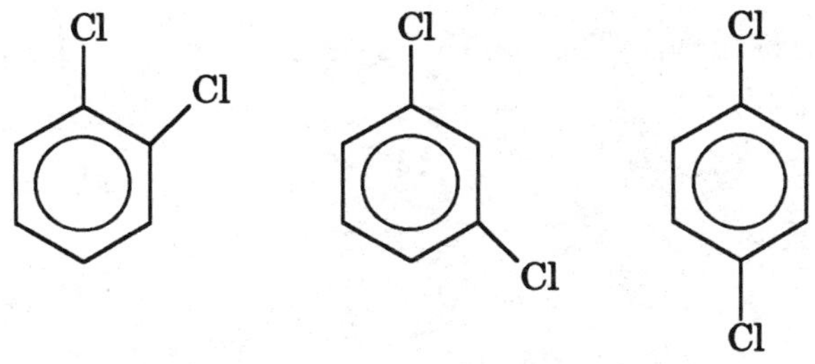

With resonance, all carbon-carbon bonds are equivalent. We can't distinguish between a single and double bond between adjacent carbons that have a chlorine attached. That only 3 isomers are observed provides evidence for the existence of resonance.

67. The Lewis structures for the various species are below:

CO (10 e ⁻): :C≡O: Triple bond between C and O

CO_2 (16 e ⁻): O=C=O Double bond between C and O

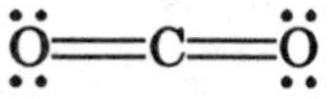

CO_3^{2-} (24 e ⁻): [resonance structures] $^{2-}$ ⟷ $^{2-}$ ⟷ $^{2-}$

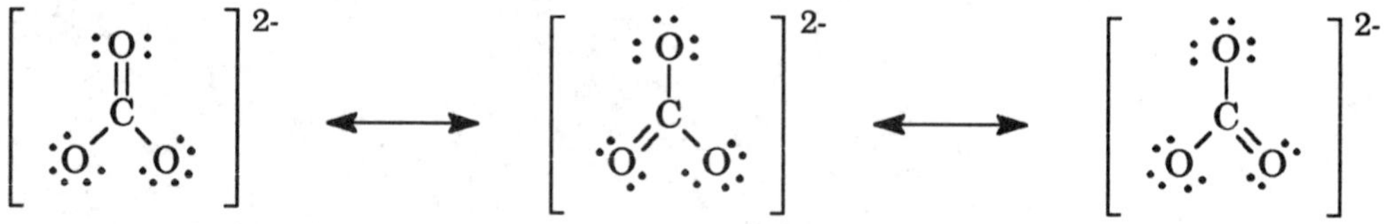

Average of 1 1/3 bond between C and O

CH_3OH (14 e ⁻): H—C(H)(H)—O—H Single bond between C and O

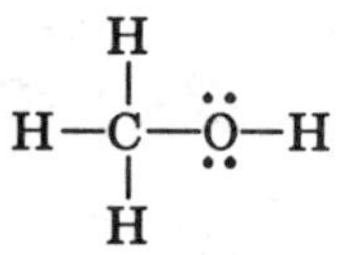

As the number of bonds increase between two atoms, bond length decreases and bond strength increases. With this in mind, then:

longest → shortest C – O bond: $CH_3OH > CO_3^{2-} > CO_2 > CO$

weakest → strongest C – O bond: $CH_3OH < CO_3^{2-} < CO_2 < CO$

Formal Charge

69. BF_3 has 3 + 3(7) = 24 valence electrons. The two Lewis structures to consider are:

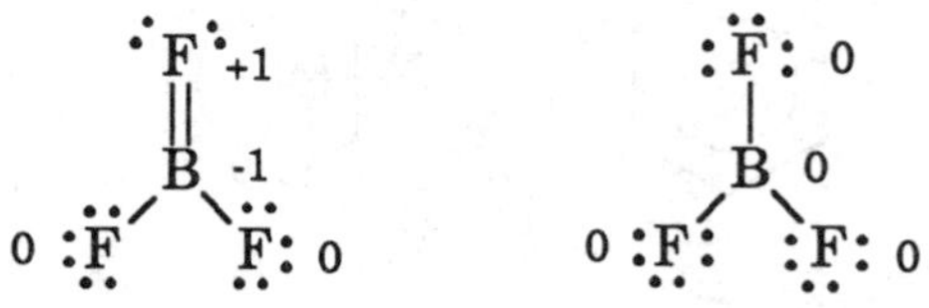

F +1, B -1, F 0, F 0   F 0, B 0, F 0, F 0

The formal charge for the various atoms are assigned in the Lewis structures. Formal charge = number of valence electrons on free atom - number of lone pair electrons on atoms - 1/2 (number of shared electrons of atom). For B in the first Lewis structure, FC = 3 - 0 - 1/2(8) = -1. For F in the first structure with the double bond, FC = 7 - 4 - 1/2(4) = +1. The others all have a formal charge equal to zero.

The first Lewis structure obeys the octet rule but has a +1 formal charge on the most electronegative element there is, fluorine, and a negative formal charge on a much less electronegative element, boron. This is just the opposite of what we expect; negative formal charge on F and positive formal charge on B. The other Lewis structure does not obey the octet rule for B but has a zero formal charge on each element in BF_3. Since structures generally want to minimize formal charge, then BF_3 with only single bonds is best from a formal charge point of view.

71. See Exercise 8.58a for the Lewis structures of $POCl_3$, SO_4^{2-}, ClO_4^- and PO_4^{3-}.

a. $POCl_3$: P, FC = 5 - 1/2(8) = +1

b. SO_4^{2-}: S, FC = 6 - 1/2(8) = +2

c. ClO_4^-: Cl, FC = 7 - 1/2(8) = +3

d. PO_4^{3-}: P, FC = 5 - 1/2(8) = +1

e. SO_2Cl_2, 6 + 2(6) + 2(7) = 32 e^-

```
        ..
       :O:
  ..    |    ..
 :Cl —  S — Cl:
  ..    |    ..
       :O:
        ..
```

S, FC = 6 - 1/2(8) = +2

f. XeO_4, 8 + 4(6) = 32 e^-

```
        ..
       :O:
  ..    |     ..
 :O —  Xe  — O:
  ..    |     ..
       :O:
        ..
```

Xe, FC = 8 - 1/2(8) = +4

g. ClO_3^-, 7 + 3(6) + 1 = 26 e^-

```
 [  ..   ..   ..  ] -
 [ :O — Cl — O:   ]
 [  ..   |    ..  ]
 [      :O:       ]
 [       ..       ]
```

Cl, FC = 7 - 2 - 1/2(6) = +2

h. NO_4^{3-}, 5 + 4(6) + 3 = 32 e^-

```
 [        ..        ] 3-
 [       :O:        ]
 [  ..    |    ..   ]
 [ :O  —  N  — O:   ]
 [  ..    |    ..   ]
 [       :O:        ]
 [        ..        ]
```

N, FC = 5 - 1/2(8) = +1

Molecular Geometry and Polarity

73. The first step always is to draw a valid Lewis strucure when predicting molecular structure. When resonance is possible, only one of the possible resonance structures is necessary to predict the correct structure since all resonance structures give the same structure. The Lewis structures are in Exercises 8.57 and 8.61. The structures and bond angles for each follow.

8.57 a. HCN: linear, 180°

b. PH_3: trigonal pyramid, < 109.5°

c. $CHCl_3$: tetrahedral, 109.5°

d. NH_4^+: tetrahedral, 109.5°

e. H_2CO: trigonal planar, 120°

f. SeF_2: V-shaped or bent, < 109.5°

g. CO_2: linear, 180°

h and i. O_2 and HBr are both linear, but there is no bond angle in either.

Note: PH_3 and SeF_2 both have lone pairs of electrons on the central atom which result in bond angles that are something less than predicted from a tetrahedral arrangement (109.5°). However, we cannot predict the exact number. For these cases, we will just insert a less than sign to show this phenomenon.

8.61 a. NO_2^-: V-shaped, < 120°; NO_3^-: trigonal planar, 120°

b. OCN^-, SCN^- and N_3^- are all linear with 180° bond angles.

75. See Exercise 8.59 for the Lewis structures.

PF_5: trigonal bipyramid, 120° and 90°
BeH_2: linear, 180°
BH_3: trigonal planar, 120°
Br_3^-: linear, 180°
SF_4: see-saw, ≈ 120° and ≈ 90°
XeF_4: square planar, 90°
ClF_5: square pyramid, ≈ 90°
SF_6: octahedral, 90°

Note: Reference Figures 19.25 and 19.26 of the text for molecular structures based on the trigonal bipyramid and octahedral geometries. We will use the term see-saw to describe the molecular structure of SF_4 type molecules instead of distorted tetrahedron.

77. a. BF_3, $3 + 3(7) = 24\ e^-$

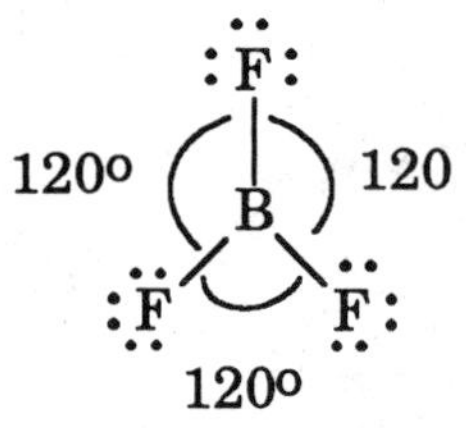

:F:
120o 120
B
:F F:
120o

Trigonal planar; All angles = 120°.

b. BeH_2^{2-}, $2 + 2(1) + 2 = 6\ e^-$

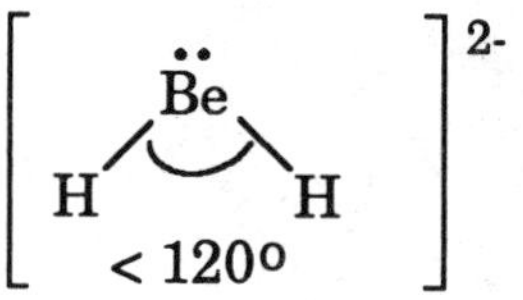

[Be]2-
H H
< 120o

V-shaped; Angle is < 120°.

Note: All of these structures have three effective pairs of electrons about the central atom. All of the structures are based on a trigonal planar geometry, but only BF_3 is described as having a trigonal planar structure. Molecular structure always describes the relative positions of the atoms.

79. a. XeF_2 has $8 + 2(7) = 22$ valence electrons.

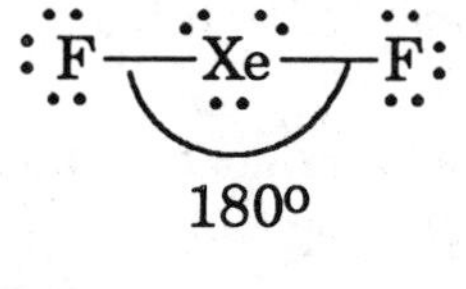

:F—Xe—F:
180o

There are 5 pairs of electrons about the central Xe atom. The structure will be based on a trigonal bipyramid geometry. The most stable arrangement of the atoms in XeF_2 is linear with a 180° bond angle.

b. IF_3 has 7 + 3(7) = 28 valence electrons.

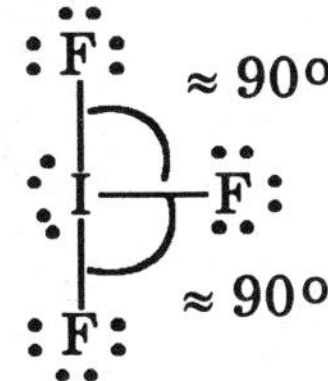

T-shaped; The FIF angles are ≈ 90°. Since the lone pairs will take up more space, the FIF bond angles will probably be slightly less than 90°.

c. IF_4^+ has 7 + 4(7) - 1 = 34 valence electrons.

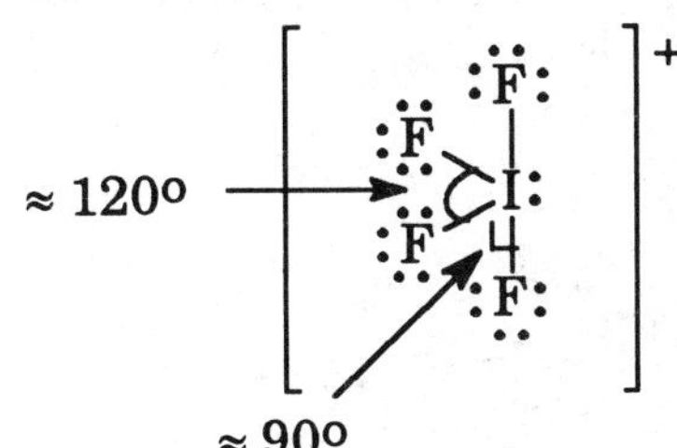

See-saw or teeter-totter or distorted tetrahedron

d. SF_5^+ has 6 + 5(7) - 1 = 40 valence electrons.

Trigonal bipyramid

All of the species in this exercise have 5 pairs of electrons around the central atom. All of the structures are based on a trigonal bipyramid geometry, but only in SF_5^+ are all of the pairs bonding pairs. Thus, SF_5^+ is the only one we describe the molecular structure as trigonal bipyramid. Still, we had to begin with the trigonal bipyramid geometry to get to the structures of the others.

Note: Reference Figures 19.25 and 19.26 of the text for molecular structures based on the trigonal bipyramid geometry. We will use the term see-saw to describe the molecular structure of IF_4^+ type species instead of distorted tetrahedron.

81. BF_3 and BeH_2^{2-} both have polar bonds but only BeH_2^{2-} has a dipole moment. The three bond dipoles from the three polar B – F bonds in BF_3 will all cancel when summed together. Hence, BF_3 is nonpolar since the overall molecule has no resulting dipole moment. In BeH_2^{2-}, the two Be – H bond dipoles do not cancel when summed together, hence BeH_2^{2-} has a dipole moment. Since H is more electronegative than Be, the negative end of the dipole moment is between the two H atoms and the positive end is around the Be atom. The arrow in the following illustration represents the overall dipole moment in BeH_2^{2-}.

83. All have polar bonds, but only IF_3 and IF_4^+ have dipole moments. The bond dipoles from the five S–F bonds in SF_5^+ cancel each other so SF_5^+ has no dipole moment. The bond dipoles in XeF_2 also cancel:

Since the bond dipoles from the two Xe – F bonds are equal in magnitude but point in opposite directions, then they cancel each other and XeF_2 has no dipole moment (is nonpolar). For IF_3 and IF_4^+, the arrangement of these molecules are such that the individual bond dipoles do <u>not</u> all cancel so each has an overall dipole moment.

85. Molecules which have an overall dipole moment are called polar molecules and molecules which do not have an overall dipole moment are called nonpolar molecules.

a. OCl_2, $6 + 2(7) = 20\ e^-$

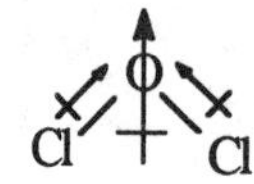

V-shaped, polar; OCl_2 is polar because the two O – Cl bond dipoles don't cancel each other. The resultant dipole moment is shown in the drawing.

KrF_2, $8 + 2(7) = 22\ e^-$

Linear, nonpolar; The molecule is nonpolar because the two Kr – F bond dipoles cancel each other.

BeH_2, $2 + 2(1) = 4\ e^-$

Linear, nonpolar; Be – H bond dipoles are equal and point in opposite directions. They cancel each other. BeH_2 is nonpolar.

SO_2, $6 + 2(6) = 18\ e^-$

V-shaped, polar; The S – O bond dipoles do not cancel so SO_2 is polar (has a dipole moment). Only one resonance structure is shown.

Note: All four species contain three atoms. They have different structures because the number of lone pairs of electrons around the central atom are different in each case.

b. SO_3, $6 + 3(6) = 24\ e^-$

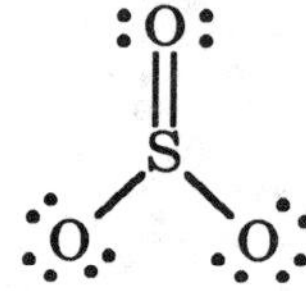

Trigonal planar, nonpolar; Bond dipoles cancel. Only one resonance structure is shown.

NF_3, $5 + 3(7) = 26\ e^-$

Trigonal pyramid, polar; Bond dipoles do not cancel.

ClF_3 has 7 + 3(7) = 28 valence electrons.

T-shaped, polar; Bond dipoles do not cancel.

Note: Each molecule has the same number of atoms, but the structures are different because of differing numbers of lone pairs around each central atom.

c. CF_4, 4 + 4(7) = 32 e^-

Tetrahedral, nonpolar;
Bond dipoles cancel.

SeF_4, 6 + 4(7) = 34 e^-

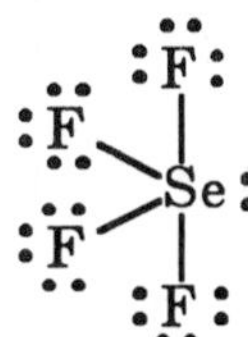

See-saw, polar;
Bond dipoles do not cancel.

XeF_4, 8 + 4(7) = 36 valence electrons

Square planar, nonpolar;
Bond dipoles cancel.

Again, each molecule has the same number of atoms, but a different structure because of differing numbers of lone pairs around the central atom.

d. IF_5, 7 + 5(7) = 42 e^-

Square pyramid, polar;
Bond dipoles do not cancel.

AsF_5, 5 + 5(7) = 40 e^-

Trigonal bipyramid, nonpolar;
Bond dipoles cancel.

Yet again, the molecules have the same number of atoms, but different structures because of the presence of differing numbers of lone pairs.

87. SbF_5, $5 + 5(7) = 40\ e^-$

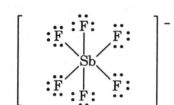

Trigonal bipyramid, nonpolar

SbF_6^-, $5 + 6(7) + 1 = 48\ e^-$

Octahedral, nonpolar (no overall dipole moment)

HF, $1 + 7 = 8\ e^-$

Linear, polar

H_2F^+, $2(1) + 7 - 1 = 8\ e^-$

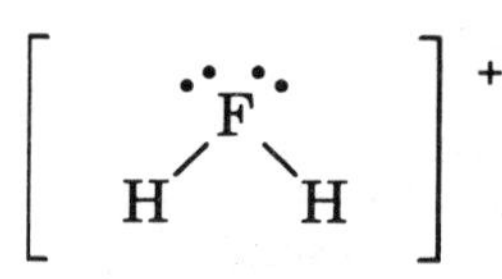

V-shaped, polar (has an overall dipole moment)

89. All these molecules have polar bonds that are symmetrically arranged about the central atoms. In each molecule, the individual bond dipoles cancel to give no net overall dipole moment.

Additional Exercises

91. All these series of ions are isoelectronic series. In each series, the smallest ion has the largest number of protons holding the constant number of electrons. The smallest ion will also have the largest ionization energy. The ion with the fewest protons will be largest and will have the smallest ionization energy.

a. radius trend: $Mg^{2+} < Na^+ < F^- < O^{2-}$; IE trend: $O^{2-} < F^- < Na^+ < Mg^{2+}$

b. radius trend: $Ca^{2+} < P^{3-}$; IE trend: $P^{3-} < Ca^{2+}$

c. radius trend: $K^+ < Cl^- < S^{2-}$; IE trend: $S^{2-} < Cl^- < K^+$

93. a. $Na^+(g) + Cl^-(g) \rightarrow NaCl(s)$ b. $NH_4^+(g) + Br^-(g) \rightarrow NH_4Br(s)$

c. $Mg^{2+}(g) + S^{2-}(g) \rightarrow MgS(s)$ d. $O_2(g) \rightarrow 2\ O(g)$

95. CO_3^{2-} has $4 + 3(6) + 2 = 24$ valence electrons.

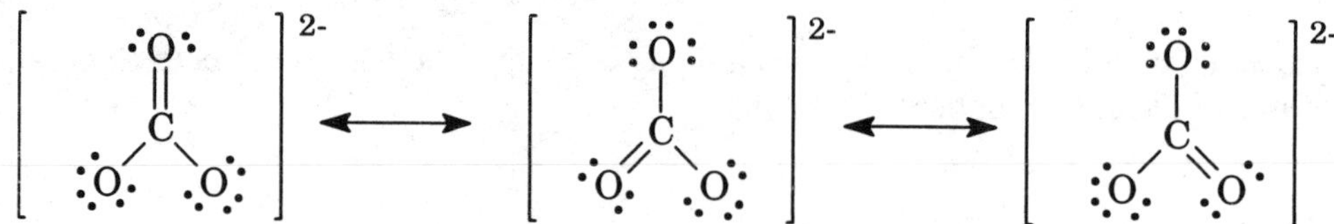

HCO_3^- has 1 + 4 + 3(6) + 1 = 24 valence electrons.

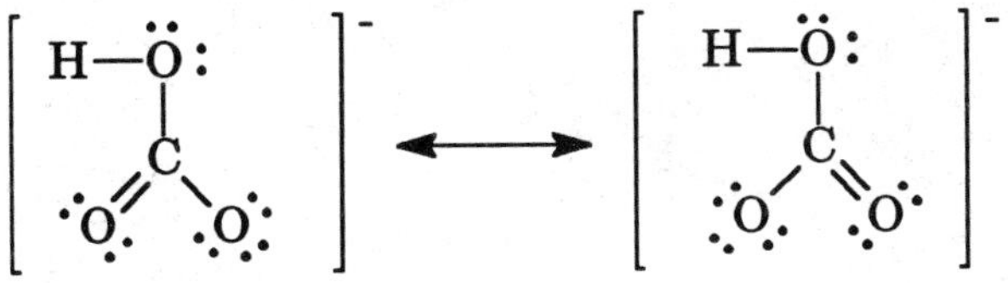

H_2CO_3 has 2(1) + 4 + 3(6) = 24 valence electrons.

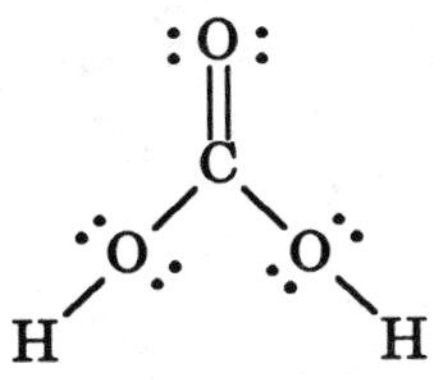

The Lewis structures for the reactants and products are:

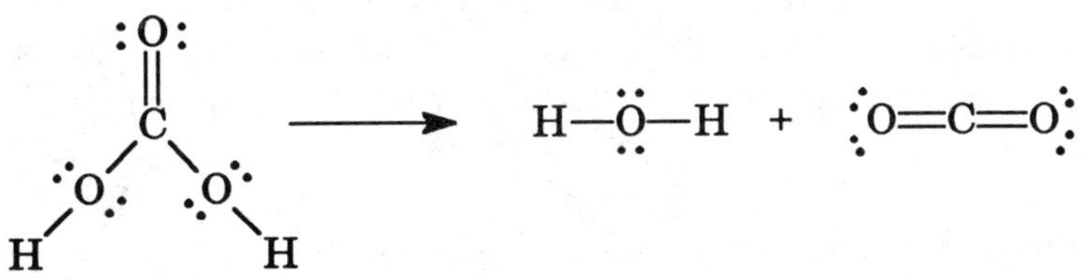

H—O—H + O=C=O

Bonds broken:

2 C – O (358 kJ/mol)
1 O – H (467 kJ/mol)

Bonds formed:

1 C = O (799 kJ/mol)
1 O – H (467 kJ/mol)

ΔH = 2(358) + 467 - [799 + 467] = -83 kJ; The carbon-oxygen double bond is stronger than two carbon-oxygen single bonds, hence CO_2 and H_2O are more stable than H_2CO_3.

97. a. NO_2, 5 + 2(6) = 17 e^-

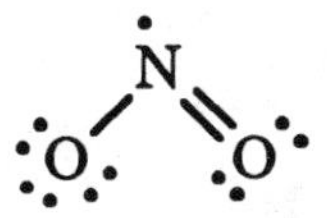

N_2O_4, 2(5) + 4(6) = 34 e^-

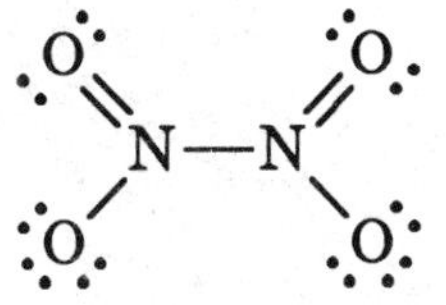

plus other resonance structures

plus other resonance structures

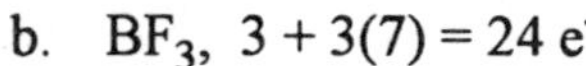

b. BF_3, $3 + 3(7) = 24\ e^-$

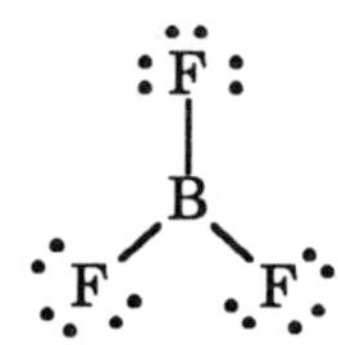

NH_3, $5 + 3(1) = 8\ e^-$

BF_3NH_3, $24 + 8 = 32\ e^-$

In reaction a, NO_2 has an odd number of electrons so it is impossible to satisfy the octet rule. By dimerizing to form N_2O_4, the odd electron on two NO_2 molecules can pair up giving a species whose Lewis structure can satisfy the octet rule. In general odd electron species and very reactive. In reaction b, BF_3 can be considered electron deficient. Boron only has six electrons around it. By forming BF_3NH_3, the boron atom satisfies the octet rule by accepting a lone pair of electrons from NH_3 to form a fourth bond.

99. The general structure of the trihalide ions is: $[X—X—X]^-$

Bromine and iodine are large enough and have low energy, empty d-orbitals to accommodate the expanded octet. Fluorine is small and its valence shell contains only 2s and 2p orbitals (4 orbitals) and cannot expand its octet. The lowest energy d orbitals in F are 3d; they are too high in energy compared to 2s and 2p to be used in bonding.

101. Yes, each structure has the same number of effective pairs around the central atom. (A multiple bond is counted as a single group of electrons.)

103. a.

Angles a and b are both ≈ 120°.

Angles c and d are both a little less than 120° (due to the lone pairs).

b. The N – F bond dipoles cancel in the first structure so it is nonpolar. The N – F bond dipoles do not cancel in the second structure so it is polar.

Challenge Problems

105. $C \equiv O$ (1072 kJ/mol) and $N \equiv N$ (941 kJ/mol); CO is polar while N_2 is nonpolar. This may lead to a great reactivity for the CO bond.

107. Let us look at the complete cycle for Li_2S.

$2\ Li(s) \rightarrow 2\ Li(g)$	$2\ \Delta H_{sub,\ Li} = 2(161)$ kJ
$2\ Li(g) \rightarrow 2\ Li^+(g) + 2\ e^-$	$2\ IE = 2(520.)$ kJ
$S(s) \rightarrow S(g)$	$\Delta H_{sub,\ S} = 277$ kJ
$S(g) + e^- \rightarrow S^-(g)$	$EA_1 = -200.$ kJ
$S^-(g) + e^- \rightarrow S^{2-}(g)$	$EA_2 = ?$
$2\ Li^+(g) + S^{2-}(g) \rightarrow Li_2S$	$LE = -2472$ kJ
$2\ Li(s) + S(s) \rightarrow Li_2S(s)$	$\Delta H_f^\circ = -500.$ kJ

$\Delta H_f^\circ = 2\ \Delta H_{sub,\ Li} + 2\ IE + \Delta H_{sub,\ S} + EA_1 + EA_2 + LE$, $-500. = -1033 + EA_2$, $EA_2 = 533$ kJ

For each salt: $\Delta H_f^\circ = 2\ \Delta H_{sub,\ M} + 2\ IE + 277 - 200. + LE + EA_2$

Na_2S: $-365 = 2(109) + 2(495) + 277 - 200. - 2203 + EA_2$; $EA_2 = 553$ kJ

K_2S: $-381 = 2(90.) + 2(419) + 277 - 200. - 2052 + EA_2$; $EA_2 = 576$ kJ

Rb_2S: $-361 = 2(82) + 2(409) + 277 - 200. - 1949 + EA_2$; $EA_2 = 529$ kJ

Cs_2S: $-360 = 2(78) + 2(382) + 277 - 200. - 1850 + EA_2$; $EA_2 = 493$ kJ

We get values from 493 to 576 kJ.

The mean value is: $\dfrac{533 + 553 + 576 + 529 + 493}{5} = 537$ kJ

We can represent the results as $EA_2 = 540 \pm 50$ kJ.

109. a. $N(NO_2)_2^-$ contains $5 + 2(5) + 4(6) + 1 = 40$ valence electrons.

The most likely structures are:

There are other possible resonance structures, but these are most likely.

b. The NNN and all ONN and ONO bond angles should be about 120°.

111. The nitrogen-nitrogen bond length of 112 pm is between a double (120 pm) and a triple (110 pm) bond. The nitrogen-oxygen bond length of 119 pm is between a single (147 pm) and a double bond (115 pm). The last resonance structure doesn't appear to be as important as the other two since there is no evidence from bond lengths for a nitrogen-oxygen triple bond or a nitrogen-nitrogen single bond as in the third resonance form. We can adequately describe the structure of N_2O using the resonance forms:

$$N{=}N{=}O \longleftrightarrow :N{\equiv}N{-}O:$$

Assigning formal charges for all 3 resonance forms:

$$\underset{-1\quad +1\quad 0}{N{=}N{=}O} \longleftrightarrow \underset{0\quad +1\quad -1}{:N{\equiv}N{-}O:} \longleftrightarrow \underset{-2\quad +1\quad +1}{:N{-}N{\equiv}O:}$$

For:

$(N{=})$, FC = 5 - 4 - 1/2(4) = -1

$({=}N{=})$, FC = 5 - 1/2(8) = +1, Same for $({\equiv}N{-})$ and $({-}N{\equiv})$

$(:\ddot{\underset{..}{N}}-)$, FC = 5 - 6 - 1/2(2) = -2; $(:N\equiv)$, FC = 5 - 2 - 1/2(6) = 0

$(=\ddot{\underset{..}{O}})$, FC = 6 - 4 - 1/2(4) = 0; $(-\ddot{\underset{..}{O}}:)$, FC = 6 - 6 - 1/2(2) = -1

$(\equiv O:)$, FC = 6 - 2 - 1/2(6) = +1

We should eliminate N – N ≡ O since it has a formal charge of +1 on the most electronegative element (O). This is consistent with the observation that the N – N bond is between a double and triple bond and that the N – O bond is between a single and double bond.

CHAPTER NINE

COVALENT BONDING: ORBITALS

Questions

7. Bond energy is directly proportional to bond order. Bond length is inversely proportional to bond order. Bond energy and bond length can be measured.

9. Paramagnetic: Unpaired electrons are present. Measure the mass of a substance in the presence and absence of a magnetic field. A substance with unpaired electrons will be attracted by the magnetic field, giving an apparent increase in mass in the presence of the field. Greater number of unpaired electrons will give greater attraction and greater observed mass increase.

Exercises

The Localized Electron Model and Hybrid Orbitals

11. H_2O has 2(1) + 6 = 8 valence electrons.

O
H H

H_2O has a tetrahedral arrangement of the electron pairs about the O atom which requires sp^3 hybridization. Two of the four sp^3 hybrid orbitals are used to form bonds to the two hydrogen atoms and the other two sp^3 hybrid orbitals hold the two lone pairs on oxygen. The two O – H bonds are formed from overlap of the sp^3 hybrid orbitals on oxygen with the 1s atomic orbitals on the hydrogen atoms.

13. H_2CO has 2(1) + 4 + 6 = 12 valence electrons.

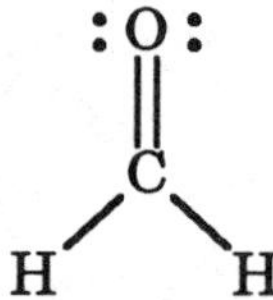

The central carbon atom has a trigonal planar arrangement of the electron pairs which requires sp^2 hybridization. The two C – H sigma bonds are formed from overlap of the sp^2 hybrid orbitals on carbon with the hydrogen 1s atomic orbitals. The double bond between carbon and oxygen consists of one σ and one π bond. The oxygen atom, like the carbon atom, also has a trigonal planar arrangement of the electrons which requires sp^2 hybridization. The σ bond in the double bond is formed from overlap of a carbon sp^2 hybrid orbital with an oxygen sp^2 hybrid orbital. The π bond in the double bond is formed from overlap of the unhybridized p atomic orbitals. Carbon and oxygen each have one unhybridized p atomic orbital which are parallel to each other. When two parallel p atomic orbitals overlap, a π bond results.

15. See Exercises 8.57 and 8.61 for the Lewis structures. To predict the hybridization, first determine the arrangement of electrons pairs about each central using the VSEPR model, then utilize the information in Figure 9.24 of the text to deduce the hybridization required for that arrangement of electron pairs.

8.57
a. HCN; C is sp hybridized.
b. PH_3; P is sp^3 hybridized.
c. $CHCl_3$; C is sp^3 hybridized.
d. NH_4^+; N is sp^3 hybridized.
e. H_2CO; C is sp^2 hybridized.
f. SeF_2; Se is sp^3 hybridized.
g. CO_2; C is sp hybridized.
h. O_2; Each O atom is sp^2 hybridized.
i. HBr; Br is sp^3 hybridized.

8.61
a. In NO_2^-, N is sp^2 hybridized and in NO_3^-, N is also sp^2 hybridized.
b. In OCN^- and SCN^-, the central carbon atoms in each ion are sp hybridized and in N_3^-, the central N atom is also sp hybridized.

17. See Exercise 8.59 for the Lewis structures.

PF_5:	P is dsp^3 hybridized.	BeH_2:	Be is sp hybridized.
BH_3:	B is sp^2 hybridized.	Br_3^-:	Br is dsp^3 hybridized.
SF_4:	S is dsp^3 hybridized.	XeF_4:	Xe is d^2sp^3 hybridized.
ClF_5:	Cl is d^2sp^3 hybridized.	SF_6:	S is d^2sp^3 hybridized.

19. All have a trigonal planar arrangement of electron pairs about the central atom so all have central atoms with sp^2 hybridization. See Exercise 8.77 for the Lewis structures.

21. All have central atoms with dsp^3 hybridization since all are based on the trigonal bipyramid arrangement of electron pairs. See Exercise 8.79 for the Lewis structures.

23. a. b.

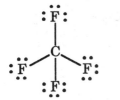

tetrahedral 109.5° sp^3 nonpolar

trigonal pyramid ≈ 109.5° sp^3 polar

Note: The angles in NF_3 should be slightly less than 109.5° because the lone pair requires more room than the bonding pairs. In Chapter 9, we will ignore this effect and say bond angles are ≈ 109.5°.

c.

V-shaped ≈ 109.5° sp^3 polar

d.

trigonal planar 120° sp^2 nonpolar

e.

H—Be—H

linear 180° sp nonpolar

f.

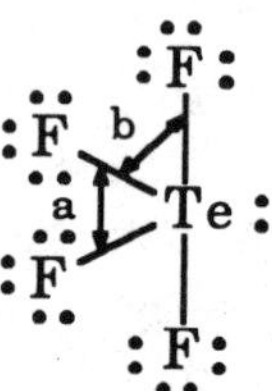

see-saw
a. ≈ 120°, b. ≈ 90°
dsp^3 polar

g.

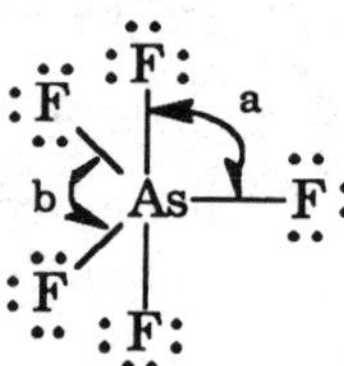

trigonal bipyramid dsp^3
a. 90°, b. 120° nonpolar

h.

F—Kr—F

linear 180° dsp^3 nonpolar

i.

square planar 90° d^2sp^3 nonpolar

j.

octahedral 90° d^2sp^3 nonpolar

k.

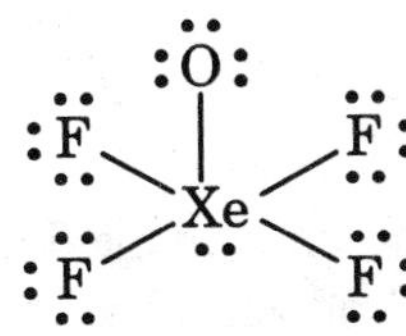

square pyramid, ≈ 90°, d^2sp^3, polar

l.

T-shaped, ≈ 90°, dsp^3, polar

m.

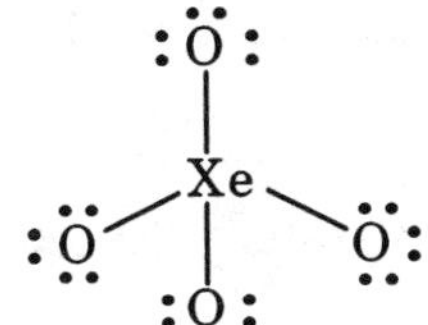

tetrahedral
109.5°
sp^3
nonpolar

25.

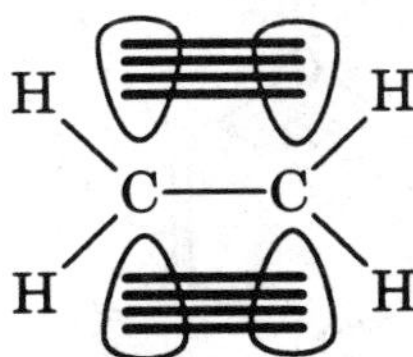

For the p-orbitals to properly line up to form the π bond, all six atoms are forced into the same plane. If the atoms are not in the same plane, then the π bond could not form since the p-orbitals would no longer be parallel to each other.

27. To complete the Lewis structures, just add lone pairs of electrons to satisfy the octet rule for the atoms with fewer than eight electrons.

Biacetyl ($C_4H_6O_2$) has 4(4) + 6(1) + 2(6) = 34 valence electrons.

All CCO angles are 120°. The six atoms are not in the same plane because of free rotation about the carbon-carbon single (sigma) bonds. There are 11 σ and 2 π bonds in biacetyl.

Acetoin ($C_4H_8O_2$) has 4(4) + 8(1) + 2(6) = 36 valence electrons.

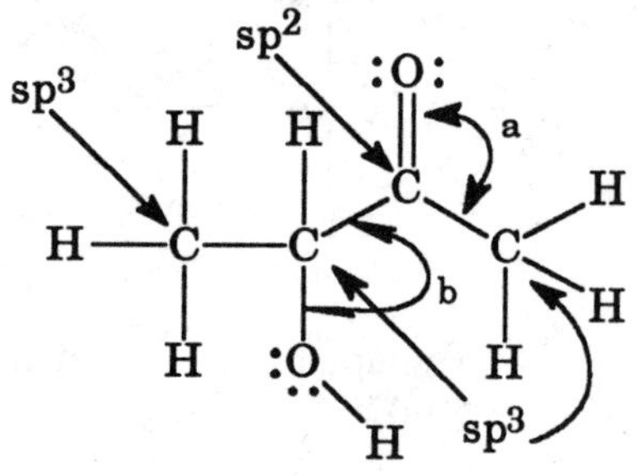

The carbon with the doubly bonded O is sp^2 hybridized. The other 3-C atoms are sp^3 hybridized. Angle a = 120° and angle b = 109.5°. There are 13 σ and 1 π bonds in acetoin.

Note: All single bonds are σ bonds, all double bonds are one σ and one π bond and all triple bonds are one σ and two π bonds.

29. To complete the Lewis structure, just add lone pairs of electrons to satisfy the octet rule for the atoms that have fewer than eight electrons.

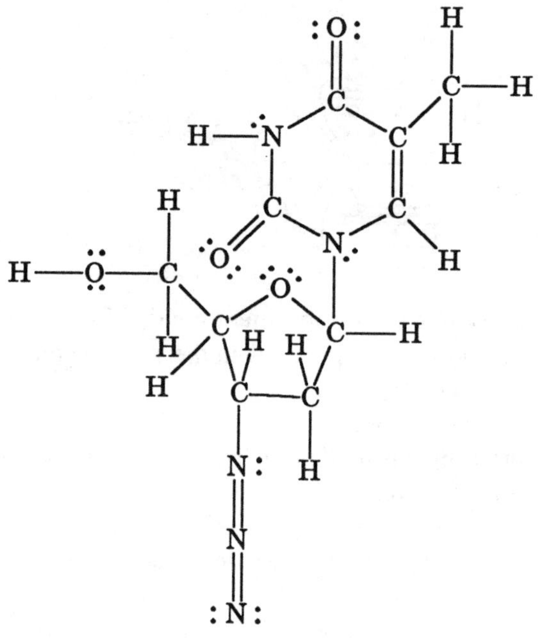

a. 6 b. 4 c. The center N in –N=N=N group

d. 33 σ e. 5 π bonds f. 180°

g. ≈ 109.5° h. sp^3

31. a. Add lone pairs to complete octets for each O and N.

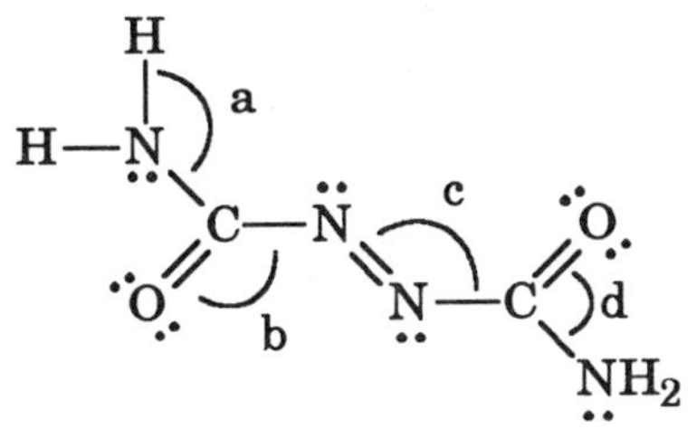

azidocarbonamide

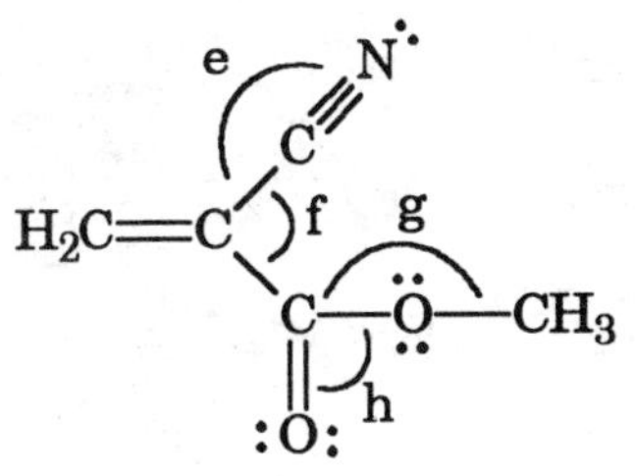

methyl cyanoacrylate

Note: NH_2, CH_2 and CH_3 are shorthand for carbon atoms singly bonded to hydrogen atoms.

b. In azidocarbonamide, the two carbon atoms are sp^2 hybridized. The two nitrogen atoms with hydrogens attached are sp^3 hybridized and the other two nitrogens are sp^2 hybridized. In methyl cyanoacrylate, the CH_3 carbon is sp^3 hybridized, the carbon with the triple bond is sp hybridized, and the other three carbons are sp^2 hybridized.

c. Azidocarbonamide contains 3π bonds and methyl cyanoacrylate contains 4π bonds.

d. a. $\approx 109.5°$ b. 120° c. $\approx 120°$ d. 120° e. 180°

f. 120° g. $\approx 109.5°$ h. 120°

The Molecular Orbital Model

33. If we calculate a non-zero bond order for a molecule, then we predict that it can exist (is stable).

a. H_2^+: $(\sigma_{1s})^1$ B.O. = (1-0)/2 = 1/2, stable
H_2: $(\sigma_{1s})^2$ B.O. = (2-0)/2 = 1, stable
H_2^-: $(\sigma_{1s})^2(\sigma_{1s}*)^1$ B.O. = (2-1)/2 = 1/2, stable
H_2^{2-}: $(\sigma_{1s})^2(\sigma_{1s}*)^2$ B.O. = (2-2)/2 = 0, not stable

b. He_2^{2+}: $(\sigma_{1s})^2$ B.O. = (2-0)/2 = 1, stable
He_2^+: $(\sigma_{1s})^2(\sigma_{1s}*)^1$ B.O. = (2-1)/2 = 1/2, stable
He_2: $(\sigma_{1s})^2(\sigma_{1s}*)^2$ B.O. = (2-2)/2 = 0, not stable

35. The electron configurations are:

a. H_2: $(\sigma_{1s})^2$ B.O. = (2-0)/2 = 1, diamagetic (0 unpaired e^-)
b. B_2: $(\sigma_{2s})^2(\sigma_{2s}*)^2(\pi_{2p})^2$ B.O. = (4-2)/2 = 1, paramagnetic (2 unpaired e^-)
c. F_2: $(\sigma_{2s})^2(\sigma_{2s}*)^2(\sigma_{2p})^2(\pi_{2p})^4(\pi_{2p}*)^4$ B.O. = (8-6)/2 = 1, diamagnetic (0 unpaired e^-)

37. The electron configurations are:

O_2^+: $(\sigma_{2s})^2(\sigma_{2s}*)^2(\sigma_{2p})^2(\pi_{2p})^4(\pi_{2p}*)^1$

O_2: $(\sigma_{2s})^2(\sigma_{2s}*)^2(\sigma_{2p})^2(\pi_{2p})^4(\pi_{2p}*)^2$

O_2^-: $(\sigma_{2s})^2(\sigma_{2s}*)^2(\sigma_{2p})^2(\pi_{2p})^4(\pi_{2p}*)^3$

O_2^{2-}: $(\sigma_{2s})^2(\sigma_{2s}*)^2(\sigma_{2p})^2(\pi_{2p})^4(\pi_{2p}*)^4$

	O_2^+	O_2	O_2^-	O_2^{2-}
Bond order	2.5	2	1.5	1
# of unpaired electrons	1	2	1	0

Bond energy: $O_2^{2-} < O_2^- < O_2 < O_2^+$; Bond length: $O_2^+ < O_2 < O_2^- < O_2^{2-}$

Bond energy is directly proportional to bond order and bond length is inversely proportional to bond order.

39. C_2^{2-} has 10 valence electrons. The Lewis structure predicts sp hybridization for each carbon with two unhybridized p orbitals on each carbon.

$[:C\equiv C:]^{2-}$ sp hybrids orbitals form the σ bond and the two unhybridized p atomic orbitals from each carbon form the two π bonds.

MO: $(\sigma_{2s})^2(\sigma_{2s}*)^2(\pi_{2p})^4(\sigma_{2p})^2$, B.O. = (8 - 2)/2 = 3

Both give the same picture, a triple bond composed of a σ and two π-bonds. Both predict the ion to be diamagnetic. Lewis structures deal well with diamagnetic (all electrons paired) species. The Lewis model cannot really predict magnetic properties.

41. The electron configurations are:

a. CN^+: $(\sigma_{2s})^2(\sigma_{2s}^*)^2(\pi_{2p})^4$ B.O. = (6-2)/2 = 2, diamagnetic
b. CN: $(\sigma_{2s})^2(\sigma_{2s}^*)^2(\pi_{2p})^4(\sigma_{2p})^1$ B.O. = (7-2)/2 = 2.5, paramagnetic
c. CN^-: $(\sigma_{2s})^2(\sigma_{2s}^*)^2(\pi_{2p})^4(\sigma_{2p})^2$ B.O. = 3, diamagnetic

43. The bond orders are: CN^+, 2; CN, 2.5; CN^-, 3; Since bond order is directly proportional to bond energy and inversely proportional to bond length, then:

shortest → longest bond length: $CN^- < CN < CN^+$

lowest → highest bond energy: $CN^+ < CN < CN^-$

45. The two types of overlap that result in bond formation for p orbitals are side to side overlap (π bond) and head to head overlap (σ bond).

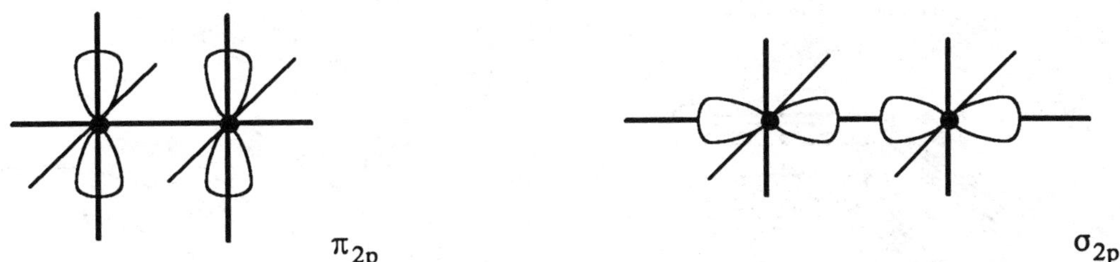

47. a. The electron density would be closer to F on the average. The F atom is more electronegative than the H atom and the 2p orbital of F is lower in energy than the 1s orbital of H.

b. The bonding MO would have more fluorine 2p character since it is closer in energy to the fluorine 2p atomic orbital.

c. The antibonding MO would place more electron density closer to H and would have a greater contribution from the higher energy hydrogen 1s atomic orbital.

49. O_3 and NO_2^- are isoelectronic, so we only need consider one of them since the same bonding ideas apply to both. The Lewis structures for O_3 are:

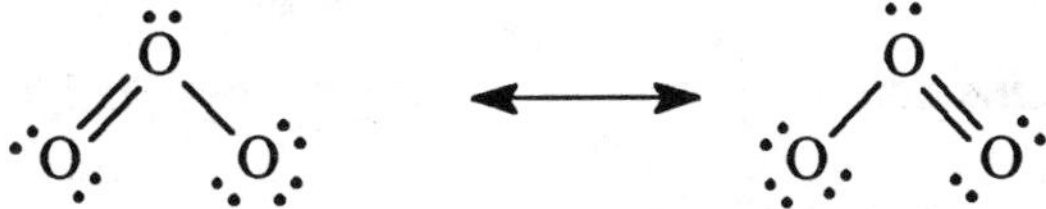

For each of the two resonance forms, the central O atom is sp^2 hybridized with one unhybridized p atomic orbital. The sp^2 hybrid orbitals are used to form the two sigma bonds to the central atom. The localized electron view of the π bond utilizes unhybridized p atomic orbitals. The π bond resonates between the two positions in the Lewis structures:

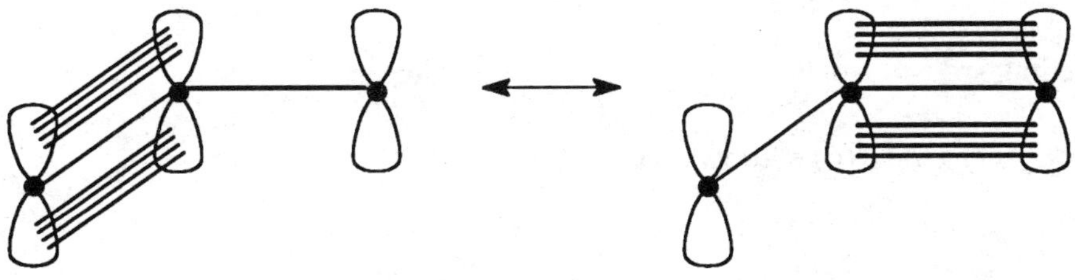

In the MO picture of the π bond, all three unhybridized p-orbitals overlap at the same time, resulting in π electrons that are delocalized over the entire surface of the molecule. This is represented as:

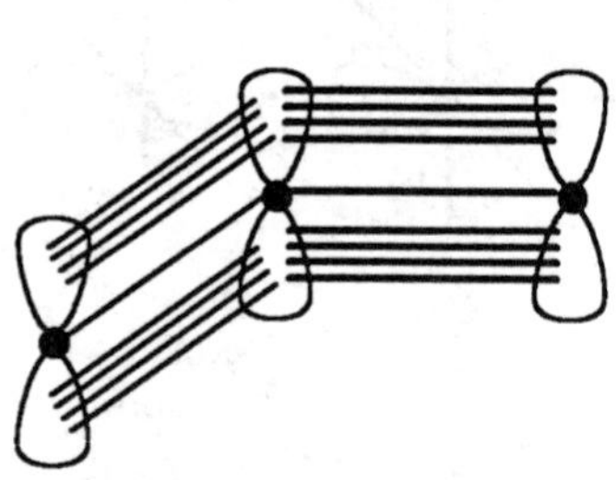

or

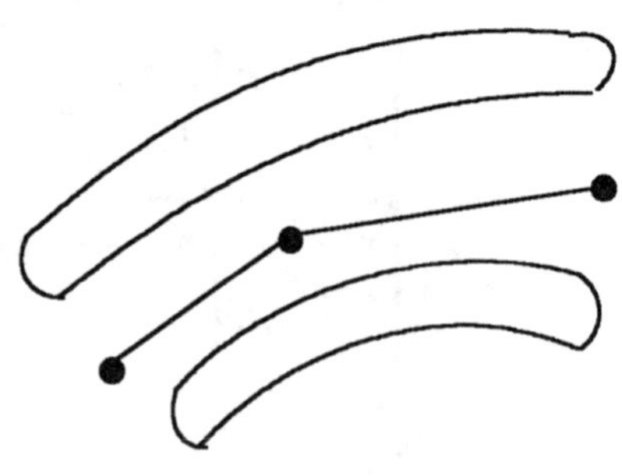

Additional Exercises

51. a. FClO, $7 + 7 + 6 = 20\ e^-$

V-shaped, sp^3 hybridization

b. $FClO_2$, $7 + 7 + 2(6) = 26\ e^-$

trigonal pyramid, sp^3

c. $FClO_3$, $7 + 7 + 3(6) = 32\ e^-$

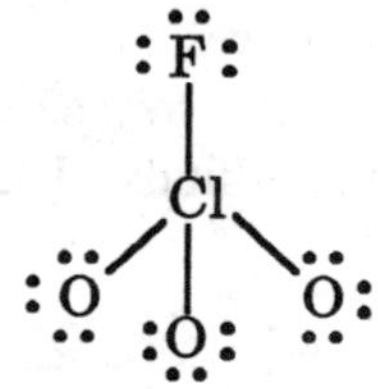

tetrahedral, sp^3

d. F_3ClO, $3(7) + 7 + 6 = 34\ e^-$

see-saw, dsp^3

e. F_3ClO_2, $3(7) + 7 + 2(6) = 40$ valence e^-

trigonal bipyramid, dsp^3

Note: Two additional Lewis structures are possible, depending on the positions of the oxygen atoms.

53. a. There are 33 σ and 9 π bonds.

b. All C atoms are sp^2 hybridized since all have a trigonal planar arrangement of the electrons.

55. a. $COCl_2$ has 4 + 6 + 2(7) = 24 valence electrons.

trigonal planar
polar
120°
sp^2

b. N_2F_2 has 2(5) + 2(7) = 24 valence electrons.

Can also be:

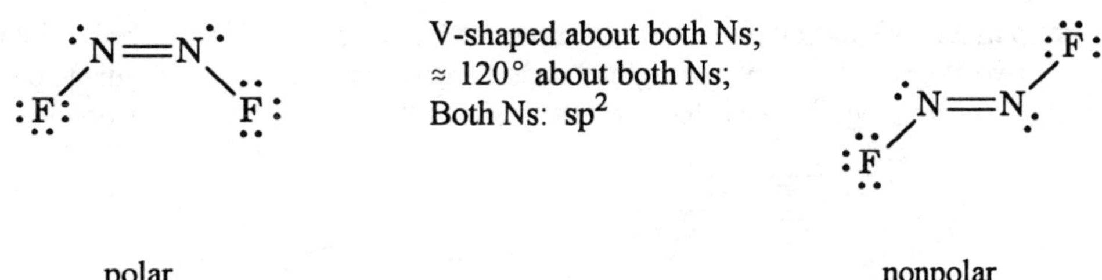

V-shaped about both Ns;
≈ 120° about both Ns;
Both Ns: sp^2

polar | nonpolar

These are distinctly different molecules.

c. COS has 4 + 6 + 6 = 16 valence electrons.

:O=C=S: linear, polar, 180°, sp

d. ICl_3 has 7 + 3(7) = 28 valence electrons.

T-shaped
polar
a. ≈ 90°
dsp^3

57. a. The Lewis structures for NNO and NON are:

:N=N=O: ⟷ :N≡N—N: ⟷ :N—N≡O:

:N=O=N: ⟷ :N≡O—N: ⟷ :N—O≡N:

The NNO structure is correct. From the Lewis structures we would predict both NNO and NON to be linear. However, we would predict NNO to be polar and NON to be nonpolar. Since experiments show N_2O to be polar, then NNO is the correct structure.

b. Formal charge = number of valence electrons of atoms - [(number of lone pair electrons) + 1/2 (number of shared electrons)].

$$:\ddot{N}=N=\ddot{O}: \longleftrightarrow :N\equiv N-\ddot{\underset{..}{O}}: \longleftrightarrow :\underset{..}{\ddot{N}}-N\equiv O:$$

-1 +1 0   0 +1 -1   -2 +1 +1

The formal charges for the atoms in the various resonance structures are below each atom. The central N is sp hybridized in all of the resonance structures. We can probably ignore the 3rd resonance structure on the basis of the relatively large formal charges as compared to the first two resonance structures.

c. The sp hybrid orbitals on the center N overlap with atomic orbitals (or hybrid orbitals) on the other two atoms to form the two sigma bonds. The remaining two unhybridized p orbitals on the center N overlap with two p orbitals on the peripheral N to form the two π bonds.

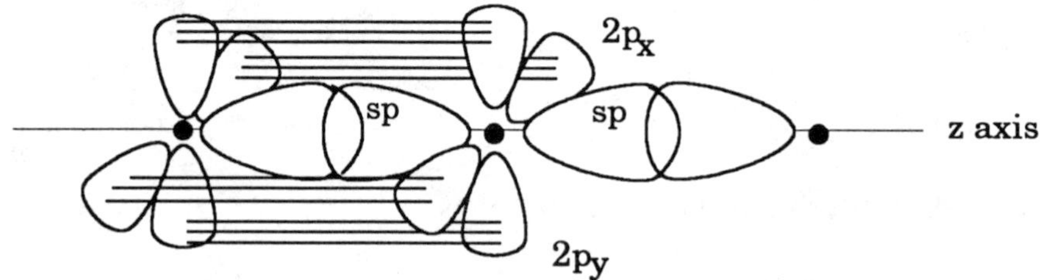

59. N_2 (ground state): $(\sigma_{2s})^2(\sigma_{2s}^*)^2(\pi_{2p})^4(\sigma_{2p})^2$, B.O. = 3, diamagnetic (0 unpaired e^-)

N_2 (1st excited state): $(\sigma_{2s})^2(\sigma_{2s}^*)^2(\pi_{2p})^4(\sigma_{2p})^1(\pi_{2p}^*)^1$

B.O. = (7-3)/2 = 2, paramagnetic (2 unpaired e^-)

The first excited state of N_2 should have a weaker bond and should be paramagnetic.

Challenge Problems

61. a. No, some atoms are in different places. Thus, these are not resonance structures; they are different compounds.

b. For the first Lewis structure, all nitrogens are sp^3 hybridized and all carbons are sp^2 hybridized. In the second Lewis structure, all nitrogens and carbons are sp^2 hybridized.

c. For the reaction:

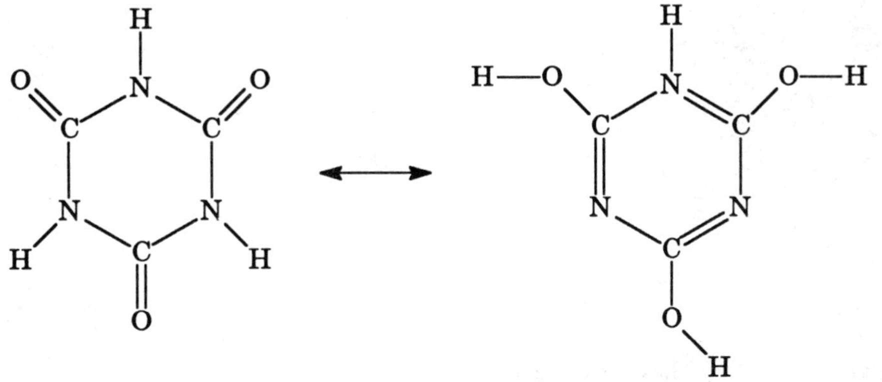

[Structures: cyanuric acid (triketo form with three C=O, three N–H) ⟷ enol form with three C=N and three C–O–H]

Bonds broken:

3 C = O (745 kJ/mol)
3 C – N (305 kJ/mol)
3 N – H (391 kJ/mol)

Bonds formed:

3 C = N (615 kJ/mol)
3 C – O (358 kJ/mol)
3 O – H (467 kJ/mol)

$\Delta H = 3(745) + 3(305) + 3(391) - [3(615) + 3(358) + 3(467)]$

$\Delta H = 4323 \text{ kJ} - 4320 \text{ kJ} = 3 \text{ kJ}$

The bonds are slightly stronger in the first structure with the carbon-oxygen double bonds since ΔH for the reaction is positive. However, the value of ΔH is so small that the best conclusion is that the bond strengths are comparable in the two structures.

63. a. NCN^{2-} has 5 + 4 + 5 + 2 = 16 valence electrons.

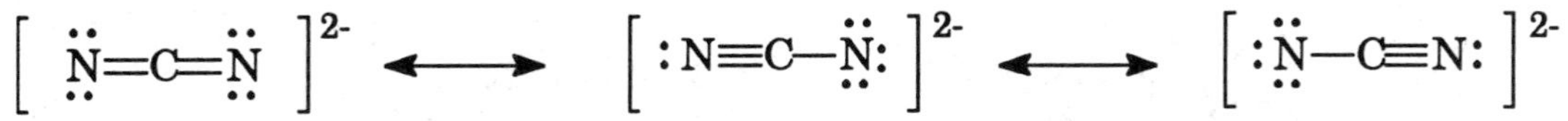

$[\,\ddot{\underset{..}{N}}=C=\ddot{\underset{..}{N}}\,]^{2-} \longleftrightarrow [\,:N\equiv C-\ddot{\underset{..}{N}}:\,]^{2-} \longleftrightarrow [\,:\ddot{\underset{..}{N}}-C\equiv N:\,]^{2-}$

H_2NCN has 2(1) + 5 + 4 + 5 = 16 valence electrons.

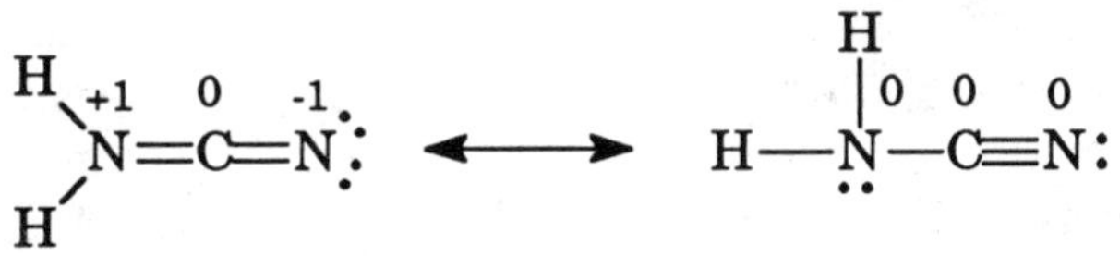

$H_2N=C=\ddot{N}:$ (formal charges +1, 0, -1) $\longleftrightarrow$ $H-\underset{..}{N}(H)-C\equiv N:$ (formal charges 0, 0, 0)

favored by formal charge

$NCNC(NH_2)_2$ has 5 + 4 + 5 + 4 + 2(5) + 4(1) = 32 valence electrons.

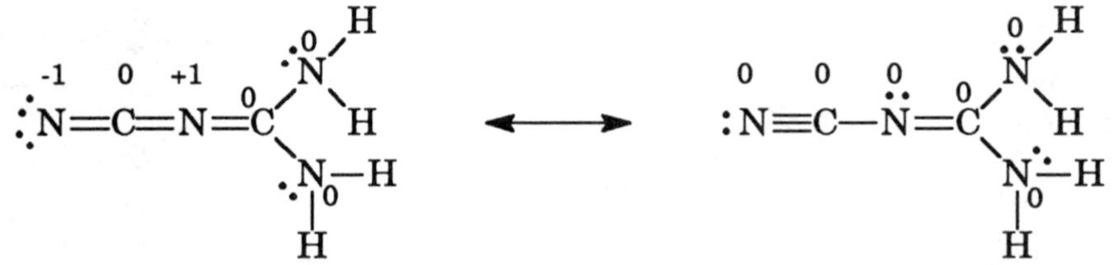

favored by formal charge

Melamine ($C_3N_6H_6$) has 3(4) + 6(5) + 6(1) = 48 valence electrons.

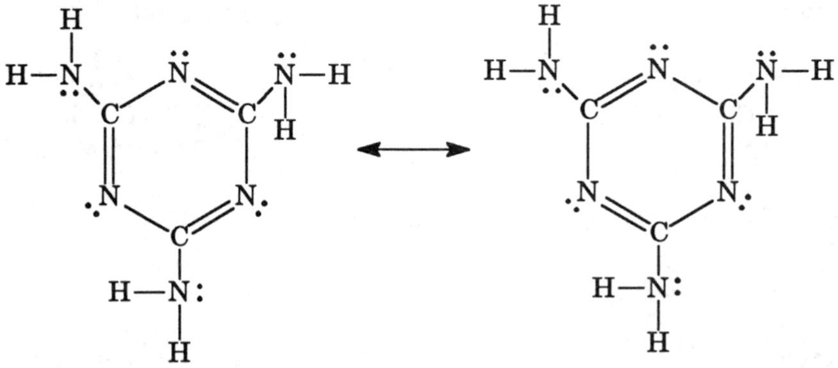

b. NCN^{2-}: C is sp hybridized. Depending on the resonance form, N can be sp, sp^2, or sp^3 hybridized. For the remaining compounds, we will give hybrids for the favored resonance structures as predicted from formal charge considerations.

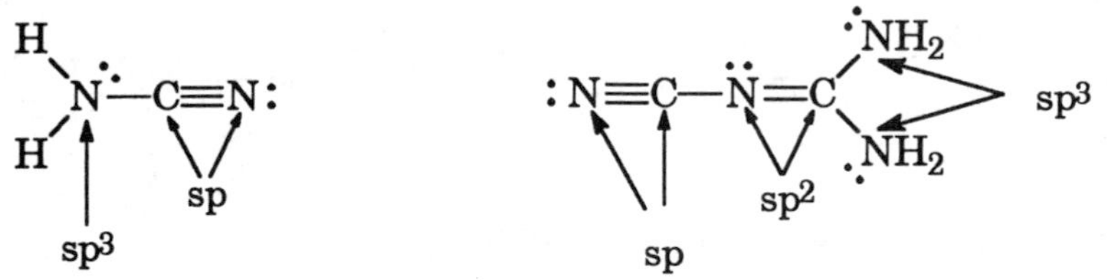

Melamine: N in NH_2 groups are all sp^3 hybridized. Atoms in ring are all sp^2 hybridized.

c. NCN^{2-}: 2 σ and 2 π bonds; H_2NCN: 4 σ and 2 π bonds; dicyandiamide: 9 σ and 3 π bonds; melamine: 15 σ and 3 π bonds

d. The π-system forces the ring to be planar just as the benzene ring is planar.

e. The structure:

```
:N≡C—N̈
       \\
        C—N̈—H
       /    |
     :N     H
     / \
    H   H
```

is the most important since it has three different CN bonds. This structure is also favored on the basis of formal charge.

65. $E = \dfrac{hc}{\lambda} = \dfrac{(6.626 \times 10^{-34}\ \text{J s})\ (2.998 \times 10^{8}\ \text{m/s})}{25 \times 10^{-9}\ \text{m}} = 7.9 \times 16^{-18}\ \text{J}$

$$7.9 \times 10^{-18}\ \text{J} \times \frac{6.022 \times 10^{23}}{\text{mol}} \times \frac{1\ \text{kJ}}{1000\ \text{J}} = 4800\ \text{kJ/mol}$$

Using ΔH values from the various reactions, 25 nm light has sufficient energy to ionize N_2 and N, and to break the triple bond. Thus, N_2, N_2^+, N, and N^+ will all be present, assuming excess N_2.

To produce atomic nitrogen but no ions, the range of energies of the light must be from 941 kJ/mol to just below 1402 kJ/mol.

$$\frac{941\ \text{kJ}}{\text{mol}} \times \frac{1\ \text{mol}}{6.022 \times 10^{23}} \times \frac{1000\ \text{J}}{\text{kJ}} = 1.56 \times 10^{-18}\ \text{J/photon}$$

$$\lambda = \frac{hc}{E} = \frac{(6.626 \times 10^{-34}\ \text{J s})\ (2.998 \times 10^{8}\ \text{m/s})}{1.56 \times 10^{-18}\ \text{J}} = 1.27 \times 10^{-7}\ \text{m} = 127\ \text{nm}$$

$$\frac{1402\ \text{kJ}}{\text{mol}} \times \frac{1\ \text{mol}}{6.0221 \times 10^{23}} \times \frac{1000\ \text{J}}{\text{kJ}} = 2.328 \times 10^{-18}\ \text{J/photon}$$

$$\lambda = \frac{hc}{E} = \frac{(6.6261 \times 10^{-34}\ \text{J s})\ (2.9979 \times 10^{8}\ \text{m/s})}{2.328 \times 10^{-18}\ \text{J}} = 8.533 \times 10^{-8}\ \text{m} = 85.33\ \text{nm}$$

Light with wavelengths in the range of $85.33\ \text{nm} < \lambda \le 127\ \text{nm}$ will produce N but no ions.

67. O=N–Cl: The bond order of the NO bond in NOCl is 2 (a double bond).

NO: The bond order of this NO bond is 2.5 (see Exercise 9.42).

Both reactions apparently only involve the breaking of the N–Cl bond. However, in the reaction: $ONCl \rightarrow NO + Cl$ some energy is released in forming the stronger NO bond, lowering the value of ΔH. Therefore, the apparent N–Cl bond energy is artifically low for this reaction. The first reaction only involves the breaking of the N–Cl bond.

CHAPTER TEN

LIQUIDS AND SOLIDS

Questions

7. London dispersion (LD) < dipole-dipole < H-bonding < metallic bonding, covalent network, ionic.

   Yes, there is considerable overlap. Consider some of the examples in Exercise 10.86. Benzene (only LD forces) has a higher boiling point than acetone (dipole-dipole forces). Also, there is even more overlap between the stronger forces (metallic, covalent, and ionic).

9. As the strengths of interparticle forces increase: surface tension, viscosity, melting point and boiling point increase, while the vapor pressure decreases.

11. a. Polarizability of an atom refers to the ease of distorting the electron cloud. It can also refer to distorting the electron clouds in molecules or ions. Polarity refers to the presence of a permanent dipole moment in a molecule.

    b. London dispersion forces are present in all substances. LD forces can be referred to as accidental dipole - induced dipole forces. Dipole - dipole forces involve the attraction of molecules with permanent dipoles for each other.

    c. inter: between; intra: within; For example, in Br_2 the covalent bond is an intramolecular force, holding the two Br-atoms together in the molecule. The much weaker London dispersion forces are the intermolecular forces of attraction which hold different molecules of Br_2 together in the liquid phase.

13. Atoms have an approximately spherical shape (on the average). It is impossible to pack spheres together without some empty space between the spheres.

15. As the intermolecular forces increase, the critical temperature increases.

17. a. Crystalline solid: Regular, repeating structure

       Amorphous solid: Irregular arrangement of atoms or molecules

    b. Ionic solid: Made up of ions held together by ionic bonding.

       Molecular solid: Made up of discrete covalently bonded molecules held together in the solid phase by weaker forces (LD, dipole or hydrogen bonds).

c. Molecular solid: Discrete, individual molecules

Covalent network solid: No discrete molecules; A covalent network solid is one large molecule. The interparticle forces are the covalent bonds between atoms.

d. Metallic solid: Completely delocalized electrons, conductor of electricity (ions in a sea of electrons)

Covalent network solid: Localized electrons; Insulator or semiconductor

19. No, an example is common glass which is primarily amorphous SiO_2 (a covalent network solid) as compared to ice (a crystalline solid held together by weaker H-bonds). The intermolecular forces in the amorphous solid in this case are stronger than those in the crystalline solid. Whether or not a solid is amorphous or crystalline depends on the long range order in the solid and not on the strengths of the intermolecular forces.

21. a. As the temperature is increased, more electrons in the filled molecular orbitals have sufficient kinetic energy to jump into the conduction bands (the unfilled molecular orbitals).

b. A photon of light is absorbed by an electron which then has sufficient energy to jump into the conduction bands.

c. An impurity either adds electrons at an energy near that of the conduction bands (n-type) or creates holes (empty energy levels) at energies in the previously filled molecular orbitals (p-type).

23. a. Condensation: vapor $\rightarrow$ liquid b. Evaporation: liquid $\rightarrow$ vapor

c. Sublimation: solid $\rightarrow$ vapor

d. A supercooled liquid is a liquid which is at a temperature below its freezing point.

25. a. As the intermolecular forces increase, the rate of evaporation decreases.

b. As temperature increases, the rate of evaporation increases.

c. As surface area increases, the rate of evaporation increases.

27. The phase change, $H_2O(g) \rightarrow H_2O(l)$, releases heat that can cause additional damage. Also steam can be at a temperature greater than 100°C.

Exercises

Intermolecular Forces and Physical Properties

29. a. ionic b. dipole, LD (LD = London dispersion) c. LD only

d. LD only; For all practical purposes, we consider a C – H bond to be nonpolar.

e. ionic f. LD only g. H-bonding, LD

31. a. OCS; OCS is polar and has dipole-dipole forces in addition to London dispersion (LD) forces. All polar molecules have dipole forces. CO_2 is nonpolar and only has LD forces. In all of the following (b-d), only one molecule is polar and, in turn, has dipole-dipole forces. To predict polarity, draw the Lewis structure and deduce if the individual bond dipoles cancel.

b. PF_3; PF_3 is polar (PF_5 is nonpolar). c. SF_2; SF_2 is polar (SF_6 is nonpolar).

d. SO_2; SO_2 is polar (SO_3 is nonpolar).

33. a. Neopentane is more compact than n-pentane. There is less surface area contact between neopentane molecules. This leads to weaker LD forces and a lower boiling point.

b. Ethanol is capable of H-bonding, dimethyl ether is not.

c. HF is capable of H-bonding, HCl is not.

d. LiCl is ionic and HCl is a molecular solid with only dipole forces and LD forces. Ionic forces are much stronger than the forces for molecular solids.

e. n-pentane is a larger molecule so has stronger LD forces.

f. Dimethyl ether is polar so has dipole forces, in addition to LD forces, unlike n-propane which only has LD forces.

35. See Question 10.9 to review the dependence of some physical properties on the strength of the intermolecular forces.

a. HCl; HCl is polar while Ar and F_2 are nonpolar. HCl has dipole forces unlike Ar and F_2.

b. NaCl; Ionic forces are much stronger than molecular forces.

c. I_2; All are nonpolar so the largest molecule (I_2) will have the strongest LD forces and the lowest vapor pressure.

d. N_2; Nonpolar and smallest, so has weakest intermolecular forces.

e. CH_4; Smallest, nonpolar molecule so has weakest LD forces.

f. HF; HF can form relatively strong H-bonding interactions unlike the others.

g. $CH_3CH_2CH_2OH$; H-bonding unlike the others so has strongest intermolecular forces.

Properties of Liquids

37. The attraction of H_2O for glass is stronger than the $H_2O - H_2O$ attraction. The miniscus is concave to increase the area of contact between glass and H_2O. The Hg – Hg attraction is greater than the Hg – glass attraction. The miniscus is convex to minimize the Hg – glass contact. Polyethylene is a nonpolar substance. The $H_2O - H_2O$ attraction is stronger than the H_2O – polyethylene attraction. Thus, the miniscus will have a convex shape.

39. The structure of H_2O_2 is H – O – O – H, which produces greater hydrogen bonding than water. Long chains of hydrogen bonded H_2O_2 molecules then get tangled together.

Structures and Properties of Solids

41. $n\lambda = 2d \sin\theta,\ d = \dfrac{n\lambda}{2 \sin\theta} = \dfrac{1 \times 1.54\ \text{Å}}{2 \times \sin 14.22°} = 3.13\ \text{Å} = 3.13 \times 10^{-10}\ \text{m} = 313\ \text{pm}$

43. A cubic closest packed structure has a face-centered cubic unit cell. In a face-centered cubic unit, there are:

$$8 \text{ corners} \times \frac{1/8 \text{ atom}}{\text{corner}} + 6 \text{ faces} \times \frac{1/2 \text{ atom}}{\text{face}} = 4 \text{ atoms}$$

The atoms in a face-centered cubic unit cell touch along the face diagonal of the cubic unit cell. Using the Pythagorean formula where l = length of the face diagonal and r = radius of the atom:

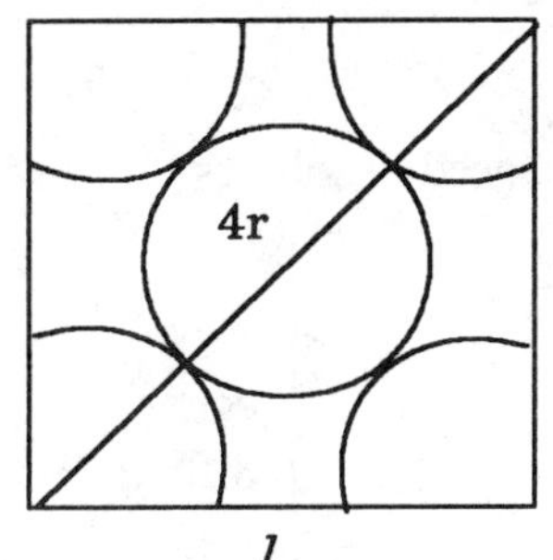

$$l^2 + l^2 = (4r)^2$$

$$2\,l^2 = 16\,r^2$$

$$l = r\sqrt{8}$$

$l = r\sqrt{8} = 125 \times 10^{-12}\ \text{m} \times \sqrt{8} = 3.54 \times 10^{-10}\ \text{m} = 3.54 \times 10^{-8}\ \text{cm}$

Volume of a unit cell $= l^3 = (3.54 \times 10^{-8}\ \text{cm})^3 = 4.44 \times 10^{-23}\ \text{cm}^3$

Mass of a unit cell $= 4 \text{ Co atoms} \times \dfrac{1 \text{ mol Co}}{6.022 \times 10^{23} \text{ atoms}} \times \dfrac{58.93 \text{ g Co}}{\text{mol Co}} = 3.914 \times 10^{-22}\ \text{g Co}$

density $= \dfrac{\text{mass}}{\text{volume}} = \dfrac{3.914 \times 10^{-22}\ \text{g}}{4.44 \times 10^{-23}\ \text{cm}^3} = 8.82\ \text{g/cm}^3$

45. The volume of a unit cell is:

$$V = l^3 = (383.3 \times 10^{-10}\ \text{cm})^3 = 5.631 \times 10^{-23}\ \text{cm}^3$$

There are 4 Ir atoms in the unit cell as is the case for all face-centered cubic unit cells. The mass of atoms in a unit cell is:

$$\text{mass} = 4\ \text{Ir atoms} \times \frac{1\ \text{mol Ir}}{6.022 \times 10^{23}\ \text{atoms}} \times \frac{192.2\ \text{g Ir}}{\text{mol Ir}} = 1.277 \times 10^{-21}\ \text{g}$$

$$\text{density} = \frac{\text{mass}}{\text{volume}} = \frac{1.277 \times 10^{-21}\ \text{g}}{5.631 \times 10^{-23}\ \text{cm}^3} = 22.68\ \text{g/cm}^3$$

47. For a body-centered unit cell: 8 corners $\times \frac{1/8\ \text{Ti}}{\text{corner}}$ + Ti at body center = 2 Ti atoms

All body-centered unit cells have 2 atoms per unit cell. For a unit cell:

$$\text{density} = 4.50\ \text{g/cm}^3 = \frac{2\ \text{atoms Ti} \times \frac{1\ \text{mol Ti}}{6.022 \times 10^{23}\ \text{atoms}} \times \frac{47.88\ \text{g Ti}}{\text{mol Ti}}}{l^3},\ l = \text{cube edge length}$$

Solving: l = edge length of unit cell = 3.28×10^{-8} cm = 328 pm

Assume Ti atoms just touch along the body diagonal of the cube, so body diagonal = 4 × radius of atoms = 4r.

The triangle we need to solve is:

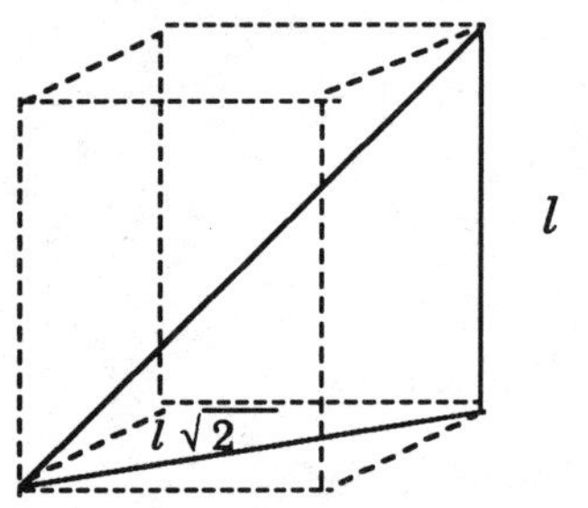

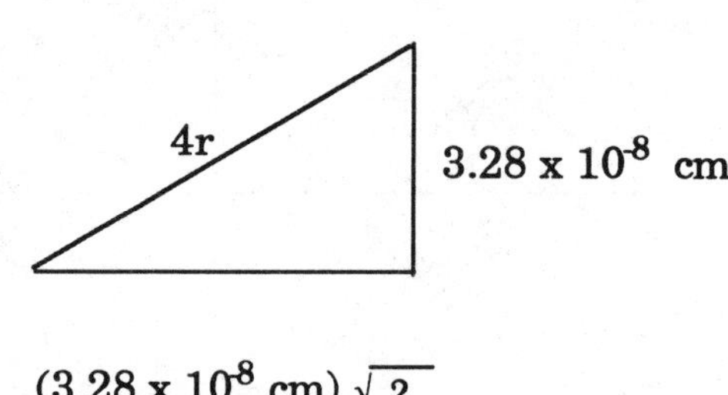

$$(4r)^2 = (3.28 \times 10^{-8}\ \text{cm})^2 + [(3.28 \times 10^{-8}\ \text{cm})\sqrt{2}\ ]^2,\ r = 1.42 \times 10^{-8}\ \text{cm} = 142\ \text{pm}$$

For a body-centered unit cell, the radius of the atom is related to the cube edge length by $4r = l\sqrt{3}$ or $l = 4r/\sqrt{3}$.

49. For the fcc unit cell:

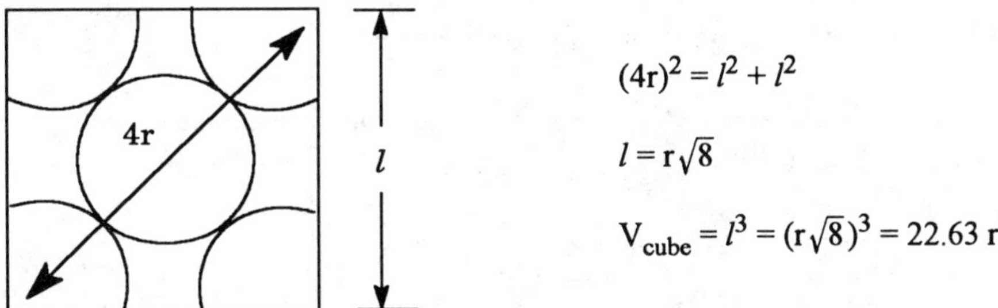

$(4r)^2 = l^2 + l^2$

$l = r\sqrt{8}$

$V_{cube} = l^3 = (r\sqrt{8})^3 = 22.63\ r^3$

There are four atoms in a face-centered cubic cell (see Exercise 10.43). Each atom has a volume of 4/3 πr^3.

$$V_{atoms} = 4 \times \frac{4}{3}\pi r^3 = 16.76\ r^3$$

So, $\frac{V_{atom}}{V_{cube}} = \frac{16.76\ r^3}{22.63\ r^3}$ = 0.7406 or 74.06% of the volume of each unit cell is occupied by atoms.

51. To produce a n-type semiconductor, dope Ge with a substance that has more than 4 valence electrons, e.g., a group 5A element. Phosphorus or arsenic are a couple substances which would produce n-type semiconductors when they are doped into germanium.

53. In has fewer valence electrons than Se, thus, Se doped with In would be a p-type semiconductor.

55. $E_{gap} = 2.5\ \text{eV} \times 1.6 \times 10^{-19}\ \text{J/eV} = 4.0 \times 10^{-19}\ \text{J}$; We want $E_{gap} = E_{light}$, so:

$$E_{light} = \frac{hc}{\lambda},\ \lambda = \frac{hc}{E} = \frac{(6.63 \times 10^{-34}\ \text{J s})(3.00 \times 10^8\ \text{m/s})}{4.0 \times 10^{-19}\ \text{J}} = 5.0 \times 10^{-7}\ \text{m} = 5.0 \times 10^2\ \text{nm}$$

57. a. $8\ \text{corners} \times \frac{1/8\ \text{Cl}}{\text{corner}} + 6\ \text{faces} \times \frac{1/2\ \text{Cl}}{\text{face}} = 4\ \text{Cl}$

$12\ \text{edges} \times \frac{1/4\ \text{Na}}{\text{edge}}$ + 1 Na at body center = 4 Na; NaCl is the formula.

b. 1 Cs at body center; $8\ \text{corners} \times \frac{1/8\ \text{Cl}}{\text{corner}} = 1\ \text{Cl}$; CsCl is the formula.

c. There are 4 Zn inside the cube.

$8\ \text{corners} \times \frac{1/8\ \text{S}}{\text{corner}} + 6\ \text{faces} \times \frac{1/2\ \text{S}}{\text{face}} = 4\ \text{S}$; ZnS is the formula.

d. $8\ \text{corners} \times \frac{1/8\ \text{Ti}}{\text{corner}}$ + 1 Ti at body center = 2 Ti

$4\ \text{faces} \times \frac{1/2\ \text{O}}{\text{face}}$ + 2 O inside cube = 4 O; TiO_2 is the formula.

59. Re at 8 corners: 8(1/8) = 1 Re; O at 12 edges: 12(1/4) = 3 O

Formula is ReO_3. If O has 2- charge, then charge on Re is +6.

61. Since magnesium oxide has the same structure as NaCl, each unit cell contains 4 Mg^{2+} ions and 4 O^{2-} ions. The mass of a unit cell is:

$$4 \text{ MgO molecules}\left(\frac{1 \text{ mol MgO}}{6.022 \times 10^{23} \text{ molecules}}\right)\left(\frac{40.31 \text{ g MgO}}{1 \text{ mol MgO}}\right) = 2.678 \times 10^{-22} \text{ g MgO}$$

$$\text{Volume of unit cell} = 2.678 \times 10^{-22} \text{ g MgO}\left(\frac{1 \text{ cm}^3}{3.58 \text{ g}}\right) = 7.48 \times 10^{-23} \text{ cm}^3$$

Volume of unit cell = l^3, l = cube edge length; $l = (7.48 \times 10^{-23} \text{ cm}^3)^{1/3} = 4.21 \times 10^{-8}$ cm = 421 pm

From the NaCl structure in Figure 10.35 of the text, Mg^{2+} and O^{2-} ions should touch along the cube edge, l:

$$l = 2\, r_{Mg^{2+}} + 2\, r_{O^{2-}} = 2\,(65 \text{ pm}) + 2\,(140. \text{ pm}) = 410. \text{ pm}$$

The two values agree within 3%. In the actual crystals the Mg^{2+} and O^{2-} ions may not touch which is assumed in calculating the 410. pm value.

63. a. CO_2: molecular b. SiO_2: covalent network c. Si: atomic, covalent network

d. CH_4: molecular e. Ru: atomic, metallic f. I_2: molecular

g. KBr: ionic h. H_2O: molecular i. NaOH: ionic

j. U: atomic, metallic k. $CaCO_3$: ionic l. PH_3: molecular

65. Al: 8 corners $\times \frac{1/8 \text{ Al}}{\text{corner}}$ = 1 Al; Ni: 6 face centers $\times \frac{1/2 \text{ Ni}}{\text{face center}}$ = 3 Ni

Composition: $AlNi_3$

67.

Structure 1	Structure 2
8 corners $\times \frac{1/8 \text{ Ca}}{\text{corner}}$ = 1 Ca atom	8 corners $\times \frac{1/8 \text{ Ti}}{\text{corner}}$ = 1 Ti atom
6 faces $\times \frac{1/2 \text{ O}}{\text{face}}$ = 3 O atoms	12 edges $\times \frac{1/4 \text{ O}}{\text{edge}}$ = 3 O atoms
1 Ti at body center. Formula = $CaTiO_3$	1 Ca at body center. Formula = $CaTiO_3$

In the extended lattice of both structures, each Ti atom is surrounded by six O atoms.

69. a. Y: 1 Y in center; Ba: 2 Ba in center

Cu: $8 \text{ corners} \times \frac{1/8 \text{ Cu}}{\text{corner}} = 1 \text{ Cu}$, $8 \text{ edges} \times \frac{1/4 \text{ Cu}}{\text{edge}} = 2 \text{ Cu}$, total = 3 Cu atoms

O: $20 \text{ edges} \times \frac{1/4 \text{ O}}{\text{edge}} = 5 \text{ oxygen}$, $8 \text{ faces} \times \frac{1/2 \text{ O}}{\text{face}} = 4 \text{ oxygen}$, total = 9 O atoms

Formula: $YBa_2Cu_3O_9$

b. The structure of this superconductor material follows the second perovskite structure described in Exercise 10.67. The $YBa_2Cu_3O_9$ structure is three of these cubic perovskite unit cells stacked on top of each other. The oxygen atoms are in the same places, Cu takes the place of Ti, two of the calcium atoms are replaced by two barium atoms and one Ca is replaced by Y.

c. Y, Ba, and Cu are the same. Some oxygen atoms are missing.

$12 \text{ edges} \times \frac{1/4 \text{ O}}{\text{edge}} = 3 \text{ O}$, $8 \text{ faces} \times \frac{1/2 \text{ O}}{\text{face}} = 4 \text{ O}$, total = 7 O atoms

Superconductor formula is $YBa_2Cu_3O_7$.

Phase Changes and Phase Diagrams

71. If we graph ln P_{vap} vs 1/T, the slope of the resulting straight line will be $-\Delta H_{vap}/R$.

P_{vap}	ln P_{vap}	T (Li)	1/T	T (Mg)	1/T
1 torr	0	1023 K	9.775×10^{-4} K^{-1}	893 K	11.2×10^{-4} K^{-1}
10.	2.3	1163	8.598×10^{-4}	1013	9.872×10^{-4}
100.	4.61	1353	7.391×10^{-4}	1173	8.525×10^{-4}
400.	5.99	1513	6.609×10^{-4}	1313	7.616×10^{-4}
760.	6.63	1583	6.317×10^{-4}	1383	7.231×10^{-4}

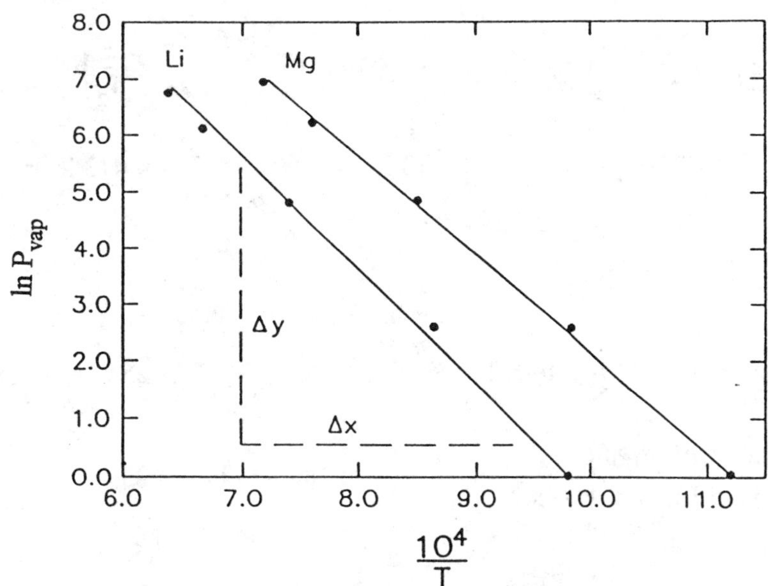

For Li:

We get the slope by taking two points (x, y) that are on the line we draw. For a line:

$$\text{slope} = \frac{\Delta y}{\Delta x} = \frac{y_2 - y_1}{x_2 - x_1}$$

or we can determine the straight line equation using a computer or calculator. The general straight line equation is y = mx + b where m = slope and b = y-intercept.

The equation of the Li line is: $\ln P_{vap} = -1.90 \times 10^4(1/T) + 18.6$, slope = -1.90×10^4 K

Slope = $-\Delta H_{vap}/R$, $\Delta H_{vap} = -\text{slope} \times R = 1.90 \times 10^4 \text{ K} \times 8.3145$ J/K•mol

$\Delta H_{vap} = 1.58 \times 10^5$ J/mol = 158 kJ/mol

For Mg:

The equation of the line is: $\ln P_{vap} = -1.67 \times 10^4(1/T) + 18.7$, slope = -1.67×10^4 K

$\Delta H_{vap} = -\text{slope} \times R = 1.67 \times 10^4 \text{ K} \times 8.3145$ J/K•mol, $\Delta H_{vap} = 1.39 \times 10^5$ J/mol = 139 kJ/mol

The bonding is stronger in Li since ΔH_{vap} is larger for Li.

73. At 100.°C (373 K), the vapor pressure of H_2O is 1.00 atm. For water, ΔH_{vap} = 40.7 kJ/mol.

$$\ln\left(\frac{P_1}{P_2}\right) = \frac{\Delta H_{vap}}{R}\left(\frac{1}{T_2} - \frac{1}{T_1}\right) \text{ or } \ln\left(\frac{P_2}{P_1}\right) = \frac{\Delta H_{vap}}{R}\left(\frac{1}{T_1} - \frac{1}{T_2}\right)$$

$$\ln\left(\frac{P_2}{1.00 \text{ atm}}\right) = \frac{40.7 \times 10^3 \text{ J/mol}}{8.3145 \text{ J/K•mol}}\left(\frac{1}{373 \text{ K}} - \frac{1}{388 \text{ K}}\right), \ \ln P_2 = 0.51, \ P_2 = e^{0.51} = 1.7 \text{ atm}$$

$$\ln\left(\frac{3.50}{1.00}\right) = \frac{40.7 \times 10^3 \text{ J/mol}}{8.3145 \text{ J/K•mol}}\left(\frac{1}{373 \text{ K}} - \frac{1}{T_2}\right), \ 2.56 \times 10^{-4} = \left(\frac{1}{373} - \frac{1}{T_2}\right)$$

$$2.56 \times 10^{-4} = 2.68 \times 10^{-3} - \frac{1}{T_2}, \ \frac{1}{T_2} = 2.42 \times 10^{-3}, \ T_2 = \frac{1}{2.42 \times 10^{-3}} = 413 \text{ K or } 140.\text{°C}$$

75. $$\ln\left(\frac{P_1}{P_2}\right) = \frac{\Delta H_{vap}}{R}\left(\frac{1}{T_2} - \frac{1}{T_1}\right)$$

At normal boiling point, P_1 = 760. torr, T_1 = 56.5°C = 329.7 K; T_2 = 25.0°C = 298.2 K, P_2 = ?

$$\ln\left(\frac{760.}{P_2}\right) = \frac{32.0 \times 10^3 \text{ J/mol}}{8.3145 \text{ J/K•mol}}\left(\frac{1}{298.2} - \frac{1}{329.7}\right), \ \ln 760. - \ln P_2 = 1.23$$

$\ln P_2 = 5.40,\ P_2 = e^{5.40} = 221$ torr

77.

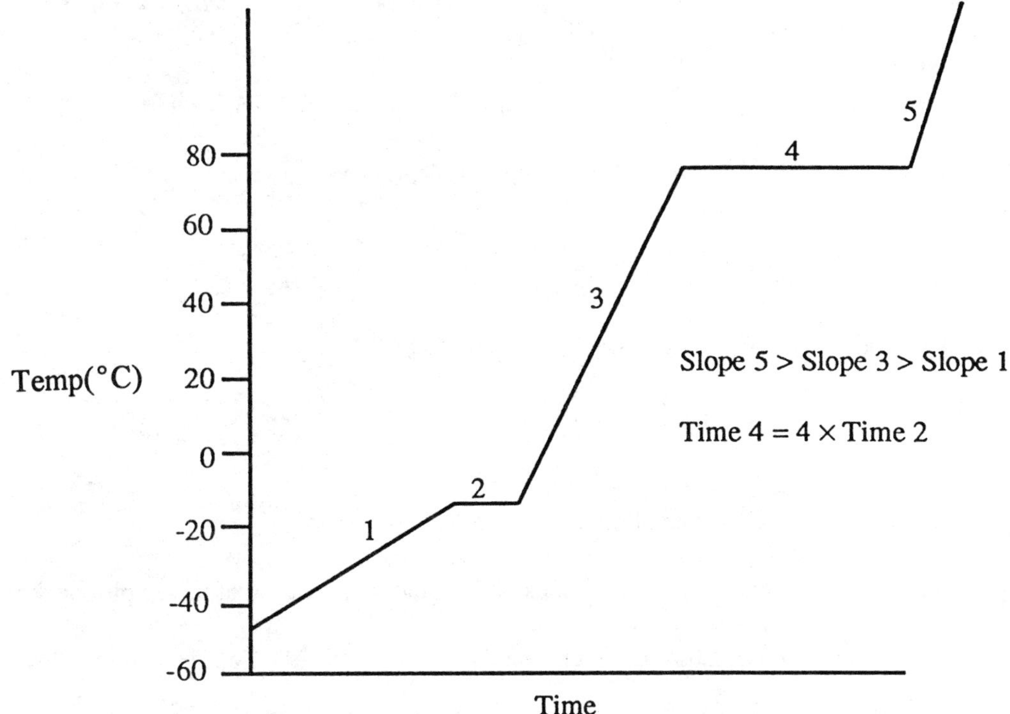

79. $H_2O(s, -20.°C) \rightarrow H_2O(s, 0°C),\ \Delta T = 20.°C$

$$q_1 = s_{ice} \times m \times \Delta T = \frac{2.1\ J}{g\ °C} \times 5.00 \times 10^2\ g \times 20.°C = 2.1 \times 10^4\ J = 21\ kJ$$

$$H_2O(s, 0°C) \rightarrow H_2O(l, 0°C),\ q_2 = 5.00 \times 10^2\ g\ H_2O \times \frac{1\ mol}{18.02\ g} \times \frac{6.02\ kJ}{mol} = 167\ kJ$$

$$H_2O(l, 0°C) \rightarrow H_2O(l, 100.°C),\ q_3 = \frac{4.2\ J}{g\ °C} \times 5.00 \times 10^2\ g \times 100.°C = 2.1 \times 10^5 J = 210\ kJ$$

$$H_2O(l, 100.°C) \rightarrow H_2O(g, 100.°C),\ q_4 = 5.00 \times 10^2\ g \times \frac{1\ mol}{18.02\ g} \times \frac{40.7\ kJ}{mol} = 1130\ kJ$$

$$H_2O(g, 100.°C) \rightarrow H_2O(g, 250.°C),\ q_5 = \frac{2.0\ J}{g\ °C} \times 5.00 \times 10^2\ g \times 150.°C = 1.5 \times 10^5\ J = 150\ kJ$$

$$q_{total} = q_1 + q_2 + q_3 + q_4 + q_5 = 21 + 167 + 210 + 1130 + 150 = 1680\ kJ$$

81. Total mass H_2O = 18 cubes $\times \frac{30.0\ g}{cube} = 540.\ g;\ 540.\ g\ H_2O \times \frac{1\ mol\ H_2O}{18.02\ g} = 30.0\ mol\ H_2O$

Heat removed to produce ice at -5.0°C:

$$\frac{4.18\ \text{J}}{\text{g °C}} \times 540.\ \text{g} \times 22.0\ \text{°C} + \frac{6.02 \times 10^3\ \text{J}}{\text{mol}} \times 30.0\ \text{mol} + \frac{2.08\ \text{J}}{\text{g °C}} \times 540.\ \text{g} \times 5.0\ \text{°C}$$

$$= 4.97 \times 10^4\ \text{J} + 1.81 \times 10^5\ \text{J} + 5.6 \times 10^3\ \text{J} = 2.36 \times 10^5\ \text{J}$$

$2.36 \times 10^5\ \text{J} \times \frac{1\ \text{g}\ CF_2Cl_2}{158\ \text{J}} = 1.49 \times 10^3\ \text{g}\ CF_2Cl_2$ must be vaporized.

83. A: solid; B: liquid; C: vapor

D: solid + vapor; E: solid + liquid + vapor

F: liquid + vapor; G: liquid + vapor; H: vapor

triple point: E; critical point: G

normal freezing point: temperature at which solid - liquid line is at 1.0 atm (see plot below).

normal boiling point: temperature at which liquid - vapor line is at 1.0 atm (see plot below).

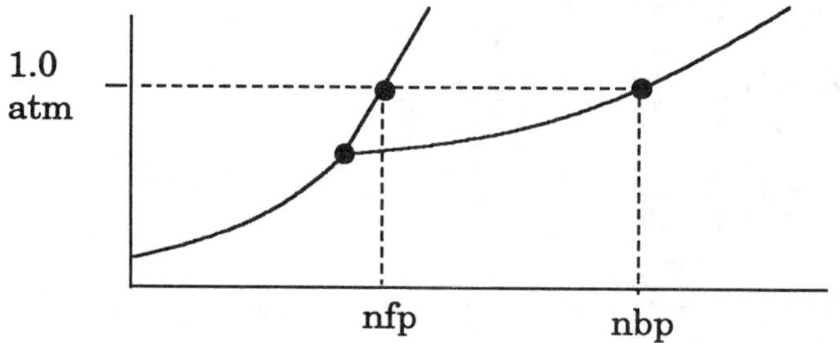

Since the solid-liquid equilibrium line has a positive slope, then the solid phase is denser than the liquid phase.

Additional Exercises

85. $C_{25}H_{52}$ has the stronger intermolecular forces because it has the higher boiling point. Even though $C_{25}H_{52}$ is nonpolar, it is so large that its London dispersion forces are much stronger than the sum of the London dispersion and hydrogen bonding interactions found in H_2O.

87.

O------H—O

H_3C—C   C—CH_3 ----- = H-bonding

O—H------O

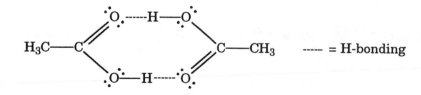

89. $n\lambda = 2d \sin\theta$, $\lambda = \dfrac{2d \sin\theta}{n} = \dfrac{2 \times 201 \text{ pm} \times \sin 34.68°}{1}$, $\lambda = 229 \text{ pm} = 2.29 \times 10^{-10} \text{ m} = 0.229 \text{ nm}$

91. If TiO_2 conducts electricity as a liquid then it is an ionic solid, if not then TiO_2 is a network solid.

93. $1.00 \text{ lb} \times \dfrac{454 \text{ g}}{\text{lb}} = 454 \text{ g } H_2O$; A change of 1.00°F is equal to a change of 5/9°C.

The amount of heat in J in 1 Btu is: $\dfrac{4.18 \text{ J}}{\text{g °C}} \times 454 \text{ g} \times \dfrac{5}{9}\text{°C} = 1.05 \times 10^3 \text{ J or } 1.05 \text{ kJ}$

It takes 40.7 kJ to vaporize 1 mol H_2O (ΔH_{vap}). Combining these:

$$\frac{1.00 \times 10^4 \text{ Btu}}{\text{hr}} \times \frac{1.05 \text{ kJ}}{\text{Btu}} \times \frac{1 \text{ mol } H_2O}{40.7 \text{ kJ}} = 258 \text{ mol/hr}$$

or: $$\frac{258 \text{ mol}}{\text{hr}} \times \frac{18.02 \text{ g } H_2O}{\text{mol}} = 4650 \text{ g/hr} = 4.65 \text{ kg/hr}$$

95. $\ln\left(\dfrac{P_1}{P_2}\right) = \dfrac{\Delta H_{vap}}{R}\left(\dfrac{1}{T_2} - \dfrac{1}{T_1}\right)$; $P_1 = 760.$ torr, $T_1 = 630.$ K; $P_2 = ?$, $T_2 = 298$ K

$$\ln\left(\frac{760.}{P_2}\right) = \frac{59.1 \times 10^3 \text{ J/mol}}{8.3145 \text{ J/K•mol}}\left(\frac{1}{298 \text{ K}} - \frac{1}{630. \text{ K}}\right) = 12.6$$

$760./P_2 = e^{12.6}$, $P_2 = 760./2.97 \times 10^5 = 2.56 \times 10^{-3}$ torr

Challenge Problems

97. A single hydrogen bond in H_2O has a strength of 21 kJ/mol. Each H_2O molecule forms two H-bonds. Thus, it should take 42 kJ/mol of energy to break all of the H-bonds in water. Consider the phase transitions:

$$\text{solid} \overset{6.0 \text{ kJ}}{\rightarrow} \text{liquid} \overset{40.7 \text{ kJ}}{\rightarrow} \text{vapor} \qquad \Delta H_{sub} = \Delta H_{fus} + \Delta H_{vap}$$

It takes a total of 46.7 kJ/mol to convert solid H_2O to vapor (ΔH_{sub}). This would be the amount of energy necessary to disrupt all of the intermolecular forces in ice. Thus, $(42 \div 46.7) \times 100 = 90\%$ of the attraction in ice can be attributed to H-bonding.

99. NaCl, $MgCl_2$, NaF, MgF_2 AlF_3 all have very high melting points indicative of strong intermolecular forces. They are all ionic solids. $SiCl_4$, SiF_4, F_2, Cl_2, PF_5 and SF_6 are nonpolar covalent molecules. Only LD forces are present. PCl_3 and SCl_2 are polar molecules. LD forces and dipole forces are present. In these 8 molecular substances, the intermolecular forces are weak and the melting points low. $AlCl_3$ doesn't seem to fit in as well. From the melting point, there are much stronger forces present than in the nonmetal halides, but they aren't as strong as we would expect for an ionic solid. $AlCl_3$ illustrates a gradual transition from ionic to covalent bonding; from an ionic solid to discrete molecules.

101. $n\lambda = 2d \sin \theta$; 100 pm = 10^{-10} m = 10^{-8} cm

$$d = \frac{n\lambda}{2 \sin \theta} = \frac{1 \times 71.2 \text{ pm}}{2 \times \sin 5.564} = 367 \text{ pm} = 3.67 \times 10^{-8} \text{ cm} = \text{cube edge length} = l$$

Mass of Hf in unit cell = volume unit cell × density; volume = $l^3 = (3.67 \times 10^{-8} \text{ cm})^3$

$$\text{Mass of Hf} = (3.67 \times 10^{-8} \text{ cm})^3 \times \frac{13.28 \text{ g Hf}}{\text{cm}^3} = 6.56 \times 10^{-22} \text{ g Hf}$$

$$\text{Atoms Hf in unit cell} = 6.56 \times 10^{-22} \text{ g Hf} \times \frac{1 \text{ mol Hf}}{178.5 \text{ g Hf}} \times \frac{6.022 \times 10^{23} \text{ atoms Hf}}{\text{mol Hf}} = 2.21$$

This is most consistent with a body-centered cubic unit cell which contains 2 atoms per unit cell. To determine the radius, the cube edge length (l) is related to the radius by the equation $l = 4r/\sqrt{3}$ (see Exercise 10.47).

$$\text{radius} = r = l\sqrt{3}/4 = 3.67 \times 10^{-8} \text{ cm} \times \sqrt{3}/4 = 1.59 \times 10^{-8} \text{ cm} = 159 \text{ pm}$$

103. A face-centered cubic unit cell contains 4 atoms. For a unit cell:

$$\text{mass of X} = \text{volume} \times \text{density} = (4.09 \times 10^{-8} \text{ cm})^3 \times 10.5 \text{ g/cm}^3 = 7.18 \times 10^{-22} \text{ g}$$

$$\text{mol X} = 4 \text{ atoms X} \times \frac{1 \text{ mol X}}{6.022 \times 10^{23} \text{ atoms}} = 6.642 \times 10^{-24} \text{ mol X}$$

$$\text{Atomic mass} = \frac{7.18 \times 10^{-22} \text{ g X}}{6.642 \times 10^{-24} \text{ mol X}} = 108 \text{ g/mol}; \text{ The metal is silver (Ag).}$$

105.

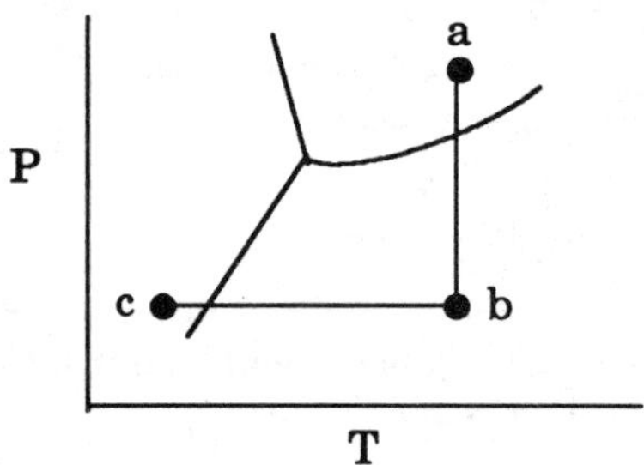

As P is lowered, we go from a to b on the phase diagram. The water boils. The evaporation of the water is endothermic and the water is cooled (b → c), forming some ice. If the pump is left on, the ice will sublime until none is left. This is the basis of freeze drying.

CHAPTER ELEVEN

PROPERTIES OF SOLUTIONS

Solution Review

9. $125 \text{ g sucrose} \times \dfrac{1 \text{ mol}}{342.30 \text{ g}} = 0.365 \text{ mol}; \; M = \dfrac{0.365 \text{ mol}}{1.00 \text{ L}} = \dfrac{0.365 \text{ mol sucrose}}{\text{L}}$

11. $25.00 \times 10^{-3} \text{ L} \times \dfrac{0.308 \text{ mol}}{\text{L}} = 7.70 \times 10^{-3} \text{ mol}; \; \dfrac{7.70 \times 10^{-3} \text{ mol}}{0.500 \text{ L}} = \dfrac{1.54 \times 10^{-2} \text{ mol NiCl}_2}{\text{L}}$

    $NiCl_2(s) \rightarrow Ni^{2+}(aq) + 2\ Cl^-(aq); \; M_{Ni^{2+}} = \dfrac{1.54 \times 10^{-2} \text{ mol}}{\text{L}}; \; M_{Cl^-} = \dfrac{3.08 \times 10^{-2} \text{ mol}}{\text{L}}$

13. $1.00 \text{ L} \times \dfrac{0.040 \text{ mol HCl}}{\text{L}} = 0.040 \text{ mol HCl}; \; 0.040 \text{ mol HCl} \times \dfrac{1 \text{ L}}{0.25 \text{ mol HCl}} = 0.16 \text{ L} = 160 \text{ mL}$

Questions

15. $\text{Molarity} = \dfrac{\text{moles solute}}{\text{L solution}}; \; \text{Molality} = \dfrac{\text{moles solute}}{\text{kg solvent}}$

    Since volume is temperature dependent and mass isn't, then molarity is temperature dependent and molality is temperature independent. In determining ΔT_f and ΔT_b, we are interested in how some temperature depends on composition. Thus, we don't want our expression of composition to also depend on temperature.

17. hydrophobic: water hating; hydrophilic: water loving

19. If solute-solvent attractions > solvent-solvent and solute-solute attractions, then there is a negative deviation from Raoult's law. If solute-solvent attractions < solvent-solvent and solute-solute attractions, then there is a positive deviation from Raoult's law.

21. No, the solution is not ideal. For an ideal solution, this strength of intermolecular forces in the solution are the same as in the pure solute and pure solvent. This results in $\Delta H_{soln} = 0$ for an ideal solution. ΔH_{soln} for methanol/water is not zero. Since $\Delta H_{soln} < 0$, then this solution shows negative deviation from Raoult's law.

23. With addition of salt or sugar, the osmotic pressure inside the fruit cells (and bacteria) is less than outside the cell. Water will leave the cells which will dehydrate bacteria present, causing them to die.

25. Both solutions and colloids have suspended particles in some medium. The major difference between the two is the size of the particles. A colloid is a suspension of relatively large particles as compared to a solution. Because of this, colloids will scatter light while solutions will not. The scattering of light by a colloidal suspension is called the Tyndall effect.

Exercises

Concentration of Solutions

27. $$\text{mass \% CsCl} = \frac{\text{mass CsCl}}{\text{mass solution}} \times 100 = \frac{50.0\text{ g CsCl}}{100.0\text{ g solution}} \times 100 = 50.0\%\text{ CsCl by mass}$$

$$\text{molarity} = M = \frac{\text{mol solute}}{\text{L solution}} = \frac{50.0\text{ g}}{63.3\text{ mL}} \times \frac{1000\text{ mL}}{\text{L}} \times \frac{1\text{ mol}}{168.4\text{ g}} = 4.69\text{ mol/L}$$

$$\text{molality} = m = \frac{\text{mol solute}}{\text{kg solvent}} = \frac{50.0\text{ g CsCl}}{50.0\text{ g solvent}} \times \frac{1000\text{ g}}{\text{kg}} \times \frac{1\text{ mol CsCl}}{168.4\text{ g}} = 5.94\text{ mol/kg}$$

$$50.0\text{ g CsCl} \times \frac{1\text{ mol}}{168.4\text{ g}} = 0.297\text{ mol CsCl};\ 50.0\text{ g }H_2O \times \frac{1\text{ mol}}{18.02\text{ g}} = 2.77\text{ mol }H_2O$$

$$\text{mole fraction CsCl} = \chi_{CsCl} = \frac{\text{mol CsCl}}{\text{total mol}} = \frac{0.297}{0.297 + 2.77} = 9.68 \times 10^{-2}$$

29. Hydrochloric acid:

$$\text{molarity} = \frac{38\text{ g HCl}}{100.\text{ g soln}} \times \frac{1.19\text{ g soln}}{\text{cm}^3\text{ soln}} \times \frac{1000\text{ cm}^3}{\text{L}} \times \frac{1\text{ mol HCl}}{36.46\text{ g}} = 12\text{ mol/L}$$

$$\text{molality} = \frac{38\text{ g HCl}}{62\text{ g solvent}} \times \frac{1000\text{ g}}{\text{kg}} \times \frac{1\text{ mol HCl}}{36.46\text{ g}} = 17\text{ mol/kg}$$

$$38\text{ g HCl} \times \frac{1\text{ mol}}{36.46\text{ g}} = 1.0\text{ mol HCl};\ 62\text{ g }H_2O \times \frac{1\text{ mol}}{18.02\text{ g}} = 3.4\text{ mol }H_2O$$

$$\text{mole fraction of HCl} = \chi_{HCl} = \frac{1.0}{3.4 + 1.0} = 0.23$$

Nitric acid:

$$\frac{70.\text{ g }HNO_3}{100.\text{ g soln}} \times \frac{1.42\text{ g soln}}{\text{cm}^3\text{ soln}} \times \frac{1000\text{ cm}^3}{\text{L}} \times \frac{1\text{ mol }HNO_3}{63.02\text{ g}} = 16\text{ mol/L}$$

$$\frac{70.\text{ g }HNO_3}{30.\text{ g solvent}} \times \frac{1000\text{ g}}{\text{kg}} \times \frac{1\text{ mol }HNO_3}{63.02\text{ g}} = 37\text{ mol/kg}$$

$$70.\ \text{g HNO}_3 \times \frac{1\ \text{mol}}{63.02\ \text{g}} = 1.1\ \text{mol HNO}_3;\ 30.\ \text{g H}_2\text{O} \times \frac{1\ \text{mol}}{18.02\ \text{g}} = 1.7\ \text{mol H}_2\text{O}$$

$$\chi_{HNO_3} = \frac{1.1}{1.7 + 1.1} = 0.39$$

Sulfuric acid:

$$\frac{95\ \text{g H}_2\text{SO}_4}{100.\ \text{g soln}} \times \frac{1.84\ \text{g soln}}{\text{cm}^3\ \text{soln}} \times \frac{1000\ \text{cm}^3}{\text{L}} \times \frac{1\ \text{mol H}_2\text{SO}_4}{98.09\ \text{g H}_2\text{SO}_4} = 18\ \text{mol/L}$$

$$\frac{95\ \text{g H}_2\text{SO}_4}{5\ \text{g H}_2\text{O}} \times \frac{1000\ \text{g}}{\text{kg}} \times \frac{1\ \text{mol}}{98.09\ \text{g}} = 194\ \text{mol/kg} \approx 200\ \text{mol/kg}$$

$$95\ \text{g H}_2\text{SO}_4 \times \frac{1\ \text{mol}}{98.09\ \text{g}} = 0.97\ \text{mol H}_2\text{SO}_4;\ 5\ \text{g H}_2\text{O} \times \frac{1\ \text{mol}}{18.02\ \text{g}} = 0.3\ \text{mol H}_2\text{O}$$

$$\chi_{H_2SO_4} = \frac{0.97}{0.97 + 0.3} = 0.76$$

Acetic Acid:

$$\frac{99\ \text{g HC}_2\text{H}_3\text{O}_2}{100.\ \text{g soln}} \times \frac{1.05\ \text{g soln}}{\text{cm}^3\ \text{soln}} \times \frac{1000\ \text{cm}^3}{\text{L}} \times \frac{1\ \text{mol}}{60.05\ \text{g}} = 17\ \text{mol/L}$$

$$\frac{99\ \text{g HC}_2\text{H}_3\text{O}_2}{1\ \text{g H}_2\text{O}} \times \frac{1000\ \text{g}}{\text{kg}} \times \frac{1\ \text{mol}}{60.05\ \text{g}} = 1600\ \text{mol/kg} \approx 2000\ \text{mol/kg}$$

$$99\ \text{g HC}_2\text{H}_3\text{O}_2 \times \frac{1\ \text{mol}}{60.05\ \text{g}} = 1.6\ \text{mol HC}_2\text{H}_3\text{O}_2;\ 1\ \text{g H}_2\text{O} \times \frac{1\ \text{mol}}{18.02\ \text{g}} = 0.06\ \text{mol H}_2\text{O}$$

$$\chi_{HC_2H_3O_2} = \frac{1.6}{1.6 + 0.06} = 0.96$$

Ammonia:

$$\frac{28\ \text{g NH}_3}{100.\ \text{g soln}} \times \frac{0.90\ \text{g}}{\text{cm}^3} \times \frac{1000\ \text{cm}^3}{\text{L}} \times \frac{1\ \text{mol}}{17.03\ \text{g}} = 15\ \text{mol/L}$$

$$\frac{28\ \text{g NH}_3}{72\ \text{g H}_2\text{O}} \times \frac{1000\ \text{g}}{\text{kg}} \times \frac{1\ \text{mol}}{17.03\ \text{g}} = 23\ \text{mol/kg}$$

$$28\ \text{g NH}_3 \times \frac{1\ \text{mol}}{17.03\ \text{g}} = 1.6\ \text{mol NH}_3;\ 72\ \text{g H}_2\text{O} \times \frac{1\ \text{mol}}{18.02\ \text{g}} = 4.0\ \text{mol H}_2\text{O}$$

$$\chi_{NH_3} = \frac{1.6}{4.0 + 1.6} = 0.29$$

31. $50.0\text{ mL toluene} \times \frac{0.867\text{ g}}{\text{mL}} = 43.4\text{ g toluene}$; $125\text{ mL benzene} \times \frac{0.874\text{ g}}{\text{mL}} = 109\text{ g benzene}$

$$\text{mass \% toluene} = \frac{\text{mass of toluene}}{\text{total mass}} \times 100 = \frac{43.4}{43.4 + 109} \times 100 = 28.5\%$$

$$\text{molarity} = \frac{43.4\text{ g toluene}}{175\text{ mL soln}} \times \frac{1000\text{ mL}}{\text{L}} \times \frac{1\text{ mol toluene}}{92.13\text{ g toluene}} = 2.69\text{ mol/L}$$

$$\text{molality} = \frac{43.4\text{ g toluene}}{109\text{ g benzene}} \times \frac{1000\text{ g}}{\text{kg}} \times \frac{1\text{ mol toluene}}{92.13\text{ g toluene}} = 4.32\text{ mol/kg}$$

$$43.4\text{ g toluene} \times \frac{1\text{ mol}}{92.13\text{ g}} = 0.471\text{ mol toluene}$$

$$109\text{ g benzene} \times \frac{1\text{ mol benzene}}{78.11\text{ g benzene}} = 1.40\text{ mol benzene};\quad \chi_{\text{toluene}} = \frac{0.471}{0.471 + 1.40} = 0.252$$

33. If we have 1.00 L of solution:

$$1.37\text{ mol citric acid} \times \frac{192.12\text{ g}}{\text{mol}} = 263\text{ g citric acid}$$

$$1.00 \times 10^3\text{ mL solution} \times \frac{1.10\text{ g}}{\text{mL}} = 1.10 \times 10^3\text{ g solution}$$

$$\text{mass \% of citric acid} = \frac{263\text{ g}}{1.10 \times 10^3\text{ g}} \times 100 = 23.9\%$$

In 1.00 L of solution, we have 263 g citric acid and $(1.10 \times 10^3 - 263) = 840$ g of H_2O.

$$\text{molality} = \frac{1.37\text{ mol citric acid}}{0.84\text{ kg } H_2O} = 1.6\text{ mol/kg}$$

$$840\text{ g } H_2O \times \frac{1\text{ mol}}{18.02\text{ g}} = 47\text{ mol } H_2O;\quad \chi_{\text{citric acid}} = \frac{1.37}{47 + 1.37} = 0.028$$

Since citric acid is a triprotic acid, then the amount of protons citric acid can provide is three times the molarity. Therefore, normality = 3 × molarity:

$$\text{normality} = 3 \times 1.37\ M = 4.11\ N$$

Energetics of Solutions and Solubility

35. Using Hess's law:

$KCl(s) \rightarrow K^+(g) + Cl^-(g)$	$\Delta H = -\Delta H_{LE} = -(-715 \text{ kJ/mol})$
$K^+(g) + Cl^-(g) \rightarrow K^+(aq) + Cl^-(aq)$	$\Delta H = \Delta H_{hyd} = -684 \text{ kJ/mol}$
$KCl(s) \rightarrow K^+(aq) + Cl^-(aq)$	$\Delta H_{soln} = 31 \text{ kJ/mol}$

ΔH_{soln} refers to the heat released or gained when a solute dissolves in a solvent. Here, an ionic compound dissolves in water.

37. Both $Al(OH)_3$ and NaOH are ionic compounds. Since the lattice energy is proportional to the charge of the ions, then the lattice energy of aluminum hydroxide is greater than that of sodium hydroxide. The attraction of water molecules for Al^{3+} and OH^- cannot overcome the larger lattice energy and $Al(OH)_3$ is insoluble. For NaOH, the favorable hydration energy is large enough to overcome the smaller lattice energy and NaOH is soluble.

39. Water is a polar solvent and dissolves polar solutes and ionic solutes. Carbon tetrachloride (CCl_4) is a nonpolar solvent and dissolves nonpolar solutes (like dissolves like).

a. Water; $Cu(NO_3)_2$ is an ionic solid.

b. CCl_4; CS_2 is a nonpolar molecule.

c. Water; CH_3CO_2H is polar.

d. CCl_4; The long nonpolar hydrocarbon chain favors a nonpolar solvent (the molecule is mostly nonpolar).

e. Water; HCl is polar.

f. CCl_4; C_6H_6 is nonpolar.

41. Water is a polar molecule capable of hydrogen bonding. Polar molecules, especially molecules capable of hydrogen bonding, and ions are all attracted to water. For covalent compounds, as polarity increases, the attraction to water increases. For ionic compounds, as the charge of the ions increase and/or the size of the ions decrease, the attraction to water increases.

a. CH_3CH_2OH; CH_3CH_2OH is polar while $CH_3CH_2CH_3$ is nonpolar.

b. $CHCl_3$; $CHCl_3$ is polar while CCl_4 is nonpolar.

c. CH_3CH_2OH; CH_3CH_2OH is much more polar than $CH_3(CH_2)_{14}CH_2OH$.

43. As the length of the hydrocarbon chain increases, the solubility decreases. The –OH end of the alcohols can hydrogen bond with water. The hydrocarbon chain, however, is basically nonpolar and interacts poorly with water. As the hydrocarbon chain gets longer, a greater portion of the molecule cannot interact with the water molecules and the solubility decreases, i.e., the effect of the –OH group decreases as the alcohols get larger.

45. $P_{gas} = kC$; $120 \text{ torr} \times \frac{1 \text{ atm}}{760 \text{ torr}} = \frac{780 \text{ atm L}}{\text{mol}} \times C$; $C = 2.0 \times 10^{-4} \text{ mol/L}$

Vapor Pressures of Solutions

47. $P_{H_2O} = \chi_{H_2O} P^{\circ}_{H_2O}$; $\chi_{H_2O} = \dfrac{\text{mol } H_2O \text{ in solution}}{\text{total mol in solution}}$

$50.0 \text{ g } C_6H_{12}O_6 \times \dfrac{1 \text{ mol } C_6H_{12}O_6}{180.16 \text{ g } C_6H_{12}O_6} = 0.278 \text{ mol glucose}$

$600.0 \text{ g } H_2O \times \dfrac{1 \text{ mol}}{18.02 \text{ g}} = 33.30 \text{ mol } H_2O$; Total mol = 0.278 + 33.30 = 33.58 mol

$\chi_{H_2O} = \dfrac{33.30}{33.58} = 0.9917$; $P_{H_2O} = \chi_{H_2O} P^{\circ}_{H_2O} = 0.9917 \times 23.8 \text{ torr} = 23.6 \text{ torr}$

49. $25.8 \text{ g } CH_4N_2O \times \dfrac{1 \text{ mol}}{60.06 \text{ g}} = 0.430 \text{ mol}$; $275 \text{ g } H_2O \times \dfrac{1 \text{ mol}}{18.02 \text{ g}} = 15.3 \text{ mol}$

$\chi_{H_2O} = \dfrac{15.3}{15.3 + 0.430} = 0.973$; $P_{H_2O} = \chi_{H_2O} P^{\circ}_{H_2O} = 0.973 \text{ (23.8 torr)} = 23.2 \text{ torr at } 25°C$

$P_{H_2O} = 0.973 \text{ (71.9 torr)} = 70.0 \text{ torr at } 45°C$

51. a. $25 \text{ mL } C_5H_{12} \times \dfrac{0.63 \text{ g}}{\text{mL}} \times \dfrac{1 \text{ mol}}{72.15 \text{ g}} = 0.22 \text{ mol } C_5H_{12}$

$45 \text{ mL } C_6H_{14} \times \dfrac{0.66 \text{ g}}{\text{mL}} \times \dfrac{1 \text{ mol}}{86.17 \text{ g}} = 0.34 \text{ mol } C_6H_{14}$; total mol = 0.22 + 0.34 = 0.56 mol

$\chi^{L}_{pen} = \dfrac{\text{mol pentane in solution}}{\text{total mol in solution}} = \dfrac{0.22 \text{ mol}}{0.56 \text{ mol}} = 0.39$, $\chi^{L}_{hex} = 1.00 - 0.39 = 0.61$

$P_{pen} = \chi^{L}_{pen} P^{\circ}_{pen} = 0.39(511 \text{ torr}) = 2.0 \times 10^2 \text{ torr}$; $P_{hex} = 0.61(150. \text{ torr}) = 92 \text{ torr}$

$P_{total} = P_{pen} + P_{hex} = 2.0 \times 10^2 + 92 = 292 \text{ torr} = 290 \text{ torr}$

b. From Chapter 5 on gases, the partial pressure of a gas is proportional to the number of moles of gas present. For the vapor phase:

$$\chi^{V}_{pen} = \frac{\text{mol pentane in vapor}}{\text{total mol vapor}} = \frac{P_{pen}}{P_{total}} = \frac{2.0 \times 10^2 \text{ torr}}{290 \text{ torr}} = 0.69$$

Note: In the Solutions Guide, we added V or L to the mole fraction symbol to emphasize which value we are solving. If the L or V is omitted, then the liquid phase is assumed.

53. $P_{total} = P_{pen} + P_{hex}$, $350. \text{ torr} = \chi^{L}_{pen}(511 \text{ torr}) + \chi^{L}_{hex}(150. \text{ torr})$; $\chi^{L}_{hex} = 1.000 - \chi^{L}_{pen}$

$350. = 511\, \chi^{L}_{pen} + (1.000 - \chi^{L}_{pen})150.$, $\dfrac{200.}{361} = \chi^{L}_{pen} = 0.554$; $\chi^{L}_{hex} = 1.000 - 0.554 = 0.446$

For the vapor composition:

$$P_{pen} = \chi^L_{pen} P^\circ_{pen} = 0.554\ (511\ \text{torr}) = 283\ \text{torr};\ \ P_{hex} = 0.446\ (150.\ \text{torr}) = 66.9\ \text{torr}$$

Since partial pressures are proportional to the moles of gas present, then:

$$\chi^V_{pen} = \frac{P_{pen}}{P_{tot}} = \frac{283\ \text{torr}}{350.\ \text{torr}} = 0.809;\ \ \chi^V_{hex} = 1.000 - 0.809 = 0.191$$

55. Compared to H_2O, solution d (methanol/water) will have the highest vapor pressure since methanol is more volatile than water. Both solution b (glucose/water) and solution c (NaCl/water) will have a lower vapor pressure than water by Raoult's law. NaCl dissolves to give Na^+ ions and Cl^- ions; glucose is a nonelectrolyte. Since there are more solute particles in solution c, the vapor pressure of solution c will be the lowest.

57. $50.0\ \text{g}\ CH_3COCH_3 \times \frac{1\ \text{mol}}{58.08\ \text{g}} = 0.861\ \text{mol acetone}$

$$50.0\ \text{g}\ CH_3OH \times \frac{1\ \text{mol}}{32.04\ \text{g}} = 1.56\ \text{mol methanol}$$

$$\chi^L_{acetone} = \frac{0.861}{0.861 + 1.56} = 0.356;\ \ \chi^L_{methanol} = 1.000 - \chi^L_{acetone} = 0.644$$

$$P_{total} = P_{methanol} + P_{acetone} = 0.644(143\ \text{torr}) + 0.356(271\ \text{torr}) = 92.1\ \text{torr} + 96.5\ \text{torr} = 188.6\ \text{torr}$$

Since partial pressures are proportional to the moles of gas present, then in the vapor phase:

$$\chi^V_{acetone} = \frac{P_{acetone}}{P_{total}} = \frac{96.5\ \text{torr}}{188.6\ \text{torr}} = 0.512;\ \ \chi^V_{methanol} = 1.000 - 0.512 = 0.488$$

The actual vapor pressure of the solution (161 torr) is less than the calculated pressure assuming ideal behavior (188.6 torr). Therefore, the solution exhibits negative deviations from Raoult's law. This occurs when the solute-solvent interactions are stronger than in pure solute and pure solvent.

Colligative Properties

59. $\text{molality} = m = \frac{\text{mol solute}}{\text{kg solvent}} = \frac{4.9\ \text{g sucrose}}{175\ \text{g solvent}} \times \frac{1000\ \text{g}}{\text{kg}} \times \frac{1\ \text{mol}\ C_{12}H_{22}O_{11}}{342.30\ \text{g}\ C_{12}H_{22}O_{11}} = 0.082\ \text{molal}$

$$\Delta T_b = K_b m = \frac{0.51°\text{C}}{\text{molal}} \times 0.082\ \text{molal} = 0.042°\text{C}$$

The boiling point is raised from 100.000°C to 100.042°C. We assumed P = 1 atm and ample significant figures in the boiling point of pure water.

61. $\Delta T_f = K_f m,\ \Delta T_f = 3.00°C = \frac{1.86°C}{\text{molal}} \times m;\ m = 1.61 \text{ mol/kg}$

$$0.150 \text{ kg } H_2O \times \frac{1.61 \text{ mol urea}}{\text{kg } H_2O} \times \frac{60.06 \text{ g urea}}{\text{mol urea}} = 14.5 \text{ g } (NH_2)_2CO$$

63. $\text{molality} = m = \frac{40.0 \text{ g } C_2H_6O_2}{60.0 \text{ g } H_2O} \times \frac{1000 \text{ g}}{\text{kg}} \times \frac{1 \text{ mol}}{62.07 \text{ g}} = 10.7 \text{ mol/kg}$

$$\Delta T_f = K_f m = 1.86°C/\text{molal} \times 10.7 \text{ molal} = 19.9°C;\ T_f = 0.0°C - 19.9°C = -19.9°C$$

$$\Delta T_b = K_b m = 0.51°C/\text{molal} \times 10.7 \text{ molal} = 5.5°C;\ T_b = 100.0°C + 5.5°C = 105.5°C$$

65. $\Delta T_f = K_f m,\ m = \frac{\Delta T_f}{K_f} = \frac{2.63°C}{40.°C \text{ kg/mol}} = \frac{6.6 \times 10^{-2} \text{ mol reserpine}}{\text{kg solvent}}$

The mol of reserpine present is:

$$0.0250 \text{ kg solvent} \times \frac{6.6 \times 10^{-2} \text{ mol reserpine}}{\text{kg solvent}} = 1.7 \times 10^{-3} \text{ mol reserpine}$$

From the problem, 1.00 g reserpine was used which must contain 1.7×10^{-3} mol reserpine. The molar mass of reserpine is:

$$\text{molar mass} = \frac{1.00 \text{ g}}{1.7 \times 10^{-3} \text{ mol}} = 590 \text{ g/mol (610 g/mol if no rounding of numbers).}$$

67. a. $M = \frac{1.0 \text{ g}}{\text{L}} \times \frac{1 \text{ mol}}{9.0 \times 10^4 \text{ g}} = 1.1 \times 10^{-5} \text{ mol/L};\ \pi = MRT$

$$\text{At 298 K: } \pi = \frac{1.1 \times 10^{-5} \text{ mol}}{\text{L}} \times \frac{0.08206 \text{ L atm}}{\text{mol K}} \times 298 \text{ K} \times \frac{760 \text{ torr}}{\text{atm}},\ \pi = 0.20 \text{ torr}$$

Since d = 1.0 g/cm^3, then 1.0 L solution has a mass of 1.0 kg. Since only 1.0 g of protein is present per liter solution, then 1.0 kg of H_2O is present and molality equals molarity.

$$\Delta T_f = K_f m = \frac{1.86°C}{\text{molal}} \times 1.1 \times 10^{-5} \text{ molal} = 2.0 \times 10^{-5}°C$$

b. Osmotic pressure is better for determining the molar mass of large molecules. A temperature change of 10^{-5}°C is very difficult to measure. A change in height of a column of mercury by 0.2 mm (0.2 torr) is not as hard to measure precisely.

69. $\pi = MRT,\ M = \frac{\pi}{RT} = \frac{15 \text{ atm}}{0.08206 \times 295 \text{ K}} = 0.62\ M;\ \frac{0.62 \text{ mol}}{\text{L}} \times \frac{342.30 \text{ g}}{\text{mol } C_{12}H_{22}O_{11}} = 212 \text{ g/L} \approx 210 \text{ g/L}$

Dissolve 210 g of sucrose in some water and dilute to 1.0 L in a volumetric flask. To get 0.62 ± 0.01 mol/L, we need 212 ± 3 g sucrose.

Properties of Electrolyte Solutions

71. $MgCl_2$ and NaCl are strong electrolytes, HOCl is a weak electrolyte and glucose is a nonelectrolyte. The effective particle concentrations are ~3.0 *m* $MgCl_2$, ~2.0 *m* NaCl, 2.0 < *m* HOCl < 1.0, and 1.0 *m* glucose. The order of freezing point depressions ($\Delta T_f = K_f m$) from lowest to highest are: glucose < HOCl < NaCl < $MgCl_2$ (a < c < b < d).

73. a. $m = \frac{5.0 \text{ g NaCl}}{0.025 \text{ kg}} \times \frac{1 \text{ mol}}{58.44 \text{ g}} = 3.4$ molal; $NaCl(aq) \rightarrow Na^+(aq) + Cl^-(aq)$, i = 2.0

$\Delta T_f = iK_f m = 2.0 \times 1.86°C/\text{molal} \times 3.4 \text{ molal} = 13°C$; $T_f = -13°C$

$\Delta T_b = iK_b m = 2.0 \times 0.51°C/\text{molal} \times 3.4 \text{ molal} = 3.5°C$; $T_b = 103.5°C$

b. $m = \frac{2.0 \text{ g Al(NO}_3)_3}{0.015 \text{ kg}} \times \frac{1 \text{ mol}}{213.01 \text{ g}} = 0.63$ mol/kg; $Al(NO_3)_3(aq) \rightarrow Al^{3+}(aq) + 3\ NO_3^-(aq)$, i = 4.0

$\Delta T_f = iK_f m = 4.0 \times 1.86°C/\text{molal} \times 0.63 \text{ molal} = 4.7°C$; $T_f = -4.7°C$

$\Delta T_b = iK_b m = 4.0 \times 0.51°C/\text{molal} \times 0.63 \text{ molal} = 1.3°C$; $T_b = 101.3°C$

75. $\Delta T_f = iK_f m,\ i = \frac{\Delta T_f}{K_f m} = \frac{0.110°C}{1.86°\ C/\text{molal} \times 0.0225 \text{ molal}} = 2.63$ for 0.0225 *m* $CaCl_2$

$i = \frac{0.440}{1.86 \times 0.0910} = 2.60$ for 0.0910 *m* $CaCl_2$; $i = \frac{1.330}{1.86 \times 0.278} = 2.57$ for 0.278 *m* $CaCl_2$

Note that i is less than the ideal value of 3.0 for $CaCl_2$. This is due to ion pairing in solution.

77. $\pi = iMRT = 3.0 \times 0.50 \text{ mol/L} \times 0.08206 \text{ L atm/K•mol} \times 298 \text{ K} = 37 \text{ atm}$

Because of ion pairing in solution, we would expect i to be less than 3.0 which results in fewer solute particles in solution which results in a lower osmotic pressure than calculated above.

Additional Exercises

79. $\text{molality} = \frac{40.0 \text{ g EG}}{60.0 \text{ g H}_2\text{O}} \times \frac{1000 \text{ g}}{\text{kg}} \times \frac{1 \text{ mol EG}}{62.07 \text{ g}} = 10.7 \text{ mol/kg}$

$\text{molarity} = \frac{40.0 \text{ g EG}}{100.0 \text{ g solution}} \times \frac{1.05 \text{ g}}{\text{cm}^3} \times \frac{1000 \text{ cm}^3}{\text{L}} \times \frac{1 \text{ mol}}{62.07 \text{ g}} = 6.77 \text{ mol/L}$

$40.0 \text{ g EG} \times \frac{1 \text{ mol}}{62.07 \text{ g}} = 0.644 \text{ mol EG}$; $60.0 \text{ g H}_2\text{O} \times \frac{1 \text{ mol}}{18.02 \text{ g}} = 3.33 \text{ mol H}_2\text{O}$

$\chi_{EG} = \frac{0.644}{3.33 + 0.644} = 0.162$ = mole fraction ethylene glycol

81. NH_3 is capable of forming hydrogen bonding interactions with water so NH_3 is very soluble in water. O_2 is a nonpolar substance and is not soluble in water.

83. $P_B = \chi_B P_B°$, $\chi_B = P_B/P_B° = 0.900\ \text{atm}/0.930\ \text{atm} = 0.968$

$$0.968 = \frac{\text{mol benzene}}{\text{total mol}};\ \text{mol benzene} = 78.11\ \text{g}\ C_6H_6 \times \frac{1\ \text{mol}}{78.11} = 1.000\ \text{mol}$$

Let x = mol solute, then: $\chi_B = 0.968 = \dfrac{1.000\ \text{mol}}{1.000 + x}$, $0.968 + 0.968\,x = 1.000$, $x = 0.033$ mol

$$\text{molar mass} = \frac{10.0\ \text{g}}{0.033\ \text{mol}} = 303\ \text{g/mol} \approx 3.0 \times 10^2\ \text{g/mol}$$

85. Out of 100.00 g, there are:

$$31.57\ \text{g C} \times \frac{1\ \text{mol C}}{12.01\ \text{g}} = 2.629\ \text{mol C};\qquad \frac{2.629}{2.629} = 1.000$$

$$5.30\ \text{g H} \times \frac{1\ \text{mol H}}{1.008\ \text{g}} = 5.26\ \text{mol H};\qquad \frac{5.26}{2.629} = 2.00$$

$$63.13\ \text{g O} \times \frac{1\ \text{mol O}}{16.00\ \text{g}} = 3.946\ \text{mol O};\qquad \frac{3.946}{2.629} = 1.501$$

empirical formula: $C_2H_4O_3$; Use the freezing point data to determine the molar mass.

$$m = \frac{\Delta T_f}{K_f} = \frac{5.20°\text{C}}{1.86°\text{C/molal}} = 2.80\ \text{molal}$$

$$\text{mol solute} = 0.0250\ \text{kg} \times \frac{2.80\ \text{mol solute}}{\text{kg}} = 0.0700\ \text{mol solute}$$

$$\text{molar mass} = \frac{10.56\ \text{g}}{0.0700\ \text{mol}} = 151\ \text{g/mol}$$

The empirical formula mass of $C_2H_4O_3$ = 76.05 g/mol. Since the molar mass is about twice the empirical mass, then the molecular formula is $C_4H_8O_6$ which has a molar mass of 152.10 g/mol.

Note: We use the experimental molar mass to determine the molecular formula. Knowing this, we calculate the molar mass precisely from the molecular formula using the periodic table.

87. a. As disccussed in Figure 11.18 of the text, the water would migrate from right to left. Initially, the level of liquid in the right arm would go down and the level in the left arm would go up. At some point, the rate of solvent transfer will be the same in both directions and the levels of the liquids in the two arms will stabilize. The height difference between the two arms will be a measure of the osmotic pressure of the NaCl solution.

b. Initially, H_2O molecules will have a net migration into the NaCl side. However, NaCl molecules can now migrate into the H_2O side. Because solute and solvent transfer are both possible, the levels of the liquids will be equal once the rate of solute and solvent transfer is equal in both directions. At this point, the concentration of NaCl will be equal in both chambers and the levels of liquid will be equal.

Challenge Problems

89. $\chi^V_{pen} = 0.15 = \frac{P_{pen}}{P_{total}}$; $P_{pen} = \chi^L_{pen} P^\circ_{pen} = \chi^L_{pen}(511 \text{ torr})$; $P_{total} = P_{pen} + P_{hex} = \chi^L_{pen}(511) + \chi^L_{hex}(150.)$

Since $\chi^L_{hex} = 1.000 - \chi^L_{pen}$, then: $P_{total} = \chi^L_{pen}(511) + (1.000 - \chi^L_{pen})(150.) = 150. + 361\ \chi^L_{pen}$

$$\chi^V_{pen} = \frac{P_{pen}}{P_{total}},\ 0.15 = \frac{\chi^L_{pen}(511)}{150. + 361\ \chi^L_{pen}},\ 0.15\,(150. + 361\ \chi^L_{pen}) = 511\,\chi^L_{pen}$$

$$23 + 54\ \chi^L_{pen} = 511\ \chi^L_{pen},\ \chi^L_{pen} = \frac{23}{457} = 0.050$$

91. $\Delta T_f = 5.51 - 2.81 = 2.70°C$; $m = \frac{\Delta T_f}{K_f} = \frac{2.7°C}{5.12°C/\text{molal}} = 0.527$ molal

Let x = mass of naphthalene (molar mass: 128.2 g/mol). Then $1.60 - x$ = mass of anthracene (molar mass = 178.2 g/mol).

$$\frac{x}{128.2} = \text{moles naphthalene and } \frac{1.60 - x}{178.2} = \text{moles anthracene}$$

$$\frac{0.527 \text{ mol solute}}{\text{kg solvent}} = \frac{\frac{x}{128.2} + \frac{1.60 - x}{178.2}}{0.0200 \text{ kg solvent}},\ 1.05 \times 10^{-2} = \frac{178.2\,x + 1.60\,(128.2) - 128.2\,x}{128.2\,(178.2)}$$

$50.0\,x + 205 = 240.$, $50.0\,x = 35$, $x = 0.70$ g naphthalene

So mixture is: $\frac{0.70 \text{ g}}{1.60 \text{ g}} \times 100 = 44\%$ naphthalene by mass and 56% anthracene by mass

93. $HCO_2H \rightarrow H^+ + HCO_2^-$; Only 4.2% of HCO_2H ionizes. The amount of H^+ or HCO_2^- produced is:

$0.042 \times 0.10\ M = 0.0042\ M$

The amount of HCO_2H remaining in solution after ionization is:

$0.10\ M - 0.0042\ M = 0.10\ M$

The total molarity of species present = $M_{HCO_2H} + M_{H^+} + M_{HCO_2^-} = 0.10 + 0.0042 + 0.0042 = 0.11\,M$.

Assuming 0.11 M = 0.11 molal and assuming ample significant figures in the freezing point and boiling point of water at P = 1 atm:

$\Delta T = K_f m = 1.86°C/molal \times 0.11\ molal = 0.20°C$; freezing point = -0.20°C

$\Delta T = K_b m = 0.51°C/molal \times 0.11\ molal = 0.056°C$; boiling point = 100.056°C

CHAPTER TWELVE

CHEMICAL KINETICS

Questions

9. a. An elementary step (reaction) is one in which the rate law can be written from the molecularity, i.e., from the coefficients in the balanced equation.

   b. The mechanism of a reaction is the series of proposed elementary reactions that may occur to give the overall reaction. The sum of all the steps in the mechanism gives the balanced chemical reaction.

   c. The rate determining step is the slowest elementary reaction in any given mechanism.

11. In a unimolecular reaction, a single reactant molecule decomposes to products. In a bimolecular reaction, two molecules collide to give products.

13. a. A homogeneous catalyst is in the same phase as the reactants.

    b. A heterogeneous catalyst is in a different phase than the reactants. The catalyst is usually a solid, although a catalyst in a liquid phase can act as a heterogeneous catalyst for some gas phase reactions.

15. No, the catalyzed reaction has a different mechanism and hence, a different rate law.

17. a. Activation energy and ΔE are independent of each other. Activation energy depends on the path reactants to take to convert to products. The overall energy change, ΔE, only depends on the initial and final energy states of the reactants and products. ΔE is path independent.

    b. The rate law can only be determined from experiment, not from the overall balanced reaction.

    c. Most reactions occur by a series of steps. The rate of the reaction is determined by the rate of the slowest step in the mechanism.

Exercises

Reaction Rates

19. The coefficients in the balanced equation tell us that the rate of consumption of $S_2O_3^{2-}$ will equal the rate of production of I^-, as well as telling us that the rate of consumption of I_2 and the rate of production of $S_4O_6^{2-}$ will be one-half the rate of consumption of $S_2O_3^{2-}$ (due to the 1:2 mol ratios).

$$\text{Rate} = \frac{-\Delta[S_2O_3^{2-}]}{\Delta t} = \frac{0.0080 \text{ mol}}{\text{L s}}$$; I_2 is consumed at half this rate. $\frac{-\Delta[I_2]}{\Delta t} = 0.0040$ mol/L•s

$$\frac{\Delta[S_4O_6^{2-}]}{\Delta t} = 0.0040 \text{ mol/L•s}; \quad \frac{\Delta[I^-]}{\Delta t} = 0.0080 \text{ mol/L•s}$$

21. a. The units for rate are always mol/L•s.

b. Rate = k; k must have units of mol/L•s

c. Rate = k[A], $\frac{\text{mol}}{\text{L s}} = k\left(\frac{\text{mol}}{\text{L}}\right)$

k must have units of s^{-1}.

d. Rate = k$[A]^2$, $\frac{\text{mol}}{\text{L s}} = k\left(\frac{\text{mol}}{\text{L}}\right)^2$

k must have units of L/mol•s.

e. L^2/mol^2•s

Rate Laws from Experimental Data: Initial Rates Method

23. a. In the first two experiments, [NO] is held constant and $[Cl_2]$ is doubled. The rate also doubled. Thus, the reaction is first order with respect to Cl_2. Or mathematically: Rate = $k[NO]^x[Cl_2]^y$

$$\frac{0.36}{0.18} = \frac{k(0.10)^x(0.20)^y}{k(0.10)^x(0.10)^y} = \frac{(0.20)^y}{(0.10)^y}, \quad 2.0 = 2.0^y, \ y = 1$$

We can get the dependence on NO from the second and third experiments. Here, as the NO concentration doubles (Cl_2 concentration is constant), the rate increases by a factor of four. Thus, the reaction is second order with respect to NO. Or mathematically:

$$\frac{1.45}{0.36} = \frac{k(0.20)^x(0.20)}{k(0.10)^x(0.20)} = \frac{(0.20)^x}{(0.10)^x}, \quad 4.0 = 2.0^x, \ x = 2; \text{ So, Rate} = k[NO]^2[Cl_2]$$

Try to examine experiments where only one concentration changes at a time. The more variables that change, the harder it is to determine the orders. Also, these types of problems can usually be solved by inspection. In general, we will solve using a mathematical approach, but keep in mind, you probably can solve for the orders by simple inspection of the data.

b. The rate constant k can be determined from the experiments. From experiment 1:

$$\frac{0.18 \text{ mol}}{\text{L min}} = k\left(\frac{0.10 \text{ mol}}{\text{L}}\right)^2\left(\frac{0.10 \text{ mol}}{\text{L}}\right), \quad k = 180 \text{ L}^2/\text{mol}^2\text{•min}$$

From the other experiments:

k = 180 L^2/mol^2•min (2nd exp.); k = 180 L^2/mol^2•min (3rd exp.)

The average rate constant is $k_{mean} = 1.8 \times 10^2$ L^2/mol^2•min.

25. a. Rate = k[NOCl]^n; Using experiments two and three:

$$\frac{2.66 \times 10^4}{6.64 \times 10^3} = \frac{k(2.0 \times 10^{16})^n}{k(1.0 \times 10^{16})^n}, \quad 4.01 = 2.0^n, \ n = 2; \ \text{Rate} = k[\text{NOCl}]^2$$

b. $$\frac{5.98 \times 10^4 \text{ molecules}}{\text{cm}^3 \text{ s}} = k\left(\frac{3.0 \times 10^{16} \text{ molecules}}{\text{cm}^3}\right)^2, \quad k = 6.6 \times 10^{-29} \text{ cm}^3/\text{molecules}\bullet\text{s}$$

The other three experiments give (6.7, 6.6 and 6.6) × 10^{-29} cm^3/molecules•s, respectively.

The mean value for k is 6.6 × 10^{-29} cm^3/molecules•s.

c. $$\frac{6.6 \times 10^{-29} \text{ cm}^3}{\text{molecules s}} \times \frac{1 \text{ L}}{1000 \text{ cm}^3} \times \frac{6.022 \times 10^{23} \text{molecules}}{\text{mol}} = \frac{4.0 \times 10^{-8} \text{ L}}{\text{mol s}}$$

27. a. Rate = k[Hb]^x[CO]^y; Comparing the first two experiments, [CO] is unchanged, [Hb] doubles, and the rate doubles. Therefore, the reaction is first order in Hb. Comparing the second and third experiments, [Hb] is unchanged, [CO] triples and the rate triples. Therefore, $y = 1$ and the reaction is first order in CO.

b. Rate = k[Hb][CO]

c. From the first experiment:

$$0.619 \ \mu\text{mol/L}\bullet\text{s} = k\ (2.21 \ \mu\text{mol/L})(1.00 \ \mu\text{mol/L}), \quad k = 0.280 \ \text{L}/\mu\text{mol}\bullet\text{s}$$

The second and third experiments give similar k values, so k_{mean} = 0.280 L/μmol•s.

d. $$\text{Rate} = k[\text{Hb}][\text{CO}] = \frac{0.280 \text{ L}}{\mu\text{mol s}} \times \frac{3.36 \ \mu\text{mol}}{\text{L}} \times \frac{2.40 \ \mu\text{mol}}{\text{L}} = 2.26 \ \mu\text{mol/L}\bullet\text{s}$$

Integrated Rate Laws

29. The first assumption to make is that the reaction is first order. For a first order reaction, a graph of ln [H_2O_2] vs time will yield a straight line. If this plot is not linear, then the reaction is not first order and we make another assumption. The data and plot for the first order assumption is below.

Time (s)	[H_2O_2] (mol/L)	ln [H_2O_2]
0	1.00	0.000
120.	0.91	-0.094
300.	0.78	-0.25
600.	0.59	-0.53
1200.	0.37	-0.99
1800.	0.22	-1.51
2400.	0.13	-2.04
3000.	0.082	-2.50
3600.	0.050	-3.00

Note: We carried extra significant figures in some of the ln values in order to reduce round off error. For the plots, we will do this most of the time when the ln function is involved.

The plot of ln $[H_2O_2]$ vs. time is linear. Thus, the reaction is first order. The rate law and integrated rate law are: Rate = $k[H_2O_2]$ and $\ln [H_2O_2] = -kt + \ln [H_2O_2]_o$.

We determine the rate constant k by determining the slope of the ln $[H_2O_2]$ vs time plot (slope = -k). Using two points on the curve gives:

$$\text{slope} = -k = \frac{\Delta y}{\Delta x} = \frac{0 - (3.00)}{0 - 3600.} = -8.3 \times 10^{-4}\ s^{-1},\ k = 8.3 \times 10^{-4}\ s^{-1}$$

To determine $[H_2O_2]$ at 4000. s, use the integrated rate law where at t = 0, $[H_2O_2]_o = 1.00\ M$.

$$\ln [H_2O_2] = -kt + \ln [H_2O_2]_o \quad \text{or} \quad \ln\left(\frac{[H_2O_2]}{[H_2O_2]_o}\right) = -kt$$

$$\ln\left(\frac{[H_2O_2]}{1.00}\right) = -8.3 \times 10^{-4}\ s^{-1} \times 4000.\ s,\ \ \ln [H_2O_2] = -3.3,\ \ [H_2O_2] = e^{-3.3} = 0.037\ M$$

31. Assume the reaction is first order and see if the plot of ln $[NO_2]$ vs. time is linear. If this isn't linear, try the second order plot of $1/[NO_2]$ vs. time. The data and plots follow.

Time (s)	$[NO_2]$ (*M*)	ln $[NO_2]$	$1/[NO_2]$ (M^{-1})
0	0.500	-0.693	2.00
1.20×10^3	0.444	-0.812	2.25
3.00×10^3	0.381	-0.965	2.62
4.50×10^3	0.340	-1.079	2.94
9.00×10^3	0.250	-1.386	4.00
1.80×10^4	0.174	-1.749	5.75

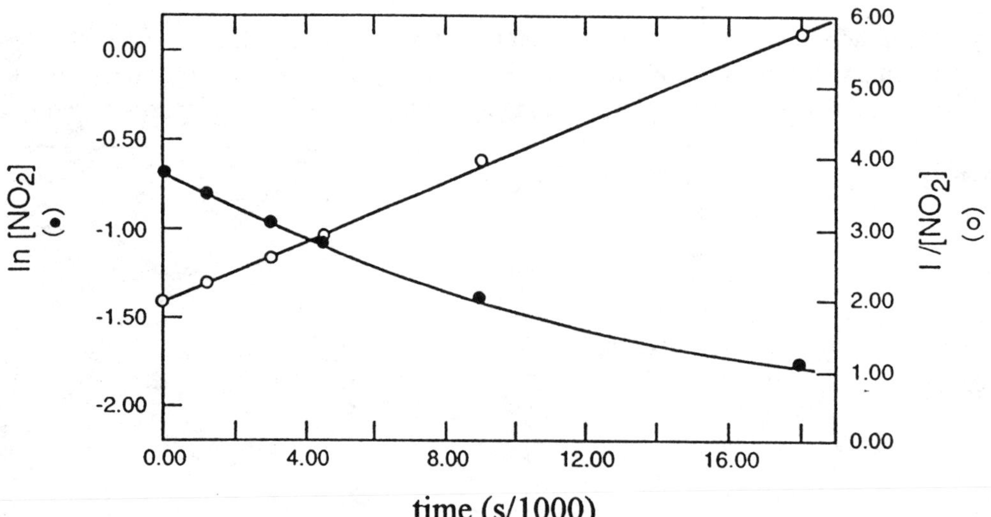

The plot of $1/[NO_2]$ vs. time is linear. The reaction is second order in NO_2. The rate law and integrated rate law are: Rate = $k[NO_2]^2$ and $\frac{1}{[NO_2]} = kt + \frac{1}{[NO_2]_o}$.

The slope of the plot $1/[NO_2]$ vs. t gives the value of k. Using a couple points on the plot:

$$\text{slope} = k = \frac{\Delta y}{\Delta x} = \frac{(5.75 - 2.00)\,M^{-1}}{(1.80 \times 10^{-4} - 0)\ \text{s}} = 2.08 \times 10^{-4}\ \text{L/mol•s}$$

To determine $[NO_2]$ at 2.70×10^4 s, use the integrated rate law where $1/[NO_2]_o = 1/0.500\,M = 2.00\,M^{-1}$.

$$\frac{1}{[NO_2]} = kt + \frac{1}{[NO_2]_o},\ \frac{1}{[NO_2]} = \frac{2.08 \times 10^{-4}\ \text{L}}{\text{mol s}} \times 2.70 \times 10^4\ \text{s} + 2.00\,M^{-1}$$

$$\frac{1}{[NO_2]} = 7.62,\ [NO_2] = 0.131\,M$$

33. a. The integrated rate law for this zero order reaction is: $[HI] = -kt + [HI]_o$. This equation is in the form of the generic straight line equation, y = mx + b. A plot of [HI] vs time will give a straight line with a negative slope equal to -k.

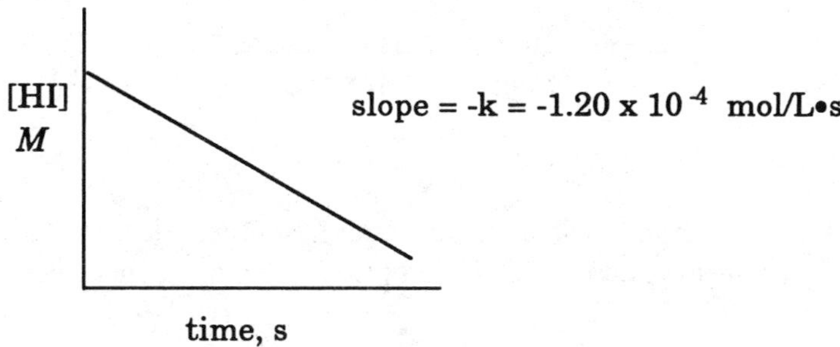

b. $[HI] = -kt + [HI]_o$, $[HI] = -\left(\frac{1.20 \times 10^{-4}\ \text{mol}}{\text{L s}}\right) \times \left(25\ \text{min} \times \frac{60\ \text{s}}{\text{min}}\right) + \frac{0.250\ \text{mol}}{\text{L}}$

$[HI] = -0.18$ mol/L + 0.250 mol/L = $0.07\,M$

c. $[HI] = 0 = -kt + [HI]_o$, $kt = [HI]_o$, $t = \frac{[HI]_o}{k}$

$$t = \frac{0.250\ \text{mol/L}}{1.20 \times 10^{-4}\ \text{mol/L•s}} = 2080\ \text{s} = 34.7\ \text{min}$$

35. The first assumption to make is that the reaction is first order. For a first order reaction, a graph of $\ln [C_4H_6]$ vs. t should yield a straight line. If this isn't linear, then try the second order plot of $1/[C_4H_6]$ vs. t. The data and the plots follow.

Time	195	604	1246	2180	6210 s
$[C_4H_6]$	1.6×10^{-2}	1.5×10^{-2}	1.3×10^{-2}	1.1×10^{-2}	0.68×10^{-2} *M*
$\ln [C_4H_6]$	-4.14	-4.20	-4.34	-4.51	-4.99
$1/[C_4H_6]$	62.5	66.7	76.9	90.9	147 M^{-1}

Note: To reduce round off error, we carried extra sig. figs. in the data points.

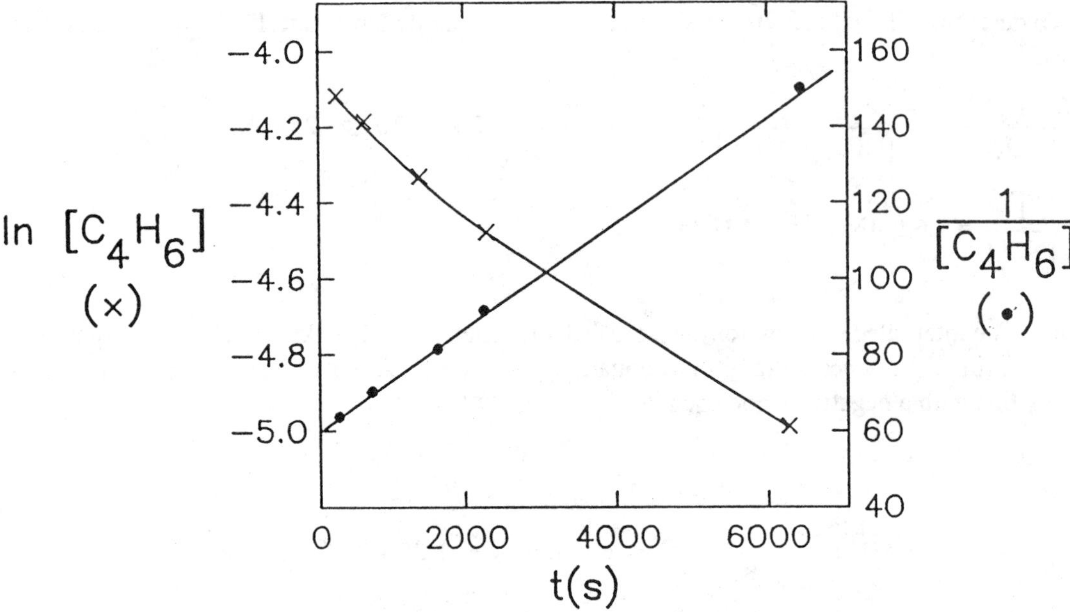

The natural log plot is not linear, so the reaction is not first order. Since the second order plot of $1/[C_4H_6]$ vs. t is linear, then we can conclude that the reaction is second order in butadiene. The rate law is:

$$\text{Rate} = k[C_4H_6]^2$$

For a second order reaction, the integrated rate law is: $\frac{1}{[C_4H_6]} = kt + \frac{1}{[C_4H_6]_o}$

The slope of the straight line equals the value of the rate constant. Using the points on the line at 1000. and 6000. s:

$$k = \text{slope} = \frac{144 \text{ L/mol} - 73 \text{ L/mol}}{6000.\text{ s} - 1000.\text{ s}} = 1.4 \times 10^{-2} \text{ L/mol•s}$$

37. If $[A]_o = 100.0$, then after 65 s, 45.0% of A has reacted or $[A] = 55.0$. For first order reactions:

$$\ln\left(\frac{[A]}{[A]_o}\right) = -kt,\ \ln\left(\frac{55.0}{100.0}\right) = -k(65\text{ s}),\ k = 9.2 \times 10^{-3}\text{ s}^{-1};\ t_{1/2} = \frac{\ln 2}{k} = \frac{0.693}{k} = 75\text{ s}$$

39. a. If the reaction is 38.5% complete, then 38.5% of the original concentration is consumed, leaving 61.5%.

$[A] = 61.5\%$ of $[A]_o$ or $[A] = 0.615\ [A]_o$; $\ln\left(\frac{[A]}{[A]_o}\right) = -kt$, $\ln\left(\frac{0.615\ [A]_o}{[A]_o}\right) = -k(480\ s)$

$\ln(0.615) = -k(480.\ s)$, $-0.486 = -k(480.\ s)$, $k = 1.01 \times 10^{-3}\ s^{-1}$

b. $t_{1/2} = (\ln 2)/k = 0.6931/1.01 \times 10^{-3}\ s^{-1} = 686\ s$

c. 25% complete: $[A] = 0.75\ [A]_o$; $\ln(0.75) = -1.01 \times 10^{-3}\ (t)$, $t = 280\ s$

75% complete: $[A] = 0.25\ [A]_o$; $\ln(0.25) = -1.01 \times 10^{-3}\ (t)$, $t = 1.4 \times 10^3\ s$

Or, we know it takes $2 \times t_{1/2}$ for reaction to be 75% complete. $t = 2 \times 686\ s = 1370\ s$

95% complete: $[A] = 0.05\ [A]_o$; $\ln(0.05) = -1.01 \times 10^{-3}\ (t)$, $t = 3 \times 10^3\ s$

41. For a second order reaction: $t_{1/2} = \frac{1}{k[A]_o}$ or $k = \frac{1}{t_{1/2}[A]_o}$

$$k = \frac{1}{143\ s(0.060\ mol/L)} = 0.12\ L/mol{\bullet}s$$

43. 100% → 50% → 25% → 12.5%; This process is 3 half-lives = 3(14 h) = 42 hours.

Reaction Mechanisms

45. For elementary reactions, the rate law can be written using the coefficients in the balanced equation to determine orders.

a. Rate = $k[CH_3NC]$ b. Rate = $k[O_3][NO]$

47. From experiment (Exercise 12.29), we know the rate law is: Rate = $k[H_2O_2]$. A mechanism consists of a series of elementary reactions where the rate law for each step can be determined using the coefficients in the balanced equations. For a plausible mechanism, the rate law derived from a mechanism must agree with the rate law determined from experiment. To derive the rate law from the mechanism, the rate of the reaction is assumed to equal the rate of the slowest step in the mechanism.

This mechanism will agree with the experimentally determined rate law only if step 1 is the slow step (called the rate determining step). If step 1 is slow, then Rate = $k[H_2O_2]$ which agrees with experiment.

Another important property of a mechanism is that the sum of all steps must give the overall balanced equation. Summing all steps gives:

$$H_2O_2 \rightarrow 2\ OH$$
$$H_2O_2 + OH \rightarrow H_2O + HO_2$$
$$HO_2 + OH \rightarrow H_2O + O_2$$
$$\overline{\quad 2\ H_2O_2 \rightarrow 2\ H_2O + O_2 \quad}$$

Temperature Dependence of Rate Constants and the Collision Model

49. In the following plot, R = reactants, P = products, E_a = activation energy and RC = reaction coordinate which is the same as reaction progress. Note for this reaction that ΔE is positive since the products are at a higher energy than the reactants.

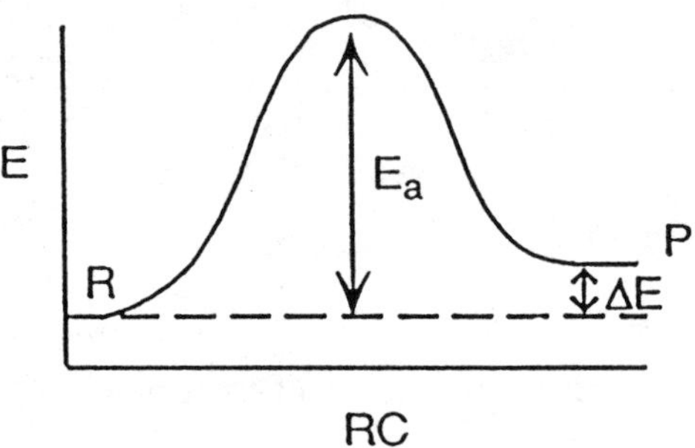

51.

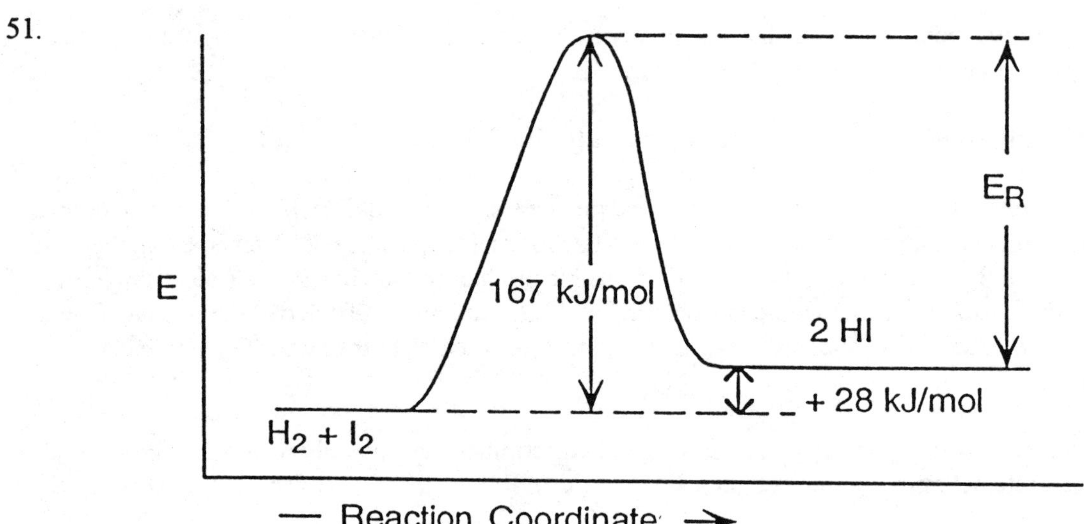

The activation energy for the reverse reaction is E_R in the diagram. $E_R = 167 - 28 = 139$ kJ/mol

53. The Arrhenius equation is: $k = A \exp(-E_a/RT)$ or in logarithmic form, $\ln k = -E_a/RT + \ln A$. Hence, a graph of ln k vs. 1/T should yield a straight line with a slope equal to $-E_a/R$ since the logarithmic form of the Arrhenius equation is in the form of a straight line equation, y = mx + b. Note: We carried one extra significant figure in the following ln k values in order to reduce round off error.

T (K)	1/T (K^{-1})	k (L/mol•s)	ln k
195	5.13×10^{-3}	1.08×10^{9}	20.80
230.	4.35×10^{-3}	2.95×10^{9}	21.81
260.	3.85×10^{-3}	5.42×10^{9}	22.41
298	3.36×10^{-3}	12.0×10^{9}	23.21
369	2.71×10^{-3}	35.5×10^{9}	24.29

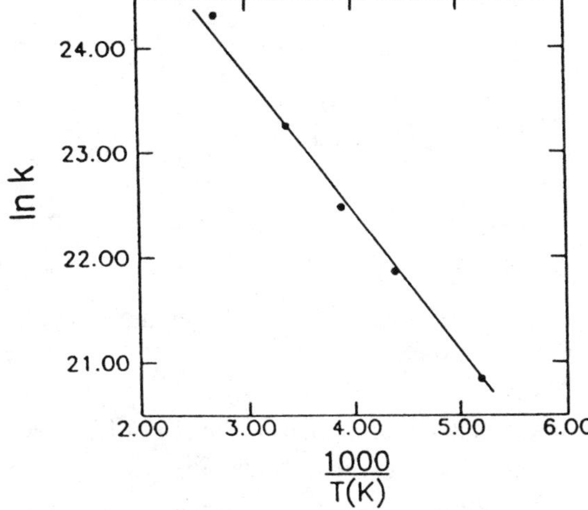

From the "eyeball" line on the graph:

$$\text{slope} = \frac{20.95 - 23.65}{5.00 \times 10^{-3} - 3.00 \times 10^{-3}} = \frac{-2.70}{2.00 \times 10^{-3}} = -1.35 \times 10^{3}\text{ K} = \frac{-E_a}{R}$$

$$E_a = 1.35 \times 10^{3}\text{ K} \times \frac{8.3145\text{ J}}{\text{K mol}} = 1.12 \times 10^{4}\text{ J/mol} = 11.2\text{ kJ/mol}$$

From the best straight line (by calculator): slope = -1.43×10^{3} K and E_a = 11.9 kJ/mol

55. $k = A \exp(-E_a/RT)$ or $\ln k = \frac{-E_a}{RT} + \ln A$ (the Arrhenius equation)

For two conditions: $\ln\left(\frac{k_2}{k_1}\right) = \frac{E_a}{R}\left(\frac{1}{T_1} - \frac{1}{T_2}\right)$ (Assuming A is temperature independent.)

Let $k_1 = 2.0 \times 10^{3}\text{ s}^{-1}$, T_1 = 298 K; k_2 = ?, T_2 = 348 K; $E_a = 15.0 \times 10^{3}$ J/mol

$$\ln\left(\frac{k_2}{2.0 \times 10^{3}\text{ s}^{-1}}\right) = \frac{15.0 \times 10^{3}\text{ J/mol}}{8.3145\text{ J/mol•K}}\left(\frac{1}{298\text{ K}} - \frac{1}{348\text{ K}}\right) = 0.87$$

$$\ln\left(\frac{k_2}{2.0 \times 10^3}\right) = 0.87, \quad \frac{k_2}{2.0 \times 10^3} = e^{0.87} = 2.4, \quad k_2 = 2.4(2.0 \times 10^3) = 4.8 \times 10^3 \text{ s}^{-1}$$

57. a. $\ln\left(\frac{k_2}{k_1}\right) = \frac{E_a}{R}\left(\frac{1}{T_1} - \frac{1}{T_2}\right)$, $\ln\left(\frac{0.950}{2.45 \times 10^{-4}}\right) = \frac{E_a}{8.3145}\left(\frac{1}{575} - \frac{1}{781}\right)$

$$8.26 = \frac{E_a}{8.3145}(4.6 \times 10^{-4}), \quad E_a = 1.5 \times 10^5 \text{ J/mol} = 150 \text{ kJ/mol}$$

$$\ln k_1 = -E_a/RT + \ln A, \quad \ln(2.45 \times 10^{-4}) = \left(\frac{-1.5 \times 10^5 \text{ J/mol}}{8.3145 \text{ J/mol•K} \times 575 \text{ K}}\right) + \ln A$$

$$-8.314 = -31.38 + \ln A, \quad \ln A = 23.07, \quad A = e^{23.07} = 1.0 \times 10^{10} \text{ L/mol•s}$$

Note: We carried extra significant figures in order to avoid large round off error.

b. $k = A \exp(-E_a/RT) = 1.0 \times 10^{10} \times \exp\left(\frac{-1.5 \times 10^5}{8.3145 \times 648}\right) = 8.1 \times 10^{-3}$ L/mol•s

Catalysts

59. a. NO is the catalyst. NO is present in the first step of the mechanism on the reactant side, but it is not a reactant since it is regenerated in the second step.

b. NO_2 is an intermediate. Intermediates also never appear in the overall balanced equation. In a mechanism, intermediates always appear first on the product side while catalysts always appear first on the reactant side. Intermediates are substances we make up in an attempt to explain the steps reactants take to get to products while catalysts are actual compounds used to speed up the reaction.

c. $k = A \exp(-E_a/RT)$; $\frac{k_{cat}}{k_{un}} = \frac{A \exp[-E_a(cat)/RT]}{A \exp[-E_a(un)/RT]} = \exp\left(\frac{E_a(un) - E_a(cat)}{RT}\right)$

$$\frac{k_{cat}}{k_{un}} = \exp\left(\frac{2100 \text{ J/mol}}{8.3145 \text{ J/mol•K} \times 298 \text{ K}}\right) = e^{0.85} = 2.3$$

The catalyzed reaction is 2.3 times faster than the uncatalyzed reaction at 25 °C.

61. The reaction at the surface of the catalyst follows the steps:

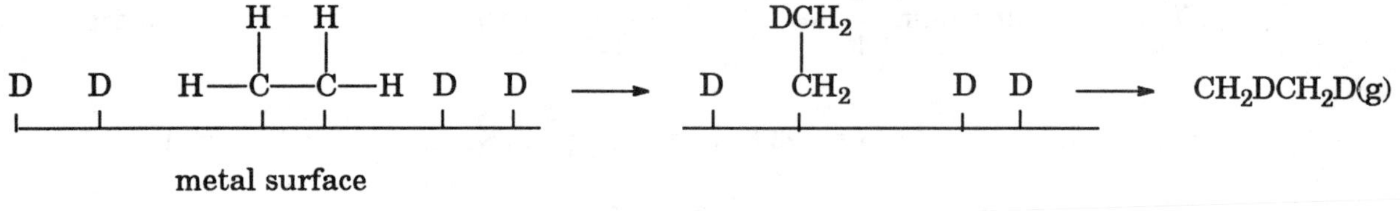

```
                  H   H                           DCH2
                  |   |                            |
D   D   H—C—C—H   D   D   ——>   D    CH2       D  D   ——>   CH2DCH2D(g)
|   |     |  |    |   |         |     |        |  |
———————————————————————         ———————————————————
         metal surface
```

Thus, $CH_2D–CH_2D$ should be the product.

Additional Exercises

63. Rate = $k[H_2SeO_3]^x[H^+]^y[I^-]^z$; Comparing the first and second experiments:

$$\frac{3.33 \times 10^{-7}}{1.66 \times 10^{-7}} = \frac{k(2.0 \times 10^{-4})^x (2.0 \times 10^{-2})^y (2.0 \times 10^{-2})^z}{k(1.0 \times 10^{-4})^x (2.0 \times 10^{-2})^y (2.0 \times 10^{-2})^z},\ 2.01 = 2.0^x,\ x = 1$$

Comparing the first and fourth experiments:

$$\frac{6.66 \times 10^{-7}}{1.66 \times 10^{-7}} = \frac{k(1.0 \times 10^{-4}) (4.0 \times 10^{-2})^y (2.0 \times 10^{-2})^z}{k(1.0 \times 10^{-4}) (2.0 \times 10^{-2})^y (2.0 \times 10^{-2})^z},\ 4.01 = 2.0^y,\ y = 2$$

Comparing the first and sixth experiments:

$$\frac{13.2 \times 10^{-7}}{1.66 \times 10^{-7}} = \frac{k(1.0 \times 10^{-4}) (2.0 \times 10^{-2})^2 (4.0 \times 10^{-2})^z}{k(1.0 \times 10^{-4}) (2.0 \times 10^{-2})^2 (2.0 \times 10^{-2})^z}$$

$$7.95 = 2.0^z,\ \log 7.95 = z \log 2.0,\ z = \frac{\log 7.95}{\log 2.0} = 2.99 \approx 3$$

Rate = $k[H_2SeO_3][H^+]^2[I^-]^3$

Experiment #1:

$$\frac{1.66 \times 10^{-7}\ \text{mol}}{\text{L s}} = k\left(\frac{1.0 \times 10^{-4}\ \text{mol}}{\text{L}}\right)\left(\frac{2.0 \times 10^{-2}\ \text{mol}}{\text{L}}\right)^2\left(\frac{2.0 \times 10^{-2}\ \text{mol}}{\text{L}}\right)^3$$

$$k = 5.19 \times 10^5\ \text{L}^5/\text{mol}^5\text{•s} = 5.2 \times 10^5\ \text{L}^5/\text{mol}^5\text{•s} = k_{mean}$$

65. The pressure of a gas is proportional to concentration. Therefore, we will use the pressure data to solve the problem since Rate = $-\Delta[SO_2Cl_2]/\Delta t = -\Delta P_{SO_2Cl_2}/\Delta t$.

Assuming a first order equation, the data and plot follow.

Time (hour)	0.00	1.00	2.00	4.00	8.00	16.00
$P_{SO_2Cl_2}$ (atm)	4.93	4.26	3.52	2.53	1.30	0.34
$\ln P_{SO_2Cl_2}$	1.595	1.449	1.258	0.928	0.262	-1.08

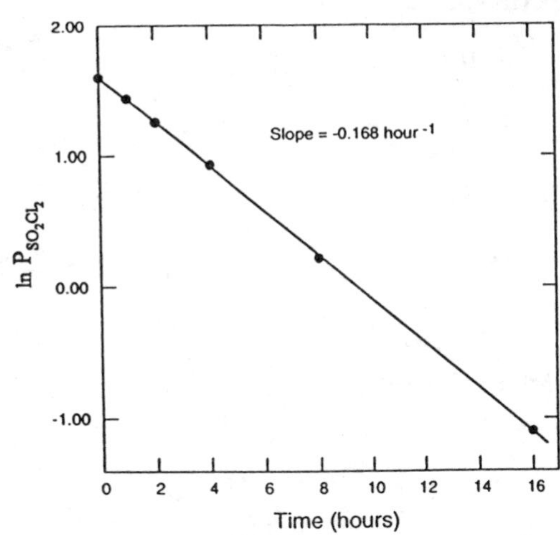

Since the ln $P_{SO_2Cl_2}$ vs. time plot is linear, then the reaction is first order in SO_2Cl_2.

a. Slope of $\ln P_{SO_2Cl_2}$ vs. t plot is -0.168 hour^{-1} = -k, k = 0.168 hour^{-1} = 4.67×10^{-5} s^{-1}

Since concentration units don't appear in first order rate constants, this value of k determined from the pressure data will be the same as if concentration data in molarity units were used.

b. $t_{1/2} = \frac{\ln 2}{k} = \frac{0.6931}{k} = \frac{0.6931}{0.168\ h^{-1}} = 4.13$ hour

c. $\ln\left(\frac{P_{SO_2Cl_2}}{P_O}\right) = -kt = -0.168\ h^{-1}\ (20.0\ hr) = -3.36,\ \left(\frac{P_{SO_2Cl_2}}{P_O}\right) = e^{-3.36} = 3.47 \times 10^{-2}$

Fraction left = 0.0347 = 3.47%

67. $\ln\left(\frac{k_2}{k_1}\right) = \frac{E_a}{R}\left(\frac{1}{T_1} - \frac{1}{T_2}\right);\ \frac{k_2}{k_1} = 7.00;\ T_1 = 275\ K;\ T_2 = 300.\ K$

$$\ln 7.00 = \frac{E_a}{8.3145\ J/mol \bullet K}\left(\frac{1}{275\ K} - \frac{1}{300.\ K}\right)$$

$$E_a = \frac{(8.3145\ J/mol \bullet K)\ (\ln 7.00)}{3.0 \times 10^{-4}\ K^{-1}} = 5.4 \times 10^4\ J/mol = 54\ kJ/mol$$

69. The rate depends on the number of reactant molecules adsorbed on the surface of the catalyst. This quantity is proportional to the concentration of reactant. However, when all of the catalyst surface sites are occupied, the rate becomes independent of the concentration of reactant.

Challenge Problems

71. Rate = $k[I^-]^x[OCl^-]^y[OH^-]^z$; Comparing the first and second experiments:

$$\frac{18.7 \times 10^{-3}}{9.4 \times 10^{-3}} = \frac{k(0.0026)^x (0.012)^y (0.10)^z}{k(0.0013)^x (0.012)^y (0.10)^z},\ 2.0 = 2.0^x,\ x = 1$$

Comparing the first and third experiments:

$$\frac{9.4 \times 10^{-3}}{4.7 \times 10^{-3}} = \frac{k(0.0013)\ (0.012)^y (0.10)^z}{k(0.0013)\ (0.006)^y (0.10)^z},\ 2.0 = 2^y,\ y = 1$$

Comparing the first and sixth experiments:

$$\frac{4.7 \times 10^{-3}}{9.4 \times 10^{-3}} = \frac{k(0.0013)\ (0.012)\ (0.20)^z}{k(0.0013)\ (0.012)\ (0.10)^z},\ 1/2 = 2.0^z,\ z = -1$$

Rate = $\frac{k[I^-][OCl^-]}{[OH^-]}$; The presence of OH^- decreases the rate of the reaction.

For the first experiment:

$$\frac{9.4 \times 10^{-3}\text{ mol}}{\text{L s}} = k\,\frac{(0.0013\text{ mol/L})\,(0.012\text{ mol/L})}{(0.10\text{ mol/L})},\quad k = 60.3\text{ s}^{-1} = 60.\text{ s}^{-1}$$

For all experiments, $k_{mean} = 60.\text{ s}^{-1}$.

73. a. We check for first order dependence by graphing ln [concentration] vs. time for each set of data. The rate dependence on NO is determined from the first set of data since the ozone concentration is relatively large compared to the NO concentration, so it is effectively constant.

time (ms)	[NO] (molecules/cm^3)	ln [NO]
0	6.0×10^8	20.21
100.	5.0×10^8	20.03
500.	2.4×10^8	19.30
700.	1.7×10^8	18.95
1000.	9.9×10^7	18.41

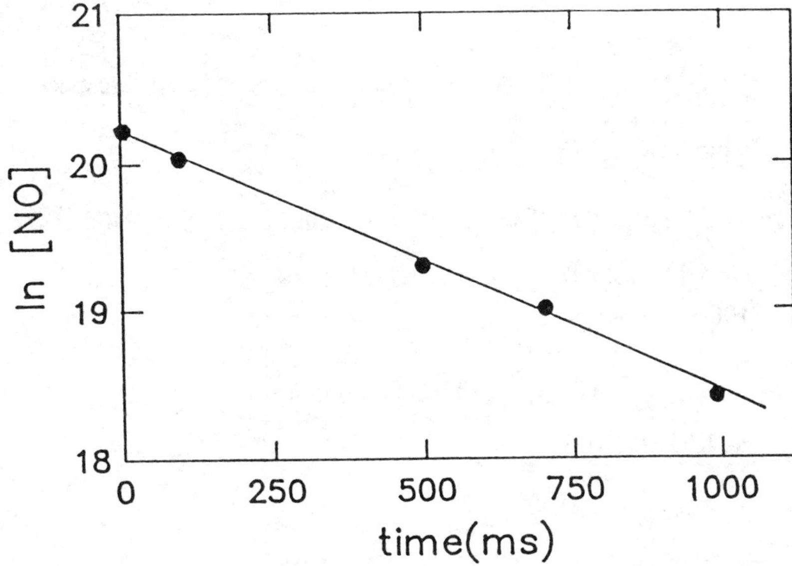

Since ln [NO] vs. t is linear, the reaction is first order with respect to NO.

We follow the same procedure for ozone using the second set of data. The data and plot are:

time (ms)	$[O_3]$ (molecules/cm^3)	ln $[O_3]$
0	1.0×10^{10}	23.03
50.	8.4×10^9	22.85
100.	7.0×10^9	22.67
200.	4.9×10^9	22.31
300.	3.4×10^9	21.95

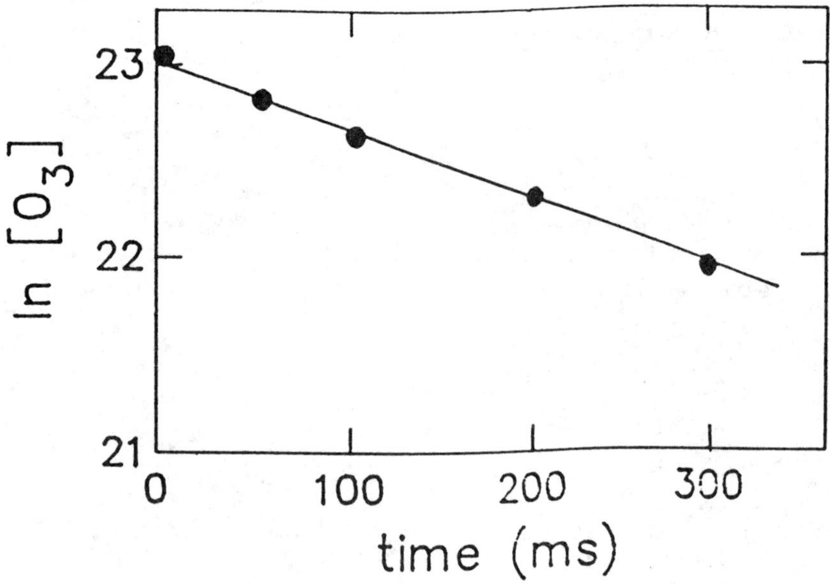

The plot of ln $[O_3]$ vs. t is linear. Hence, the reaction is first order with respect to ozone.

b. Rate = $k[NO][O_3]$ is the overall rate law.

c. For NO experiment, Rate = $k'[NO]$ and k' = -(slope from graph of ln [NO] vs. t).

$$k' = -\text{slope} = -\frac{18.41 - 20.21}{(1000. - 0) \times 10^{-3}\ \text{s}} = 1.8\ \text{s}^{-1}$$

For ozone experiment, Rate = $k''[O_3]$ and k'' = -(slope from ln $[O_3]$ vs. t).

$$k'' = -\text{slope} = -\frac{(21.95 - 23.03)}{(300. - 0) \times 10^{-3}\ \text{s}} = 3.6\ \text{s}^{-1}$$

d. From NO experiment, Rate = $k[NO][O_3] = k'[NO]$ where $k' = k[O_3]$.

$k' = 1.8\ \text{s}^{-1} = k(1.0 \times 10^{14}\ \text{molecules/cm}^3)$, $k = 1.8 \times 10^{-14}\ \text{cm}^3/\text{molecules}\bullet\text{s}$

We can check this from the ozone data. Rate = $k''[O_3] = k[NO][O_3]$ where $k'' = k[NO]$.

$k'' = 3.6\ \text{s}^{-1} = k(2.0 \times 10^{14}\ \text{molecules/cm}^3)$, $k = 1.8 \times 10^{-14}\ \text{cm}^3/\text{molecules}\bullet\text{s}$

Both values of k agree.

75. a.

T (K)	1/T (K^{-1})	k (min^{-1})	ln k
298.2	3.353×10^{-3}	178	5.182
293.5	3.407×10^{-3}	126	4.836
290.5	3.442×10^{-3}	100.	4.605

The plot of ln k vs. 1/T gives a straight line. The equation for the straight line is:

$\ln k = -6.48 \times 10^3 (1/T) + 26.9$

For a ln k vs. 1/T plot, the slope = $-E_a/R = -6.48 \times 10^3$ K

-6.48×10^3 K = $-E_a/8.3145$ J/mol•K, $E_a = 5.39 \times 10^4$ J/mol = 53.9 kJ/mol

b. $\ln k = -6.48 \times 10^3(1/288.2) + 26.9 = 4.42$, $k = e^{4.42} = 83\ min^{-1}$

About 83 chirps per minute per insect. Note: We carried extra sig. figs.

c. k gives the number of chirps per minute. The number or chirps in 15 sec is k/4.

T (C°)	T (°F)	k (min^{-1})	42 + 0.80 (k/4)
25.0	77.0	178	78° F
20.3	68.5	126	67°F
17.3	63.1	100.	62°F
15.0	59.0	83	59°F

The rule of thumb appears to be fairly accurate, about ± 1 °F.

CHAPTER THIRTEEN

CHEMICAL EQUILIBRIUM

Questions

9. a. The rates of the forward and reverse reactions are equal at equilibrium.

   b. There is no net change in the composition (as long as temperature is constant).

11. The equilibrium constant is a number that tells us the relative concentrations (pressures) of reactants and products at equilibrium. An equilibrium position is a set of concentrations that satisfy the equilibrium constant expression. More than one equilibrium position can satisfy the same equilibrium constant expression.

13. Table 13.1 of the text illustrates this nicely. Each of the three experiments in Table 13.1 have different equilibrium positions, that is, each experiment has different equilibrium concentrations. However, when these equilibrium concentrations are inserted into the equilibrium constant expression, each experiment gives the same value for K. The equilibrium position depends on the initial concentrations that one starts with. Since there are an infinite number of initial conditions, then there are an infinite number of equilibrium positions. However, each one of these infinite equilibrium positions will always give the same value for the equilibrium constant (assuming temperature is constant).

15. When we change the pressure by adding an unreactive gas, we do not change the partial pressures of any of the substances in equilibrium with each other. In this case the equilibrium will not shift. If we change the pressure by changing the volume, we will change the partial pressures of all the substances in equilibrium by the same factor. If there are unequal number of gaseous particles on the two sides of the equation, then the reaction is no longer at equilibrium and must shift to either products or reactants to return to equilibrium.

Exercises

Characteristics of Chemical Equilibrium

17. No, equilibrium is a dynamic process. Both reactions:

$$H_2O + CO \rightarrow H_2 + CO_2 \text{ and } H_2 + CO_2 \rightarrow H_2O + CO$$

are occurring, but at equal rates. Thus, ^{14}C atoms will be distributed between CO and CO_2.

The Equilibrium Constant

19. a. $K = \frac{[NO_2][O_2]}{[NO][O_3]}$ b. $K = \frac{[O_2][O]}{[O_3]}$

c. $K = \frac{[ClO][O_2]}{[Cl][O_3]}$ d. $K = \frac{[O_2]^3}{[O_3]^2}$

21. $K = 278 = \frac{[SO_3]^2}{[SO_2]^2[O_2]}$ for $2\ SO_2(g) + O_2(g) \rightleftharpoons 2\ SO_3(g)$

When a reaction is reversed, then $K_{new} = 1/K_{original}$. When a reaction is multiplied through by a value of n, then $K_{new} = (K_{original})^n$.

a. $SO_2(g) + 1/2\ O_2(g) \rightleftharpoons SO_3(g),\ K' = \frac{[SO_3]}{[SO_2][O_2]^{1/2}} = K^{1/2} = (278)^{1/2} = 16.7$

b. $2\ SO_3(g) \rightleftharpoons 2\ SO_2(g) + O_2(g),\ K'' = \frac{[SO_2]^2[O_2]}{[SO_3]^2} = \frac{1}{K} = \frac{1}{278} = 3.60 \times 10^{-3}$

c. $SO_3(g) \rightleftharpoons SO_2(g) + 1/2\ O_2(g),\ K''' = \frac{[SO_2][O_2]^{1/2}}{[SO_3]} = \left(\frac{1}{K}\right)^{1/2} = 6.00 \times 10^{-2}$

d. $4\ SO_2(g) + 2\ O_2(g) \rightleftharpoons 4\ SO_3(g),\ K'''' = \frac{[SO_3]^4}{[SO_2]^4[O_2]^2} = K^2 = 7.73 \times 10^4$

23. $K = \frac{[NO]^2}{[N_2][O_2]} = \frac{(4.7 \times 10^{-4}\ M)^2}{(0.041\ M)(0.0078\ M)} = 6.9 \times 10^{-4}$

25. $H_2(g) + I_2(g) \rightleftharpoons 2\ HI(g)$

$$K = \frac{[HI]^2}{[H_2][I_2]} = \frac{\left(\frac{3.50\ \text{mol}}{3.00\ \text{L}}\right)^2}{\left(\frac{4.10\ \text{mol}}{3.00\ \text{L}}\right)\left(\frac{0.30\ \text{mol}}{3.00\ \text{L}}\right)} = 10.$$

27. $K_p = \frac{P_{NO}^2 \times P_{Cl_2}}{P_{NOCl}^2} = \frac{(1.25 \times 10^{-2}\ \text{atm})^2(0.300\ \text{atm})}{(1.20\ \text{atm})^2} = 3.26 \times 10^{-5}$ atm

29. $K_p = K(RT)^{\Delta n}$ where Δn = sum of gaseous product coefficients - sum of gaseous reactant coefficients. For this reaction, $\Delta n = 4 - 2 = 2$.

$$K_p = \frac{2.6 \times 10^{-5}\ \text{mol}^2}{\text{L}^2} \times \left(\frac{0.08206\ \text{L atm}}{\text{mol K}} \times 400.\ \text{K}\right)^2 = 2.8 \times 10^{-2}\ \text{atm}^2$$

31. Solids and liquids do not appear in the equilibrium expression. Only gases and dissolved solutes appear in the equilibrium expression.

a. $K = \frac{1}{[O_2]^5}$

b. $K = [N_2O][H_2O]^2$

c. $K = \frac{1}{[CO_2]}$

d. $K = \frac{[SO_2]^8}{[O_2]^8}$

33. Since solids do not appear in the equilibrium constant expression, then $K = [CO_2] = 2.1 \times 10^{-3}$ mol/L.

35. $K_p = K(RT)^{\Delta n}$ where Δn equals the difference in the sum of the coefficients between gaseous products and gaseous reactants (Δn = mol gaseous products - mol gaseous reactants). When $\Delta n = 0$, then $K_p = K$. In Exercise 13.19, only reactions a and c have $\Delta n = 0$, so only these reactions have $K_p = K$. Reaction b has $\Delta n = 2 - 1 = 1$ and reaction d has $\Delta n = 3 - 2 = 1$.

Equilibrium Calculations

37. $H_2O(g) + Cl_2O(g) \rightarrow 2\ HOCl(g) \quad K = \frac{[HOCl]}{[H_2O][Cl_2O]} = 0.0900$

Use the reaction quotient Q to determine which way the reaction shifts to reach equilibrium. If Q < K, then the reaction shifts right to reach equilibrium. If Q > K, then the reaction shifts left to reach equilibrium. If Q = K, then the reaction does not have a shift in either direction since the reaction is at equilibrium.

a. $$Q = \frac{[HOCl]_o^2}{[H_2O]_o[Cl_2O]_o} = \frac{\left(\frac{1.0\ \text{mol}}{1.0\ \text{L}}\right)^2}{\left(\frac{0.10\ \text{mol}}{1.0\ \text{L}}\right)\left(\frac{0.10\ \text{mol}}{1.0\ \text{L}}\right)} = 1.0 \times 10^2$$

Q > K so the reaction shifts left to produce more reactants at equilibrium.

b. $$Q = \frac{\left(\frac{0.084\ \text{mol}}{2.0\ \text{L}}\right)^2}{\left(\frac{0.98\ \text{mol}}{2.0\ \text{L}}\right)\left(\frac{0.080\ \text{mol}}{2.0\ \text{L}}\right)} = 0.090 = K; \quad \text{at equilibrium}$$

c. $$Q = \frac{\left(\frac{0.25\ \text{mol}}{3.0\ \text{L}}\right)^2}{\left(\frac{0.56\ \text{mol}}{3.0\ \text{L}}\right)\left(\frac{0.0010\ \text{mol}}{3.0\ \text{L}}\right)} = 110 > K$$

Reaction will proceed to the left to reach equilibrium.

39. $CaCO_3(s) \rightleftharpoons CaO(s) + CO_2(g) \quad K_p = P_{CO_2} = 1.04$ atm

a. $Q = P_{CO_2}$; We only need the partial pressure of CO_2 to determine Q since solids do not appear in equilibrium expressions (or Q expressions). At this temperature all CO_2 will be in the gas phase. Q = 2.55 atm so $Q > K_p$; Reaction will shift to the left to reach equilibrium; the mass of CaO will decrease.

b. Q = 1.04 atm = K_p so the reaction is at equilibrium; mass of CaO will not change.

c. Q = 1.04 atm = K_p so the reaction is at equilibrium; mass of CaO will not change.

d. Q = 0.211 atm < K_p; The reaction will shift to the right to reach equilibrium; the mass of CaO will increase.

41. $K = \frac{[HF]^2}{[H_2][F_2]} = 2.1 \times 10^3;\ 2.1 \times 10^3 = \frac{[HF]^2}{(0.0021\,M)\,(0.0021\,M)}$, $[HF] = 0.096\,M$

43. $SO_2(g) + NO_2(g) \rightleftharpoons SO_3(g) + NO(g) \quad K = \frac{[SO_3][NO]}{[SO_2][NO_2]}$

To determine K, we must calculate the equilibrium concentrations. The initial concentrations are:

$$[SO_3]_o = [NO]_o = 0;\ \ [SO_2]_o = [NO_2]_o = \frac{2.00\text{ mol}}{1.00\text{ L}} = 2.00\,M$$

Next, we determine the change required to reach equilibrium. At equilibrium, [NO] = 1.30 mol/1.00 L = 1.30 M. Since there was zero amount of NO present initially, then 1.30 M of SO_2 and 1.30 M NO_2 must have reacted to produce 1.30 M NO as well as 1.30 M SO_3, all required by the balanced reaction. The equilibrium concentration for each substance is the sum of the initial concentration plus the change concentration necessary to reach equilibrium. The equilibrium concentrations are:

$$[SO_3] = [NO] = 0 + 1.30\,M = 1.30\,M;\ \ [SO_2] = [NO_2] = 2.00\,M - 1.30\,M = 0.70\,M$$

We now use these equilibrium concentrations to calculate K:

$$K = \frac{[SO_3]\,[NO]}{[SO_2]\,[NO_2]} = \frac{(1.30\,M)\,(1.30\,M)}{(0.70\,M)\,(0.70\,M)} = 3.4$$

45. When solving equilibrium problems, a common method to summarize all the information in the problem is to set up a table. We commonly call this table the ICE table since it summarizes initial concentrations, changes that must occur to reach equilibrium and equilibrium concentrations (the sum of the initial and change columns). For the change column, we will generally use the variable x which will be defined as the amount of reactant (or product) that must react to reach equilibrium. In this problem, the reaction must shift right since there are no products present initially. The general ICE table for this problem is:

	$2\,NO_2(g)$	$\rightleftharpoons$	$2\,NO(g)$	+	$O_2(g)$	$K = \dfrac{[NO]^2[O_2]}{[NO_2]^2}$
Initial	8.0 mol/1.0 L		0		0	
	Let x mol/L of NO_2 react to reach equilibrium					
Change	$-x$	$\rightarrow$	$+x$		$+x/2$	
Equil.	$8.0 - x$		x		$x/2$	

Note that we must use the coefficients in the balanced equation to determine how much products are produced letting x mol/L of NO_2 react to reach equilibrium. In the problem, we are told that $[NO]_e = 2.0\,M$. From the set-up, $[NO]_e = x = 2.0\,M$. Solving for the other concentrations: $[NO]_e = 8.0 - x = 8.0 - 2.0 = 6.0\,M$; $[O_2]_e = x/2 = 2.0/2 = 1.0\,M$. Calculating K:

$$K = \frac{[NO]^2[O_2]}{[NO_2]^2} = \frac{(2.0\,M)^2(1.0\,M)}{(6.0\,M)^2} = 0.11 \text{ mol/L}$$

Alternate Method: Fractions in the change column can be avoided (if you want) be defining x differently. If we were to let $2x$ mol/L of NO_2 react to reach equilibrium then the ICE table set-up is:

	$2\,NO_2(g)$	$\rightleftharpoons$	$2\,NO(g)$	+	$O_2(g)$	$K = \dfrac{[NO]^2[O_2]}{[NO_2]^2}$
Initial	$8.0\,M$		0		0	
	Let $2x$ mol/L of NO_2 react to reach equilibrium					
Change	$-2x$	$\rightarrow$	$+2x$		$+x$	
Equil.	$8.0 - 2x$		$2x$		x	

Solving: $2x = [NO]_e = 2.0\,M, x = 1.0\,M$; $[NO_2]_e = 8.0 - 2(1.0) = 6.0\,M$; $[O_2]_e = x = 1.0\,M$

These are exactly the same equilibrium concentrations as solved for previously, thus K will be the same (as it must be). The moral of the story is define x in a manner that is most comfortable for you. Your final answer is independent of how you define x initially.

47. Q = 1.00 which is less than K. Reaction shifts to the right to reach equilibrium. Summarizing the equilibrium problem in a table:

	$SO_2(g)$	+	$NO_2(g)$	$\rightleftharpoons$	$SO_3(g)$	+	$NO(g)$	K = 2.50
Initial	$1.00\,M$		$1.00\,M$		$1.00\,M$		$1.00\,M$	
	x mol/L of SO_2 reacts to reach equilibrium							
Change	$-x$		$-x$	$\rightarrow$	$+x$		$+x$	
Equil.	$1.00 - x$		$1.00 - x$		$1.00 + x$		$1.00 + x$	

Plug the equilibrium concentrations into the equilibrium constant expression:

$$K = \frac{[SO_3][NO]}{[SO_2][NO_2]} = 2.50 = \frac{(1.00 + x)^2}{(1.00 - x)^2};$$ Take the square root of both sides and solve for x:

$$\frac{1.00 + x}{1.00 - x} = 1.58, \; 1.00 + x = 1.58 - 1.58\,x, \; 2.58\,x = 0.58, \; x = 0.22\,M$$

The equilibrium concentrations are:

$[SO_3] = [NO] = 1.00 + x = 1.00 + 0.22 = 1.22\,M$; $[SO_2] = [NO_2] = 1.00 - x = 0.78\,M$

49. Since only reactants are present initially, the reaction must proceed to the right to reach equilibrium. Summarizing the problem in a table:

	$N_2(g)$	+ $O_2(g)$	$\rightleftharpoons$	$2\,NO(g)$	$K_p = 0.050$
Initial	0.80 atm	0.20 atm		0	
	x atm of N_2 reacts to reach equilibrium				
Change	$-x$	$-x$	$\rightarrow$	$+2x$	
Equil.	$0.80 - x$	$0.20 - x$		$2x$	

$$K = 0.050 = \frac{P_{NO}^2}{P_{N_2} \times P_{O_2}} = \frac{(2x)^2}{(0.80 - x)(0.20 - x)},\ 0.050(0.16 - 1.00\,x + x^2) = 4\,x^2$$

$4\,x^2 = 8.0 \times 10^{-3} - 0.050\,x + 0.050\,x^2$, $3.95\,x^2 + 0.050\,x - 8.0 \times 10^{-3} = 0$

Solving using the quadratic equation (see Appenxix 1.4 of the text):

$$x = \frac{-b \pm (b^2 - 4ac)^{1/2}}{2a} = \frac{-0.050 \pm [(0.050)^2 - 4(3.95)(-8.0 \times 10^{-3})]^{1/2}}{2(3.95)}$$

$x = 3.9 \times 10^{-2}$ atm or $x = -5.2 \times 10^{-2}$ atm; Only $x = 3.9 \times 10^{-2}$ atm makes sense (x cannot be negative), so the equilibrium NO concentration is:

$P_{NO} = 2x = 2(3.9 \times 10^{-2}\text{ atm}) = 7.8 \times 10^{-2}$ atm

51. a. The reaction must shift to reactants to reach equilibrium. Summarize the problem in a table:

	$N_2O_4(g)$	$\rightleftharpoons$	$2\,NO_2(g)$	$K_p = 0.25$
Initial	0		0.050 atm	
	$2x$ atm of NO_2 reacts to reach equilibrium			
Change	$+x$	$\leftarrow$	$-2x$	
Equil.	x		$0.050 - 2x$	

$$K_p = \frac{P_{NO_2}^2}{P_{N_2O_4}},\ 0.25 = \frac{(0.050 - 2x)^2}{x},\ 0.25\,x = 2.5 \times 10^{-3} - 0.20\,x + 4\,x^2$$

$4\,x^2 - 0.45\,x + 2.5 \times 10^{-3} = 0$; Solving using the quadratic equation:

$$x = \frac{-(-0.45) \pm [(-0.45)^2 - 4(4)(2.5 \times 10^{-3})]^{1/2}}{2(4)} = 5.9 \times 10^{-3}$$ (other value, 0.11, makes no sense)

$P_{N_2O_4} = x = 5.9 \times 10^{-3}$ atm; $P_{NO_2} = 0.050 - 2(0.0059) = 3.8 \times 10^{-2}$ atm

b. This reaction must proceed to products to reach equilibrium.

	$N_2O_4(g)$	$\rightleftharpoons$	$2\ NO_2(g)$ $K_p = 0.25$
Initial	0.040 atm		0
	x atm of N_2O_4 reacts to reach equilibrium		
Change	$-x$	$\rightarrow$	$+2x$
Equil.	$0.040 - x$		$2x$

$$K_p = \frac{(2x)^2}{0.040 - x} = 0.25, \quad 4x^2 = 0.010 - 0.25\,x, \quad 4x^2 + 0.25\,x - 0.010 = 0$$

$$x = \frac{-0.25 \pm [(0.25)^2 - 4(4)(-0.010)]^{1/2}}{2(4)} = 0.028 \text{ (other value is negative)}$$

$P_{N_2O_4} = 0.040 - x = 0.012$ atm; $P_{NO_2} = 2x = 0.056$ atm

c. Q = 1.00, so some NO_2 must react to form N_2O_4 ($Q > K_p$).

	$N_2O_4(g)$	$\rightleftharpoons$	$2\ NO_2(g)$
Initial	1.00 atm		1.00 atm
	$2x$ atm of NO_2 reacts to reach equilibrium		
Change	$+x$	$\leftarrow$	$-2x$
Equil.	$1.00 + x$		$1.00 - 2x$

$$K_p = \frac{(1.00 - 2x)^2}{1.00 + x} = 0.25, \quad 1.00 - 4.00\,x + 4x^2 = 0.25 + 0.25\,x, \quad 4x^2 - 4.25\,x + 0.75 = 0$$

$$x = \frac{-(-4.25) \pm [(-4.25)^2 - 4(4)(0.75)]^{1/2}}{2(4)} = 0.22 \quad \text{(other value, 0.84, makes no sense)}$$

$P_{N_2O_4} = 1.00 + x = 1.22$ atm; $P_{NO_2} = 1.00 - 2x = 0.56$ atm

53. a. The reaction must proceed to products to reach equilibrium since only reactants are present initially. Summarizing the problem in a table:

	$2\ NOCl(g)$	$\rightleftharpoons$	$2\ NO(g)$	+	$Cl_2(g)$	$K = 1.6 \times 10^{-5}$
Initial	$\frac{2.0\text{ mol}}{2.0\text{ L}} = 1.0\,M$		0		0	
	$2x$ mol/L of NOCl reacts to reach equilibrium					
Change	$-2x$	$\rightarrow$	$+2x$		$+x$	
Equil.	$1.0 - 2x$		$2x$		x	

$$K = 1.6 \times 10^{-5} = \frac{[NO]^2[Cl_2]}{[NOCl]^2} = \frac{(2x)^2(x)}{(1.0 - 2x)^2}$$

If we assume that $1.0 - 2x \approx 1.0$ (from the size of K, we know that not much reaction will occur so x is small), then:

$$1.6 \times 10^{-5} = \frac{4x^3}{1.0^2}, \quad x = 1.6 \times 10^{-2}; \text{ Now we must check the assumption.}$$

$$1.0 - 2x = 1.0 - 2(0.016) = 0.97 = 1.0 \text{ (to proper significant figures)}$$

Our error is about 3%, i.e., $2x$ is 3.2% of 1.0 M. Generally, if the error we introduce by making simplifying assumptions is less than 5%, we go no further, the assumption is said to be valid. We call this the 5% rule. Solving for the equilibrium concentrations:

$$[NO] = 2x = 0.032\ M;\ [Cl_2] = x = 0.016\ M;\ [NOCl] = 1.0 - 2x = 0.97\ M \approx 1.0\ M$$

Note: If we were to solve this cubic equation exactly (a long and tedious process), we get $x = 0.016$. This is the exact same answer we determined by making a simplifying assumption. We saved time and energy. Whenever K is a very small value, always make the assumption that x is small. If the assumption introduces an error of less than 5%, then the answer you calculated making the assumption will be considered the correct answer.

b. $2\ NOCl(g) \rightleftharpoons 2\ NO(g) + Cl_2(g)$

	$2\ NOCl(g)$		$2\ NO(g)$	$Cl_2(g)$
Initial	1.0 M		1.0 M	0
	$2x$ mol/L of NOCl reacts to reach equilibrium			
Change	$-2x$	$\rightarrow$	$+2x$	$+x$
Equil.	$1.0 - 2x$		$1.0 + 2x$	x

$$1.6 \times 10^{-5} = \frac{(1.0 + 2x)^2(x)}{(1.0 - 2x)^2} \approx \frac{(1.0)^2(x)}{(1.0)^2} \quad \text{(assuming } 2x << 1.0\text{)}$$

$x = 1.6 \times 10^{-5}$; Assumptions are great ($2x$ is 3.2×10^{-3}% of 1.0).

$[Cl_2] = 1.6 \times 10^{-5}\ M$ and $[NOCl] = [NO] = 1.0\ M$

c. $2\ NOCl(g) \rightleftharpoons 2\ NO(g) + Cl_2(g)$

	$2\ NOCl(g)$		$2\ NO(g)$	$Cl_2(g)$
Initial	2.0 M		0	1.0 M
	$2x$ mol/L of NOCl reacts to reach equilibrium			
Change	$-2x$	$\rightarrow$	$+2x$	$+x$
Equil.	$2.0 - 2x$		$2x$	$1.0 + x$

$$1.6 \times 10^{-5} = \frac{(2x)^2(1.0 + x)}{(2.0 - 2x)^2} \approx \frac{4x^2}{4.0} \quad \text{(assuming } x << 1.0\text{)}$$

Solving: $x = 4.0 \times 10^{-3}$; Assumptions good (x is 0.4% of 1.0 and $2x$ is 0.4% of 2.0).

$[Cl_2] = 1.0 + x = 1.0\ M$; $[NO] = 2(4.0 \times 10^{-3}) = 8.0 \times 10^{-3}$; $[NOCl] = 2.0\ M$

55. $2\,CO_2(g) \rightleftharpoons 2\,CO(g) + O_2(g)$ $\quad K = \frac{[CO]^2[O_2]}{[CO_2]^2} = 2.0 \times 10^{-6}$

	$2\,CO_2(g)$	$\rightleftharpoons$	$2\,CO(g)$	+	$O_2(g)$
Initial	2.0 mol/5.0 L		0		0
	$2x$ mol/L of CO_2 reacts to reach equilibrium				
Change	$-2x$	$\rightarrow$	$+2x$		x
Equil.	$0.40 - 2x$		$2x$		x

$$K = 2.0 \times 10^{-6} = \frac{[CO]^2[O_2]}{[CO_2]^2} = \frac{(2x)^2(x)}{(0.40 - 2x)^2};\ \text{Assuming } 2x << 0.40:$$

$$2.0 \times 10^{-6} \approx \frac{4x^3}{(0.40)^2},\ 2.0 \times 10^{-6} = \frac{4x^3}{0.16},\ x = 4.3 \times 10^{-3}\,M$$

Checking assumption: $\frac{2(4.3 \times 10^{-3})}{0.40} \times 100 = 2.2\%$; Assumption valid by the 5% rule.

$[CO_2] = 0.40 - 2x = 0.40 - 2(4.3 \times 10^{-3}) = 0.39\,M$

$[CO] = 2x = 2(4.3 \times 10^{-3}) = 8.6 \times 10^{-3}\,M$; $[O_2] = x = 4.3 \times 10^{-3}\,M$

57. This is a typical equilibrium problem except that the reaction contains a solid. Whenever solids and liquids are present, we basically ignore them in the equilibrium problem.

	$NH_4Cl(s)$	$\rightleftharpoons$	$NH_3(g)$	+	$HCl(g)$
Initial	-		0		0
	$NH_4Cl(s)$ decomposes to produce x atm each of NH_3 and HCl at equilibrium				
Change	-	$\rightarrow$	$+x$		$+x$
Equil.	-		x		x

$K_p = P_{NH_3} \times P_{HCl} = (x)(x)$; From the problem, $P_{NH_3} = x = 2.2$ atm. Solving for K_p:

$$K_p = x^2 = (2.2\ \text{atm})^2 = 4.8\ \text{atm}^2$$

Le Chatelier's Principle

59. a. No effect; Adding more of a pure solid or pure liquid has no effect on the equilibrium position.

b. Shifts left; HF(g) will be removed by reaction with the glass. As HF(g) is removed, the reaction will shift left to produce more HF(g).

c. Shifts right; As $H_2O(g)$ is removed, the reaction will shift right to produce more $H_2O(g)$.

61. a. right b. right c. no effect; He(g) is neither a reactant nor a product.

d. left; Since the reaction is exothermic, heat is a product:

$$CO(g) + H_2O(g) \rightarrow H_2(g) + CO_2(g) + \text{Heat}$$

Increasing T will add heat. The equilibrium shifts to the left to use up the added heat.

e. no effect; Since there are equal numbers of gas molecules on both sides of the reaction, then a change in volume has no effect on the equilibrium position.

63. a. left b. right c. left

d. no effect (reactant and product concentrations are unchanged)

e. no effect; Since there are equal numbers of product and reactant gas molecules, then a change in volume has no effect on the equilibrium position.

f. right; A decrease in temperature will shift the equilibrium to the right since heat is a product in this reaction (as is true in all exothermic reactions).

65. In an exothermic reaction, heat is a product. To maximize yield of products, one would want as low a temperature as possible since high temperatures would shift the reaction left (away from products). Since temperature changes also change the value of K, then at low temperatures the value of K will be largest which maximizes yield of products.

Additional Exercises

67.

$$O(g) + NO(g) \rightleftharpoons NO_2(g) \qquad K = 1/6.8 \times 10^{-49} = 1.5 \times 10^{48}$$

$$NO_2(g) + O_2(g) \rightleftharpoons NO(g) + O_3(g) \qquad K = 1/5.8 \times 10^{-34} = 1.7 \times 10^{33}$$

$$O_2(g) + O(g) \rightleftharpoons O_3(g) \qquad K = (1.5 \times 10^{48})(1.7 \times 10^{33}) = 2.6 \times 10^{81}\ \text{L/mol}$$

69. $H_2O(g) + Cl_2O(g) \rightleftharpoons 2\ HOCl(g) \qquad K = 0.090 = \dfrac{[HOCl]^2}{[H_2O]\,[Cl_2O]}$

a. The initial concentrations of H_2O and Cl_2O are:

$$\frac{1.0\ \text{g}\ H_2O}{1.0\ \text{L}} \times \frac{1\ \text{mol}}{18.02\ \text{g}} = 5.5 \times 10^{-2}\ \text{mol/L}; \qquad \frac{2.0\ \text{g}\ Cl_2O}{1.0\ \text{L}} \times \frac{1\ \text{mol}}{86.90\ \text{g}} = 2.3 \times 10^{-2}\ \text{mol/L}$$

	$H_2O(g)$ +	$Cl_2O(g)$	$\rightleftharpoons$	$2\ HOCl(g)$
Initial	$5.5 \times 10^{-2}\ M$	$2.3 \times 10^{-2}\ M$		0
	x mol/L of H_2O reacts to reach equilibrium			
Change	$-x$	$-x$	$\rightarrow$	$+2x$
Equil.	$5.5 \times 10^{-2} - x$	$2.3 \times 10^{-2} - x$		$2x$

$$K = 0.090 = \frac{(2x)^2}{(5.5 \times 10^{-2} - x)\,(2.3 \times 10^{-2} - x)}, \quad 1.14 \times 10^{-4} - 7.02 \times 10^{-3}\,x + 0.090\,x^2 = 4\,x^2$$

$3.91\,x^2 + 7.02 \times 10^{-3}\,x - 1.14 \times 10^{-4} = 0$ (We carried extra significant figures.)

Solving using the quadratic equation:

$$x = \frac{-7.02 \times 10^{-3} \pm (4.93 \times 10^{-5} + 1.78 \times 10^{-3})^{1/2}}{7.82} = 4.6 \times 10^{-3} \text{ or } -6.4 \times 10^{-3}$$

A negative answer makes no physical sense; we can't have less than nothing.
So $x = 4.6 \times 10^{-3}\,M$.

$[HOCl] = 2x = 9.2 \times 10^{-3}\,M$; $[Cl_2O] = 2.3 \times 10^{-2} - x = 0.023 - 0.0046 = 1.8 \times 10^{-2}\,M$

$[H_2O] = 5.5 \times 10^{-2} - x = 0.055 - 0.0046 = 5.0 \times 10^{-2}\,M$

b. $H_2O(g) + Cl_2O(g) \rightleftharpoons 2\,HOCl(g)$

	$H_2O(g)$	$Cl_2O(g)$		$2\,HOCl(g)$
Initial	0	0		1.0 mol/2.0 L = 0.50 M
	$2x$ mol/L of HOCl reacts to reach equilibrium			
Change	$+x$	$+x$	$\leftarrow$	$-2x$
Equil.	x	x		$0.50 - 2x$

$$K = 0.090 = \frac{[HOCl]^2}{[H_2O]\,[Cl_2O]} = \frac{(0.50 - 2x)^2}{x^2}$$

The expression is a perfect square, so we can take the square root of each side:

$$0.30 = \frac{0.50 - 2x}{x}, \quad 0.30\,x = 0.50 - 2x, \quad 2.30\,x = 0.50$$

$x = 0.217$ (We carried extra significant figures.)

$x = [H_2O] = [Cl_2O] = 0.217 = 0.22\,M$; $[HOCl] = 0.50 - 2x = 0.50 - 0.434 = 0.07\,M$

71. a. $$P_{PCl_5} = \frac{nRT}{V} = \frac{\dfrac{2.450\text{ g }PCl_5}{208.22\text{ g/mol}} \times \dfrac{0.08206\text{ L atm}}{\text{mol K}} \times 600.\text{ K}}{0.500\text{ L}} = 1.16\text{ atm}$$

b. $PCl_5(g) \rightleftharpoons PCl_3(g) + Cl_2(g)$ $\quad K_p = \dfrac{P_{PCl_3} \times P_{Cl_2}}{P_{PCl_5}} = 11.5$

	$PCl_5(g)$		$PCl_3(g)$	$Cl_2(g)$
Initial	1.16 atm		0	0
	x atm of PCl_5 reacts to reach equilibrium			
Change	$-x$	$\rightarrow$	$+x$	$+x$
Equil.	$1.16 - x$		x	x

$K_p = \dfrac{x^2}{1.16 - x} = 11.5$, $x^2 + 11.5\,x - 13.3 = 0$; Using the quadratic formula: $x = 1.06$ atm

$P_{PCl_5} = 1.16 - 1.06 = 0.10$ atm

c. $P_{PCl_3} = P_{Cl_2} = 1.06$ atm; $P_{PCl_5} = 0.10$ atm

$P_{tot} = P_{PCl_5} + P_{PCl_3} + P_{Cl_2} = 0.10 + 1.06 + 1.06 = 2.22$ atm

d. Percent dissociation $= \frac{x}{1.16} \times 100 = \frac{1.06}{1.16} \times 100 = 91.4\%$

73. a. $N_2O_4(g) \rightleftharpoons 2NO_2(g)$

	$N_2O_4(g)$		$2NO_2(g)$
Initial	4.5 atm		0
	x atm of N_2O_4 reacts to reach equilibrium		
Change	$-x$	$\rightarrow$	$+2x$
Equil.	$4.5 - x$		$2x$

$$K_p = \frac{P^2_{NO_2}}{P_{N_2O_4}} = \frac{(2x)^2}{4.5-x} = 0.25$$

Solving using successive approximations or the quadratic formula: $x = 0.50$ atm

$P_{NO_2} = 2x = 1.0$ atm; $P_{N_2O_4} = 4.5 - x = 4.0$ atm

b. $N_2O_4(g) \rightleftharpoons 2\,NO_2(g)$

	$N_2O_4(g)$		$2\,NO_2(g)$
Initial	0		9.0 atm
	$2x$ atm of NO_2 reacts to reach equilibrium		
Change	$+x$	$\leftarrow$	$-2x$
Equil.	x		$9.0 - 2x$

$\frac{(9.0 - 2x)^2}{x} = 0.25$; Solving using quadratic formula: $x = 4.0$ (the other value makes no sense).

$P_{N_2O_4} = x = 4.0$ atm; $P_{NO_2} = 9.0 - 2x = 1.0$ atm

c. No, we get the same equilibrium position starting with either pure N_2O_4 or pure NO_2 in stoichiometric amounts.

75. $H^+ + OH^- \rightarrow H_2O$; Sodium hydroxide (NaOH) will react with the H^+ on the product side of the reaction. This effectively removes H^+ from the equilibrium which will shift the reaction to the right to produce more H^+ and CrO_4^{2-}. Since more CrO_4^{2-} is produced, the solution turns yellow.

Challenge Problems

77. There is a little trick we can use to solve this problem in order to avoid solving a cubic equation. Since K for this reaction is very small, then the dominant reaction is the reverse reaction. We will let the products react to completion by the reverse reaction, then we will solve the forward equilibrium problem to determine the equilibrium concentrations. Summarizing these steps to solve in a table:

	$2\,NOCl(g)$	$\rightleftharpoons$	$2\,NO(g)$	+	$Cl_2(g)$	$K = 1.6 \times 10^{-5}$
Before	0		2.0 *M*		1.0 *M*	
	Let 1.0 mol/L Cl_2 react completely.					(K is small, reactants dominate.)
Change	+2.0	←	-2.0		-1.0	React completely
After	2.0		0		0	New initial conditions
	$2x$ mol/L of NOCl reacts to reach equilibrium					
Change	$-2x$	→	$+2x$		$+x$	
Equil.	$2.0 - 2x$		$2x$		x	

$$K = 1.6 \times 10^{-5} = \frac{(2x)^2\,(x)}{(2.0 - 2x)^2} \approx \frac{4x^3}{2.0^2} \quad \text{(assuming } 2.0 - 2x \approx 2.0\text{)}$$

$x^3 = 1.6 \times 10^{-5}$, $x = 2.5 \times 10^{-2}$ Assumption good by the 5% rule ($2x$ is 2.5% of 2.0).

[NOCl] = 2.0 - 0.050 = 1.95 *M* = 2.0 *M*; [NO] = 0.050 *M*; [Cl_2] = 0.025 *M*

Note: If we do not break this problem into two parts (a stoichiometric part and an equilibrium part), then we are faced with solving a cubic equation. The set-up would be:

	$2\,NOCl$	$\rightleftharpoons$	$2\,NO$	+	Cl_2
Initial	0		2.0 *M*		1.0 *M*
Change	$+2y$	←	$-2y$		$-y$
Equil.	$2y$		$2.0 - 2y$		$1.0 - y$

$$1.6 \times 10^{-5} = \frac{(2.0 - 2y)^2\,(1.0 - y)}{(2y)^2};$$ If we say that y is small to simplify the problem, then:

$$1.6 \times 10^{-5} = \frac{2.0^2}{4y^2};$$ We get $y = 250$. This is impossible!

To solve this equation we cannot make any simplifying assumptions; we have to find a way to solve a cubic equation. Or, we can use some chemical common sense and solve the problem the easier way.

79.

	$N_2(g)$	+	$3\,H_2(g)$	$\rightleftharpoons$	$2\,NH_3(g)$	
Initial	0		0		P_0	P_0 = initial pressure of NH_3
	$2x$ atm of NH_3 reacts to reach equilibrium					
Change	$+x$		$+3x$	←	$-2x$	
Equil.	x		$3x$		$P_0 - 2x$	

From problem, $P_0 - 2x = \dfrac{P_0}{2.00}$, so $P_0 = 4.00\,x$

$$K_p = \frac{(4.00\,x - 2x)^2}{(x)\,(3x)^3} = \frac{(2.00\,x)^2}{(x)\,(3x)^3} = \frac{4.00\,x^2}{27x^4} = \frac{4.00}{27x^2} = 5.3 \times 10^5,\ x = 5.3 \times 10^{-4}\ \text{atm}$$

$P_0 = 4.00\,x = 4.00 \times (5.3 \times 10^{-4})$ atm $= 2.1 \times 10^{-3}$ atm

81. $N_2O_4(g) \rightleftharpoons 2\,NO_2(g)$ $\qquad K_p = \dfrac{P^2_{NO_2}}{P_{N_2O_4}} = \dfrac{(1.2\text{ atm})^2}{0.33\text{ atm}} = 4.4$ atm

Doubling the volume decreases each partial pressure by a factor of 2 (P = nRT/V).

$P_{NO_2} = 0.60$ atm and $P_{N_2O_4} = 0.17$ atm are the new partial pressures.

$Q = \dfrac{(0.60)^2}{0.17} = 2.1$, so Q < K; Equilibrium will shift to the right.

	$N_2O_4(g)$	$\rightleftharpoons$	$2\,NO_2(g)$
Initial	0.17 atm		0.60 atm
	x atm of N_2O_4 reacts to reach equilibrium		
Change	$-x$	$\rightarrow$	$+2x$
Equil.	$0.17 - x$		$0.60 + 2x$

$K_p = 4.4 = \dfrac{(0.60 + 2x)^2}{(0.17 - x)}$, $4x^2 + 6.8\,x - 0.39 = 0$

Solving using the quadratic formula: $x = 0.056$

$P_{NO_2} = 0.60 + 2(0.056) = 0.71$ atm; $P_{N_2O_4} = 0.17 - 0.056 = 0.11$ atm

CHAPTER FOURTEEN

ACIDS AND BASES

Questions

13. The Arrhenius definitions are: acids produce H^+ in water and bases produce OH^- in water. The difference between strong and weak acids and bases is the amount of H^+ and OH^- produced.

    a. A strong acid is 100% dissociated in water.
    b. A strong base is 100% dissociated in water.
    c. A weak acid is much less than 100% dissociated in water (typically 1-10% dissociated).
    d. A weak base is one that only a small percentage of the molecules react with water to form OH^-.

15. $H_2O \rightleftharpoons H^+ + OH^-$ $\quad K_w = [H^+][OH^-] = 1.0 \times 10^{-14}$

    Neutral solution: $[H^+] = [OH^-]$; $[H^+] = 1.0 \times 10^{-7}$ *M*; pH = $-\log(1.0 \times 10^{-7}) = 7.00$

17. a. Arrhenius acid: produce H^+ in water
    b. Brönsted-Lowry acid: proton (H^+) donor
    c. Lewis acid: electron pair acceptor

19. An organic compound containing nitrogen with a lone pair of electrons will most likely produce a basic solution.

21. In general, the weaker the acid, the stronger the conjugate base and vice versa.

    a. Since acid strength increases as the X – H bond strength decreases, then conjugate base strength will increase as the strength of the X – H bond increases.
    b. Since acid strength increases as the electronegativity of neighboring atoms increase, then conjugate base strength will decrease as the electronegativity of neighboring atoms increase.
    c. Since acid strength increases as the number of oxygen atoms increase, then conjugate base strength decreases as the number of oxygen atoms increase.

23. a. H_2O and $CH_3CO_2^-$

    b. An acid-base reaction can be thought of as a competition between two opposing bases. Since this equilibrium lies far to the left ($K_a < 1$), then $CH_3CO_2^-$ is a stronger base than H_2O.

    c. The acetate ion is a better base than water and produces basic solutions in water. When we put acetate ion into solution as the only major basic species, the reaction is:

$$CH_3CO_2^- + H_2O \rightleftharpoons CH_3CO_2H + OH^-$$

Now the competition is between $CH_3CO_2^-$ and OH^- for the proton. Hydroxide ion is the strongest base possible in water. The above equilibrium lies far to the left resulting in a K_b value less than one. Those species we specifically call weak bases ($10^{-14} < K_b < 1$) lie between H_2O and OH^- in base strength. Weak bases are stronger bases than water but are weaker bases than OH^-.

Exercises

Nature of Acids and Bases

25. a. $HClO_4(aq) + H_2O(l) \rightarrow H_3O^+(aq) + ClO_4^-(aq)$. Only the forward reaction is indicated since $HClO_4$ is a strong acid and is basically 100% dissociated in water. For acids, the dissociation reaction is commonly written without water as a reactant. The common abbreviation for this reaction is: $HClO_4(aq) \rightarrow H^+(aq) + ClO_4^-(aq)$. This reaction is also called the K_a reaction as the equilibrium constant for this reaction is called K_a.

    b. Propanoic acid is a weak acid, so it is only partially dissociated in water. The dissociation reaction is: $CH_3CH_2CO_2H(aq) + H_2O(l) \rightleftharpoons H_3O^+(aq) + CH_3CH_2CO_2^-(aq)$ or $CH_3CH_2CO_2H(aq) \rightleftharpoons H^+(aq) + CH_3CH_2CO_2^-(aq)$.

    c. NH_4^+ is a weak acid. Similar to propanoic acid, the dissociation reaction is:

    $$NH_4^+(aq) + H_2O(l) \rightleftharpoons H_3O^+(aq) + NH_3(aq) \text{ or } NH_4^+(aq) \rightleftharpoons H^+(aq) + NH_3(aq)$$

27. An acid is a proton (H^+) donor and the base is a proton acceptor. A conjugate acid-base pair only differs by a proton (H^+).

	Acid	Base	Conjugate Base of Acid	Conjugate Acid of Base
a.	HF	H_2O	F^-	H_3O^+
b.	H_2SO_4	H_2O	HSO_4^-	H_3O^+
c.	HSO_4^-	H_2O	SO_4^{2-}	H_3O^+

29. Strong acids have a $K_a >> 1$ and weak acids have $K_a < 1$. Table 14.2 in the text lists some K_a values for weak acids. K_a values for strong acids are hard to determine so they are not listed in the text. However, there are only a few common strong acids so if you memorize the strong acids, then all other acids will be weak acids. The strong acids to memorize are HCl, HBr, HI, HNO_3, $HClO_4$ and H_2SO_4.

    a. HNO_2 is a weak acid ($K_a = 4.0 \times 10^{-4}$).
    b. HNO_3 is a strong acid.
    c. HCl is a strong acid.
    d. HF is a weak acid ($K_a = 7.2 \times 10^{-4}$).

31. The K_a value is directly related to acid strength. As K_a increases, acid strength increases. For water, use K_w when comparing the acid strength of water to other species. The K_a values are:

HNO_3: strong acid ($K_a >> 1$); HOCl: $K_a = 3.5 \times 10^{-8}$

NH_4^+: $K_a = 5.6 \times 10^{-10}$; H_2O: $K_a = K_w = 1.0 \times 10^{-14}$

From the K_a values, the ordering is: $HNO_3 > HOCl > NH_4^+ > H_2O$.

33. a. $HClO_4$ is a strong acid and water is a weak acid with $K_a = K_w = 1.0 \times 10^{-14}$. $HClO_4$ is a stronger acid than H_2O.

b. H_2O, $K_a = K_w = 1.0 \times 10^{-14}$; $HClO_2$, $K_a = 1.2 \times 10^{-2}$; $HClO_2$ is a stronger acid than H_2O since K_a for $HClO_2 > K_a$ for H_2O.

c. HF, $K_a = 7.2 \times 10^{-4}$; HCN, $K_a = 6.2 \times 10^{-10}$; HF is a stronger acid than HCN since K_a for HF > K_a for HCN.

Autoionization of Water and the pH Scale

35. At 25°C, the relationship: $[H^+][OH^-] = K_w = 1.0 \times 10^{-14}$ always holds for aqueous solutions. When $[H^+]$ is greater than 1.0×10^{-7} *M*, then the solution is acidic; when $[H^+]$ is less than 1.0×10^{-7} *M*, then the solution is basic; when $[H^+] = 1.0 \times 10^{-7}$ *M*, then the solution is neutral. In terms of $[OH^-]$, an acidic solution has $[OH^-] < 1.0 \times 10^{-7}$ *M*, a basic solution has $[OH^-] > 1.0 \times 10^{-7}$ *M* and a neutral solution has $[OH^-] = 1.0 \times 10^{-7}$ *M*.

a. $[OH^-] = \frac{K_w}{[H^+]} = \frac{1.0 \times 10^{-14}}{1.0 \times 10^{-7}} = 1.0 \times 10^{-7}\ M$; The solution is neutral.

b. $[OH^-] = \frac{1.0 \times 10^{-14}}{1.4 \times 10^{-3}} = 7.1 \times 10^{-12}\ M$; The solution is acidic.

c. $[OH^-] = \frac{1.0 \times 10^{-14}}{2.5 \times 10^{-10}} = 4.0 \times 10^{-5}\ M$; The solution is basic.

d. $[OH^-] = \frac{1.0 \times 10^{-14}}{6.1} = 1.6 \times 10^{-15}\ M$; The solution is acidic.

37. a. Since the value of the equilibrium constant increases as the temperature increases, then the reaction is endothermic. In endothermic reactions, heat is a reactant so an increase in temperature (heat) shifts the reaction to produce more products and increases K in the process.

b. $H_2O(l) \rightleftharpoons H^+(aq) + OH^-(aq)$ $K_w = 5.47 \times 10^{-14} = [H^+][OH^-]$;

In pure water $[H^+] = [OH^-]$, so $5.47 \times 10^{-14} = [H^+]^2$, $[H^+] = 2.34 \times 10^{-7}\ M = [OH^-]$

39. pH = -log [H^+]; pOH = -log [OH^-]; At 25°C, pH + pOH = 14.00

a. pH = -log [H^+] = -log (1.0×10^{-7}) = 7.00; pOH = 14.00 - pH = 14.00 - 7.00 = 7.00

b. pH = -log (1.4×10^{-3}) = 2.85; pOH = 14.00 - 2.85 = 11.15

c. pH = -log (2.5×10^{-10}) = 9.60; pOH = 14.00 - 9.60 = 4.40

d. pH = -log (6.1) = -0.79; pOH = 14.00 - (-0.79) = 14.79

Note that pH is less than zero when [H^+] is greater than 1.0 *M* (an extremely acidic solution).

41. a. [H^+] = 10^{-pH}, [H^+] = $10^{-7.41}$ = 3.9×10^{-8} *M*

pOH = 14.00 - pH = 14.00 - 7.41 = 6.59; [OH^-] = 10^{-pOH} = $10^{-6.59}$ = 2.6×10^{-7} *M*

or $[OH^-] = \frac{K_w}{[H^+]} = \frac{1.0 \times 10^{-14}}{3.9 \times 10^{-8}} = 2.6 \times 10^{-7}\ M$

b. [H^+] = $10^{-15.3}$ = 5×10^{-16} *M*; pOH = 14.00 - 15.3 = -1.3; [OH^-] = $10^{-(-1.3)}$ = 20 *M*

c. [H^+] = $10^{-(-1.0)}$ = 10 *M*; pOH = 14.0 - (-1.0) = 15.0; [OH^-] = $10^{-15.0}$ = 1×10^{-15} *M*

d. [H^+] = $10^{-3.2}$ = 6×10^{-4} *M*; pOH = 14.0 - 3.2 = 10.8; [OH^-] = $10^{-10.8}$ = 2×10^{-11} *M*

e. [OH^-] = $10^{-5.0}$ = 1×10^{-5} *M*; pH = 14.0 - pOH = 14.0 - 5.0 = 9.0; [H^+] = $10^{-9.0}$ = 1×10^{-9} *M*

f. [OH^-] = $10^{-9.6}$ = 3×10^{-10} *M*; pH = 14.0 - 9.6 = 4.4; [H^+] = $10^{-4.4}$ = 4×10^{-5} *M*

43. pOH = 14.00 - pH = 14.00 - 6.77 = 7.23; [H^+] = 10^{-pH} = $10^{-6.77}$ = 1.7×10^{-7} *M*

$[OH^-] = \frac{K_w}{[H^+]} = \frac{1.0 \times 10^{-14}}{1.7 \times 10^{-7}} = 5.9 \times 10^{-8}\ M$ or [OH^-] = 10^{-pOH} = $10^{-7.23}$ = 5.9×10^{-8} *M*

Milk is slightly acidic since the pH is less than 7.00 at 25°C.

Solutions of Acids

45. All the acids in this problem are strong acids which are always assumed to completely dissociate in water. The general dissociation reaction for a strong acid is: HA(aq) → H^+(aq) + A^-(aq) where A^- is the conjugate base of the strong acid HA. For 0.250 *M* solutions of these strong acids, 0.250 *M* H^+ and 0.250 *M* A^- are present when the acids completely dissociate. The amount of H^+ donated from water will be insignificant in this problem since H_2O is a very weak acid.

a. Major species present after dissociation = H^+(aq), Cl^-(aq) and H_2O (always present),

pH = -log [H^+] = - log (0.250) = 0.602

b. Major species = $H^+(aq)$, $Br^-(aq)$ and H_2O; pH = 0.602

47. Strong acids are assumed to completely dissociate in water: $HCl(aq) \rightarrow H^+(aq) + Cl^-(aq)$

a. A 0.10 *M* HCl solution gives 0.10 *M* H^+ and 0.10 *M* Cl^- since HCl completely dissociates. The amount of H^+ from H_2O will be insignificant.

$$\text{pH} = -\log[H^+] = -\log(0.10) = 1.00$$

b. 5.0 *M* H^+ is produced when 5.0 *M* HCl completely dissociates. The amount of H^+ from H_2O will be insignificant. pH = -log (5.0) = -0.70 (Negative pH values just indicate very concentrated acid solutions).

c. 1.0×10^{-11} *M* H^+ is produced when 1.0×10^{-11} *M* HCl completely dissociates. If you take the negative log of 1.0×10^{-11} this gives pH = 11.00. This is impossible! We dissolved an acid in water and got a basic pH. What we must consider in this problem is that water by itself donates 1.0×10^{-7} *M* H^+. We can normally ignore the small amount of H^+ from H_2O except when we have a very dilute solution of an acid (as in the case here). Therefore, the pH is that of neutral water (pH = 7.00) since the amount of HCl present is insignificant.

49. Both are strong acids.

$$0.0500\text{ L} \times 0.050\text{ mol/L} = 2.5 \times 10^{-3}\text{ mol HCl} = 2.5 \times 10^{-3}\text{ mol H}^+ + 2.5 \times 10^{-3}\text{ mol Cl}^-$$

$$0.1500\text{ L} \times 0.10\text{ mol/L} = 1.5 \times 10^{-2}\text{ mol HNO}_3 = 1.5 \times 10^{-2}\text{ mol H}^+ + 1.5 \times 10^{-2}\text{ mol NO}_3^-$$

$$[H^+] = \frac{(2.5 \times 10^{-3} + 1.5 \times 10^{-2})\text{ mol}}{0.2000\text{ L}} = 0.088\ M;\ [OH^-] = \frac{K_w}{[H^+]} = 1.1 \times 10^{-13}\ M$$

$$[Cl^-] = \frac{2.5 \times 10^{-3}\text{ mol}}{0.2000\text{ L}} = 0.013\ M;\ [NO_3^-] = \frac{1.5 \times 10^{-2}\text{ mol}}{0.2000\text{ L}} = 0.075\ M$$

51. $[H^+] = 10^{-pH} = 10^{-2.50} = 3.2 \times 10^{-3}$ *M*. Since HCl is a strong acid, then a 3.2×10^{-3} *M* HCl solution will produce 3.2×10^{-3} *M* H^+ giving a pH = 2.50.

53. a. HNO_2 ($K_a = 4.0 \times 10^{-4}$) and H_2O ($K_a = K_w = 1.0 \times 10^{-14}$) are the major species. HNO_2 is much stronger acid than H_2O so it is the major source of H^+. However, HNO_2 is a weak acid ($K_a < 1$) so it only partially dissociates in water. We must solve an equilibrium problem to determine $[H^+]$. In the Solutions Guide, we will summarize the initial, change and equilibrium concentrations into one table called the ICE table. Solving the weak acid problem:

	HNO_2	$\rightleftharpoons$	H^+	+	NO_2^-
Initial	0.250 *M*		~0		0
	x mol/L HNO_2 dissociates to reach equilibrium				
Change	$-x$	$\rightarrow$	$+x$		$+x$
Equil.	$0.250 - x$		x		x

$$K_a = \frac{[H^+][NO_2^-]}{[HNO_2]} = 4.0 \times 10^{-4} = \frac{x^2}{0.250 - x};$$ If we assume $x \ll 0.250$, then:

$$4.0 \times 10^{-4} \approx \frac{x^2}{0.250},\ x = \sqrt{4.0 \times 10^{-4}(0.250)} = 0.010\,M$$

We must check the assumption: $\frac{x}{0.250} \times 100 = \frac{0.010}{0.250} \times 100 = 4.0\%$

All the assumptions are good. The H^+ contribution from water (10^{-7} *M*) is negligible and x is small compared to 0.250 (percent error = 4.0%). If the percent error is less than 5% for an assumption, we will consider it a valid assumption (called the 5% rule). Finishing the problem: $x = 0.010\,M = [H^+]$; pH = -log(0.010) = 2.00

b. CH_3CO_2H ($K_a = 1.8 \times 10^{-5}$) and H_2O ($K_a = K_w = 1.0 \times 10^{-14}$) are the major species. CH_3CO_2H is the major source of H^+. Solving the weak acid problem:

	CH_3CO_2H	$\rightleftharpoons$	H^+	+	$CH_3CO_2^-$
Initial	0.250 *M*		~0		0
	x mol/L CH_3CO_2H dissociates to reach equilibrium				
Change	$-x$	→	$+x$		$+x$
Equil.	$0.250 - x$		x		x

$$K_a = \frac{[H^+][CH_3CO_2^-]}{[CH_3CO_2H]} = 1.8 \times 10^{-5} = \frac{x^2}{0.250 - x} \approx \frac{x^2}{0.250} \quad (\text{assuming } x << 0.250)$$

$x = 2.1 \times 10^{-3}$ *M*; Checking assumption: $\frac{2.1 \times 10^{-3}}{0.250} \times 100 = 0.84\%$. Assumptions good.

$[H^+] = x = 2.1 \times 10^{-3}$ *M*; pH = -log (2.1×10^{-3}) = 2.68

55. Major species: $HC_2H_3O_2$ ($K_a = 1.8 \times 10^{-5}$) and water; Major source of $H^+ = HC_2H_3O_2$. Since K_a for $HC_2H_3O_2$ is less than one, then $HC_2H_3O_2$ is a weak acid and we must solve an equilibrium problem to determine $[H^+]$. The set-up is:

	$HC_2H_3O_2$	$\rightleftharpoons$	H^+ +	$C_3H_3O_2^-$
Initial	0.20 *M*		~0	0
	x mol/L $HC_2H_3O_2$ dissociates to reach equilibrium			
Change	$-x$	→	$+x$	$+x$
Equl.	$0.20 - x$		x	x

$$K_a = 1.8 \times 10^{-5} = \frac{[H^+][C_2H_3O_2^-]}{[HC_2H_3O_2]} = \frac{x^2}{0.20 - x} \approx \frac{x^2}{0.20} \quad (\text{assuming } x << 0.20)$$

$x = [H^+] = 1.9 \times 10^{-3}$ *M*

We have made two assumptions which we must check.

1. $0.20 - x \approx 0.20$

   $(x/0.20) \times 100 = (1.9 \times 10^{-3}/0.20) \times 100 = 0.95\%$. Good assumption (1% error). If the percent error in the assumption is < 5%, then the assumption is valid.

2. Acetic acid is the major source of H^+, i.e., we can ignore $10^{-7}\ M\ H^+$ already present in neutral H_2O.

   $[H^+]$ from $HC_2H_3O_2 = 1.9 \times 10^{-3} >> 10^{-7}$; This assumption is valid.

   In future problems we will always begin the problem solving process by making these assumptions and we will always check them. However, we may not explicitly state that the assumptions are valid. We will always state when the assumptions are not valid and we have to use other techniques to solve the problem. Remember, anytime we make an assumption, we must check its validity before the solution to the problem is complete. Answering the question:

   $[H^+] = [C_2H_3O_2^-] = 1.9 \times 10^{-3}\ M$; $[OH^-] = 5.3 \times 10^{-12}\ M$

   $[HC_2H_3O_2] = 0.20 - x = 0.198 \approx 0.20\ M$; $pH = -\log(1.9 \times 10^{-3}) = 2.72$

57. Boric acid is a weak acid; it is the major source of H^+.

$$B(OH)_3 + H_2O \rightleftharpoons B(OH)_4^- + H^+$$

	$B(OH)_3$		$B(OH)_4^-$	H^+
Initial	$0.50\ M$		0	~0
	x mol/L $B(OH)_3$ reacts to reach equilibrium			
Change	$-x$	$\rightarrow$	$+x$	$+x$
Equil.	$0.50 - x$		x	x

$$K_a = 5.8 \times 10^{-10} = \frac{[B(OH)_4^-][H^+]}{[B(OH)_3]} = \frac{x^2}{0.50 - x} \approx \frac{x^2}{0.50} \quad \text{(assuming } x << 0.50\text{)}$$

$x = [H^+] = 1.7 \times 10^{-5}\ M$; pH = 4.77 Assumptions good (x is $3.4 \times 10^{-3}\%$ of 0.50).

59. This is a weak acid in water. Solving the weak acid problem:

$$HF \rightleftharpoons H^+ + F^- \qquad K_a = 7.2 \times 10^{-4}$$

	HF		H^+	F^-
Initial	$0.020\ M$		~0	0
	x mol/L HF dissociates to reach equilibrium			
Change	$-x$	$\rightarrow$	$+x$	$+x$
Equl.	$0.020 - x$		x	x

$$K_a = 7.2 \times 10^{-4} = \frac{[H^+][F^-]}{[HF]} = \frac{x^2}{0.020 - x} \approx \frac{x^2}{0.020} \quad \text{(assuming } x << 0.020\text{)}$$

$x = [H^+] = 3.8 \times 10^{-3}\ M$; Check assumptions: $\frac{x}{0.020} \times 100 = \frac{3.8 \times 10^{-3}}{0.020} \times 100 = 19\%$

The assumption $x << 0.020$ is not good (x is more than 5% of 0.020). We must solve $x^2/0.20 - x = 7.2 \times 10^{-4}$ exactly by using either the quadratic formula or by the method of successive approximations (see Appendix 1.4 of text). Using successive approximations, we let 0.016 M be a new approximation for [HF]. That is, in the denominator try $x = 0.0038$ (the value of x we calculated making the normal assumption), so 0.020 - 0.0038 = 0.016, then solve for a new value of x in the numerator.

$$\frac{x^2}{0.020 - x} \approx \frac{x^2}{0.016} = 7.2 \times 10^{-4}, \; x = 3.4 \times 10^{-3}$$

We use this new value of x to further refine our estimate of [HF], i.e., $0.020 - x = 0.020 - 0.0034 = 0.0166$ (carry extra significant figure).

$$\frac{x^2}{0.020 - x} \approx \frac{x^2}{0.0166} = 7.2 \times 10^{-4}, \; x = 3.5 \times 10^{-3}$$

We repeat, until we get a self-consistent answer. This would be the same answer we would get solving exactly using the quadratic equation. In this case it is: $x = 3.5 \times 10^{-3}$

So: $[H^+] = [F^-] = x = 3.5 \times 10^{-3}\ M$; $[OH^-] = K_w/[H^+] = 2.9 \times 10^{-12}\ M$

$[HF] = 0.020 - x = 0.020 - 0.0035 = 0.017\ M$; pH = 2.46

Note: When the 5% assumption fails, use whichever method you are most comfortable with to solve exactly. The method of successive approximations is probably fastest when the percent error is less than ~25%.

61. This is a weak acid in water. Solving the weak acid problem:

	HCO_2H	$\rightleftharpoons$	H^+	+	HCO_2^- $K_a = 1.8 \times 10^{-4}$
Initial	0.025 M		~0		0
	x mol/L HCO_2H dissociates to reach equilibrium				
Change	$-x$	$\rightarrow$	$+x$		$+x$
Equil.	$0.025 - x$		x		x

$$K_a = 1.8 \times 10^{-4} = \frac{[H^+][HCO_2^-]}{[HCO_2H]} = \frac{x^2}{0.025 - x} \approx \frac{x^2}{0.025}$$

$x = [H^+] = 2.1 \times 10^{-3}\ M$; Check assumptions: $\frac{2.1 \times 10^{-3}}{0.025} \times 100 = 8.4\%$

Assumption that $x << 0.025$ is not good (fails the 5% rule). Solving using the method of successive approximations (see Appendix 1.4 in text):

$$\frac{x^2}{0.025 - x} = \frac{x^2}{0.025 - 0.0021} = \frac{x^2}{0.023} = 1.8 \times 10^{-4}, \; x = 2.0 \times 10^{-3}$$ which we get consistently.

$x = [H^+] = 2.0 \times 10^{-3}\ M$; pH = 2.70

63. HCl is a strong acid. It will produce 0.10 M H^+. HOCl is a weak acid. Let's consider the equilibrium:

	HOCl	$\rightleftharpoons$	H^+	+	OCl^-
Initial	0.10 M		0.10 M		0
	x mol/L HOCl dissociates to reach equilibrium				
Change	$-x$	$\rightarrow$	$+x$		$+x$
Equil.	$0.10 - x$		$0.10 + x$		x

$$K_a = 3.5 \times 10^{-8} = \frac{[H^+][OCl^-]}{[HOCl]} = \frac{(0.10 + x)(x)}{0.10 - x} \approx x, x = 3.5 \times 10^{-8}$$

Assumptions are great (x is 3.5×10^{-5}% of 0.10). We are really assuming that HCl is the only important source of H^+, which it is. The $[H^+]$ contribution from the HOCl, x, is negligible. Therefore, $[H^+] = 0.10\ M$ and pH = 1.00

65. In all parts of this problem, acetic acid ($HC_2H_3O_2$) is the best weak acid present. We must solve a weak acid problem.

a.

	$HC_2H_3O_2$	$\rightleftharpoons$	H^+	+	$C_2H_3O_2^-$
Initial	0.50 M		~0		0
	x mol/L $HC_2H_3O_2$ dissociates to reach equilibrium				
Change	$-x$	$\rightarrow$	$+x$		$+x$
Equil.	$0.50 - x$		x		x

$$K_a = 1.8 \times 10^{-5} = \frac{[H^+][C_2H_3O_2^-]}{[HC_2H_3O_2]} = \frac{x^2}{0.50 - x} \approx \frac{x^2}{0.50}$$

$x = [H^+] = [C_2H_3O_2^-] = 3.0 \times 10^{-3}\ M$ Assumptions good.

$$\text{Perecent dissociation} = \frac{[H^+]}{[HC_2H_3O_2]_o} \times 100 = \frac{3.0 \times 10^{-3}}{0.50} \times 100 = 0.60\%$$

b. The set-up for solutions b and c are similar to solution a except the final equation is slightly different, reflecting the new concentration of $HC_2H_3O_2$.

$$K_a = 1.8 \times 10^{-5} = \frac{x^2}{0.050 - x} \approx \frac{x^2}{0.050}$$

$x = [H^+] = [C_2H_3O_2^-] = 9.5 \times 10^{-4}\ M$ Assumptions good.

$$\%\ \text{dissociation} = \frac{9.5 \times 10^{-4}}{0.050} \times 100 = 1.9\%$$

c. $$K_a = 1.8 \times 10^{-5} = \frac{x^2}{0.0050 - x} \approx \frac{x^2}{0.0050}$$

$x = [H^+] = [C_2H_3O_2^-] = 3.0 \times 10^{-4}\ M$; Check assumptions.

Assumption that x is negligible is borderline (6.0% error). We should solve exactly. Using the method of successive approximations (see Appendix 1.4 of text):

$$1.8 \times 10^{-5} = \frac{x^2}{0.0050 - 3.0 \times 10^{-4}} = \frac{x^2}{0.0047},\quad x = 2.9 \times 10^{-4}$$

Next trial also gives $x = 2.9 \times 10^{-4}$.

$$\%\ \text{dissociation} = \frac{2.9 \times 10^{-4}}{5.0 \times 10^{-3}} \times 100 = 5.8\%$$

d. As we dilute a solution, all concentrations are decreased. Dilution will shift the equilibrium to the side with the greater number of particles. For example, suppose we double the volume of an equilibrium mixture of a weak acid by adding water, then:

$$Q = \frac{\left(\frac{[H^+]_{eq}}{2}\right)\left(\frac{[X^-]_{eq}}{2}\right)}{\left(\frac{[HX]_{eq}}{2}\right)} = \frac{1}{2}K_a$$

$Q < K_a$, so the equilibrium shifts to the right or towards a greater percent dissociation.

e. $[H^+]$ depends on the initial concentration of weak acid and on how much weak acid dissociates. For solutions a-c the initial concentration of acid decreases more rapidly than the percent dissociation increases. Thus, $[H^+]$ decreases.

67. $HC_3H_5O_2$ ($K_a = 1.3 \times 10^{-5}$) and H_2O ($K_a = K_w = 1.0 \times 10^{-14}$) are the major species present. $HC_3H_5O_2$ will be the dominant producer of H^+ since $HC_3H_5O_2$ is a stronger acid than H_2O. Solving the weak acid problem:

	$HC_3H_5O_2$	$\rightleftharpoons$	H^+	+	$C_3H_5O_2^-$
Initial	$0.10\ M$		~0		0
	x mol/L $HC_3H_5O_2$ dissociates to reach equilibrium				
Change	$-x$	$\rightarrow$	$+x$		$+x$
Equil.	$0.10 - x$		x		x

$$K_a = 1.3 \times 10^{-5} = \frac{[H^+][C_3H_5O_2^-]}{[HC_3H_5O_2]} = \frac{x^2}{0.10 - x} \approx \frac{x^2}{0.10}$$

$x = [H^+] = 1.1 \times 10^{-3}\ M$; $pH = -\log(1.1 \times 10^{-3}) = 2.96$

Assumption follows the 5% rule (x is 1.1% of 0.10).

$$\text{Percent dissociation} = \frac{[H^+]}{[HC_3H_5O_2]_o} \times 100 = \frac{1.1 \times 10^{-3}}{0.10} \times 100 = 1.1\%$$

69. Let HX symbolize the weak acid. Set-up the problem like a typical weak acid equilibrium problem.

	HX	$\rightleftharpoons$	H^+	+	X^-
Initial	0.15		~0		0
	x mol/L HX dissociates to reach equilibrium				
Change	$-x$	$\rightarrow$	$+x$		$+x$
Equil.	$0.15 - x$		x		x

If the acid is 3.0% dissociated, then $x = [H^+]$ is 3.0% of 0.15: $x = 0.030 \times (0.15\ M) = 4.5 \times 10^{-3}\ M$
Now that we know the value of x, we can solve for K_a.

$$K_a = \frac{[H^+][X^-]}{[HX]} = \frac{x^2}{0.15 - x} = \frac{(4.5 \times 10^{-3})^2}{0.15 - 4.5 \times 10^{-3}} = 1.4 \times 10^{-4}$$

71. Set-up the problem using the K_a equilibrium reaction for HOBr.

	$HOBr$	$\rightleftharpoons$	H^+	+	OBr^-
Initial	0.063 M		~0		0
	x mol/L HOBr dissociates to reach equilibrium				
Change	$-x$	$\rightarrow$	$+x$		$+x$
Equil.	$0.063 - x$		x		x

$$K_a = \frac{[H^+][OBr^-]}{[HOBr]} = \frac{x^2}{0.063 - x};\ \text{Since pH} = 4.95\text{, then: } x = [H^+] = 10^{-pH} = 10^{-4.95} = 1.1 \times 10^{-5}\ M$$

$$K_a = \frac{(1.1 \times 10^{-5})^2}{0.063 - 1.1 \times 10^{-5}} = 1.9 \times 10^{-9}$$

73. Major species: $HC_2H_3O_2$ (acetic acid) and H_2O; Major source of H^+: $HC_2H_3O_2$

	$HC_2H_3O_2$	$\rightleftharpoons$	H^+	+	$C_2H_3O_2^-$	
Initial	C		~0		0	where $C = [HC_2H_3O_2]_o$
	x mol/L $HC_2H_3O_2$ dissociates to reach equilibrium					
Change	$-x$	$\rightarrow$	$+x$		$+x$	
Equil.	$C - x$		x		x	

$$K_a = 1.8 \times 10^{-5} = \frac{[H^+][C_2H_3O_2^-]}{[HC_2H_3O_2]} = \frac{x^2}{C - x} \text{ where } x = [H^+]$$

$$1.8 \times 10^{-5} = \frac{[H^+]^2}{C - [H^+]};\ \text{Since pH} = 3.0\text{, then: } [H^+] = 10^{-3.0} = 1 \times 10^{-3}\ M$$

$$1.8 \times 10^{-5} = \frac{(1 \times 10^{-3})^2}{C - 1 \times 10^{-3}},\ C - 1 \times 10^{-3} = \frac{1 \times 10^{-6}}{1.8 \times 10^{-5}},\ C = 5.7 \times 10^{-2} \approx 6 \times 10^{-2}\ M$$

A $6 \times 10^{-2}\ M$ acetic acid solution will have pH = 3.0.

Solutions of Bases

75. a. $NH_3(aq) + H_2O(l) \rightleftharpoons NH_4^+(aq) + OH^-(aq)$ $K_b = \frac{[NH_4^+][OH^-]}{[NH_3]}$

b. $C_5H_5N(aq) + H_2O(l) \rightleftharpoons C_5H_5NH^+(aq) + OH^-(aq)$ $K_b = \frac{[C_5H_5NH^+][OH^-]}{[C_5H_5N]}$

77. NO_3^-: $K_b << K_w$ since HNO_3 is a strong acid. All conjugate bases of strong acids have no base strength. H_2O: $K_b = K_w = 1.0 \times 10^{-14}$; NH_3: $K_b = 1.8 \times 10^{-5}$; CH_3NH_2: $K_b = 4.38 \times 10^{-4}$

$CH_3NH_2 > NH_3 > H_2O > NO_3^-$ (As K_b increases, base strength increases.)

79. a. NH_3 b. NH_3 c. OH^- d. CH_3NH_2

The base with the largest K_b value is the strongest base. OH^- is the strongest base possible in water.

81. $NaOH(aq) \rightarrow Na^+(aq) + OH^-(aq)$; NaOH is a strong base which completely dissociates into Na^+ and OH^-. The initial concentration of NaOH will equal the concentration of OH^- donated by NaOH.

a. $[OH^-] = 0.10\ M$; pOH = -log[OH^-] = -log(0.10) = 1.00

pH = 14.00 - pOH = 14.00 - 1.00 = 13.00

Note that H_2O is also present but the amount of OH^- produced by H_2O will be insignificant as compared to 0.10 M OH^- produced from the NaOH.

b. The [OH^-] concentration donated by the NaOH is $1.0 \times 10^{-10}\ M$. Water by itself donates $1.0 \times 10^{-7}\ M$. In this problem, water is the major OH^- contributor and $[OH^-] = 1.0 \times 10^{-7}\ M$.

pOH = -log (1.0×10^{-7}) = 7.00; pH = 14.00 - 7.00 = 7.00

c. $[OH^-] = 2.0\ M$; pOH = -log (2.0) = -0.30; pH = 14.00 - (-0.30) = 14.30

83. a. Major species: K^+, OH^-, H_2O (KOH is a strong base.)

$[OH^-] = 0.150\ M$, pOH = -log(0.150) = 0.824; pH = 14.000 - pOH = 13.176

b. Major species: Sr^{2+}, OH^-, H_2O; $Sr(OH)_2(aq) \rightarrow Sr^{2+}(aq) + 2\ OH^-(aq)$; Since the strong base $Sr(OH)_2$ dissolves in water to produce two mol OH^-, then $[OH^-] = 2(0.150\ M) = 0.300\ M$.

pOH = -log(0.300) = 0.523; pH =14.000 - 0.523 = 13.477

85. pH = 10.50; pOH = 14.00 - 10.50 = 3.50; $[OH^-] = 10^{-3.50} = 3.2 \times 10^{-4}\ M$

$KOH(aq) \rightleftharpoons K^+(aq) + OH^-(aq)$; Since KOH is a strong base, then a $3.2 \times 10^{-4}\ M$ KOH solution will produce a pH = 10.50 solution.

87. NH_3 is a weak base with $K_b = 1.8 \times 10^{-5}$. The major species present will be NH_3 and H_2O ($K_b = K_w = 1.0 \times 10^{-14}$). Since NH_3 has a much larger K_b value as compared to H_2O, then NH_3 is the stronger base present and will be the major producer of OH^-. To determine the amount of OH^- produced from NH_3, we must perform an equilibrium calculation.

	$NH_3(aq)$ + $H_2O(l)$	$\rightleftharpoons$	$NH_4^+(aq)$	+	$OH^-(aq)$
Initial	0.150 M		0		~0
	x mol/L NH_3 reacts with H_2O to reach equilibrium				
Change	$-x$	$\rightarrow$	$+x$		$+x$
Equil.	$0.150 - x$		x		x

$$K_b = 1.8 \times 10^{-5} = \frac{[NH_4^+][OH^-]}{[NH_3]} = \frac{x^2}{0.150 - x} \approx \frac{x^2}{0.150} \quad \text{(assuming } x \ll 0.150\text{)}$$

$x = [OH^-] = 1.6 \times 10^{-3}$ M; Check assumptions: x is 1.1% of 0.150 so the assumption $0.150 - x \approx 0.150$ is valid by the 5% rule. Also, the contribution of OH^- from water will be insignificant (which will usually be the case). Finishing the problem, pOH = -log [OH^-] = -log (1.6×10^{-3} M) = 2.80; pH = 14.00 - pOH = 14.00 - 2.80 = 11.20.

89. These are solutions of weak bases in water. We must solve the equilibrium weak base problem.

a. $(C_2H_5)_3N + H_2O \rightleftharpoons (C_2H_5)_3NH^+ + OH^-$ $\quad K_b = 4.0 \times 10^{-4}$

	$(C_2H_5)_3N$ + H_2O	$\rightleftharpoons$	$(C_2H_5)_3NH^+$	+	OH^-
Initial	0.20 M		0		~0
	x mol/L of $(C_2H_5)_3N$ reacts with H_2O to reach equilibrium				
Change	$-x$	$\rightarrow$	$+x$		$+x$
Equil.	$0.20 - x$		x		x

$$K_b = 4.0 \times 10^{-4} = \frac{[(C_2H_5)_3NH^+][OH^-]}{[(C_2H_5)_3N]} = \frac{x^2}{0.20 - x} \approx \frac{x^2}{0.20}, \quad x = [OH^-] = 8.9 \times 10^{-3}\ M$$

Assumptions good (x is 4.5% of 0.20). $[OH^-] = 8.9 \times 10^{-3}$ M

$$[H^+] = \frac{K_w}{[OH^-]} = \frac{1.0 \times 10^{-14}}{8.9 \times 10^{-3}} = 1.1 \times 10^{-12}\ M; \quad pH = 11.96$$

b. $HONH_2 + H_2O \rightleftharpoons HONH_3^+ + OH^-$ $\quad K_b = 1.1 \times 10^{-8}$

	$HONH_2$ + H_2O	$HONH_3^+$	OH^-
Initial	0.20 M	0	~0
Equil.	$0.20 - x$	x	x

$$K_b = 1.1 \times 10^{-8} = \frac{x^2}{0.20 - x} \approx \frac{x^2}{0.20}, \quad x = [OH^-] = 4.7 \times 10^{-5}\ M; \text{ Assumptions good.}$$

$[H^+] = 2.1 \times 10^{-10}$ M; pH = 9.68

91. This is a solution of a weak base in water. We must solve the weak base equilibrium problem.

$$C_2H_5NH_2 + H_2O \rightleftharpoons C_2H_5NH_3^+ + OH^- \quad K_b = 5.6 \times 10^{-4}$$

	$C_2H_5NH_2$		$C_2H_5NH_3^+$	OH^-
Initial	0.20 M		0	~0
	x mol/L $C_2H_5NH_2$ reacts with H_2O to reach equilibrium			
Change	$-x$	$\rightarrow$	$+x$	$+x$
Equil.	$0.20 - x$		x	x

$$K_b = \frac{[C_2H_5NH_3^+][OH^-]}{[C_2H_5NH_2]} = \frac{x^2}{0.20 - x} \approx \frac{x^2}{0.20} \quad \text{(assuming } x << 0.20\text{)}$$

$x = 1.1 \times 10^{-2}$; Checking assumption: $\frac{1.1 \times 10^{-2}}{0.20} \times 100 = 5.5\%$

Assumption fails the 5% rule. We must solve exactly using either the quadratic equation or the method of successive approximations (see Appendix 1.4 of the text). Using successive approximations and carrying extra significant figures:

$$\frac{x^2}{0.20 - 0.011} = \frac{x^2}{0.189} = 5.6 \times 10^{-4}, \; x = 1.0 \times 10^{-2} \, M \text{ (consistent answer)}$$

$$x = [OH^-] = 1.0 \times 10^{-2} \, M; \; [H^+] = \frac{K_w}{[OH^-]} = \frac{1.0 \times 10^{-14}}{1.0 \times 10^{-2}} = 1.0 \times 10^{-12} \, M; \; pH = 12.00$$

93. To solve for percent ionization, just solve the weak base equilibrium problem.

a. $$NH_3 + H_2O \rightleftharpoons NH_4^+ + OH^- \quad K_b = 1.8 \times 10^{-5}$$

	NH_3	NH_4^+	OH^-
Initial	0.10 M	0	~0
Equil.	$0.10 - x$	x	x

$$K_b = 1.8 \times 10^{-5} = \frac{x^2}{0.10 - x} \approx \frac{x^2}{0.10}, \; x = [OH^-] = 1.3 \times 10^{-3} \, M; \text{ Assumptions good.}$$

$$\text{Percent ionization} = \frac{[OH^-]}{[NH_3]_o} \times 100 = \frac{1.3 \times 10^{-3} \, M}{0.10 \, M} \times 100 = 1.3\%$$

b. $$NH_3 + H_2O \rightleftharpoons NH_4^+ + OH^-$$

	NH_3	NH_4^+	OH^-
Initial	0.010 M	0	~0
Equil.	$0.010 - x$	x	x

$$1.8 \times 10^{-5} = \frac{x^2}{0.010 - x} \approx \frac{x^2}{0.010}, \; x = [OH^-] = 4.2 \times 10^{-4} \, M; \text{ Assumptions good.}$$

$$\text{Percent ionization} = \frac{4.2 \times 10^{-4}}{0.010} \times 100 = 4.2\%$$

Note: For the same base, the percent ionization increases as the initial concentration of base decreases.

95. Using the K_b reaction to solve where PT = p-toluidine, $CH_3C_6H_4NH_2$:

$$PT + H_2O \rightleftharpoons PTH^+ + OH^-$$

	PT		PTH^+	OH^-
Initial	0.016 M		0	~0
	x mol/L of PT reacts with H_2O to reach equilibrium			
Change	$-x$	$\rightarrow$	$+x$	$+x$
Equil.	$0.016 - x$		x	x

$$K_b = \frac{[PTH^+][OH^-]}{[PT]} = \frac{x^2}{0.016 - x}$$

Since pH = 8.60: pOH = 14.00 - 8.60 = 5.40 and $[OH^-] = x = 10^{-5.40} = 4.0 \times 10^{-6}\ M$

$$K_b = \frac{(4.0 \times 10^{-6})^2}{0.016 - 4.0 \times 10^{-6}} = 1.0 \times 10^{-9}$$

Polyprotic Acids

97. $H_2SO_3(aq) \rightleftharpoons HSO_3^-(aq) + H^+(aq)$ $\qquad K_{a_1} = \frac{[HSO_3^-][H^+]}{[H_2SO_3]}$

$HSO_3^-(aq) \rightleftharpoons SO_3^{2-}(aq) + H^+(aq)$ $\qquad K_{a_2} = \frac{[SO_3^{2-}][H^+]}{[HSO_3^-]}$

99. In both these polyprotic acid problems, the dominate equilibrium is the K_{a_1} reaction. The amount of H^+ produced from the subsequent K_a reactions will be minimal since they are all have much smaller K_a values.

a. $$H_3AsO_4 \rightleftharpoons H^+ + H_2AsO_4^- \qquad K_{a_1} = 5 \times 10^{-3}$$

	H_3AsO_4		H^+	$H_2AsO_4^-$
Initial	0.10 M		~0	0
	x mol/L H_3AsO_4 dissociates to reach equilibrium			
Change	$-x$	$\rightarrow$	$+x$	$+x$
Equil.	$0.10 - x$		x	x

$$K_{a_1} = 5 \times 10^{-3} = \frac{[H^+][H_2AsO_4^-]}{[H_3AsO_4]} = \frac{x^2}{0.10 - x} \approx \frac{x^2}{0.10},\ x = 2 \times 10^{-2}$$

Assumption is bad (x is 20% of 0.10). Using successive approximations:

$$\frac{x^2}{0.10 - 0.02} = \frac{x^2}{0.08} = 5 \times 10^{-3},\ x = 2 \times 10^{-2} \quad \text{(consistent answer)}$$

$x = [H^+] = 2 \times 10^{-2}\ M$; pH = $-\log(2 \times 10^{-2}) = 1.7$

b. $H_2CO_3 \rightleftharpoons H^+ + HCO_3^-$ $K_{a_1} = 4.3 \times 10^{-7}$

	H_2CO_3	H^+	HCO_3^-
Initial	0.10 *M*	~0	0
Equil.	0.10 - x	x	x

$$K_{a_1} = 4.3 \times 10^{-7} = \frac{[H^+][HCO_3^-]}{[H_2CO_3]} = \frac{x^2}{(0.10 - x)} \approx \frac{x^2}{0.10}$$

$x = [H^+] = 2.1 \times 10^{-4}$ *M*; pH = 3.68; Assumptions good.

101. The dominant H^+ producer is the strong acid H_2SO_4. A 2.0 *M* H_2SO_4 solution produces 2.0 *M* HSO_4^- and 2.0 *M* H^+. However, HSO_4^- is a weak acid which could also add H^+ to the solution.

$$HSO_4^- \rightleftharpoons H^+ + SO_4^{2-}$$

	HSO_4^-		H^+	SO_4^{2-}
Initial	2.0 *M*		2.0 *M*	0
	x mol/L HSO_4^- dissociates to reach equilibrium			
Change	$-x$	$\rightarrow$	$+x$	$+x$
Equil.	2.0 - x		2.0 + x	x

$$K_{a_2} = 1.2 \times 10^{-2} = \frac{[H^+][SO_4^{2-}]}{[HSO_4^-]} = \frac{(2.0 + x)(x)}{2.0 - x} \approx \frac{2.0\,(x)}{2.0},\quad x = 1.2 \times 10^{-2}$$

Since x is 0.60% of 2.0, then the assumption is valid by the 5% rule. The amount of additional H^+ from HSO_4^- is 1.2×10^{-2}. The total amount of H^+ present is:

$[H^+] = 2.0 + 1.2 \times 10^{-2} = 2.0$ *M*; pH = -log (2.0) = -0.30

Note: In this problem, H^+ from HSO_4^- could have been ignored. However, this is not usually the case, especially in more dilute solutions of H_2SO_4.

Acid-Base Properties of Salts

103. One difficult aspect of acid-base chemistry is recognizing what types of species are present in solution, i.e., whether a species is a strong acid, strong base, weak acid, weak base or a neutral species. Below are some ideas and generalizations to keep in mind that will help in recognizing types of species present.

a. Memorize the following strong acids: HCl, HBr, HI, HNO_3, $HClO_4$ and H_2SO_4
b. Memorize the following strong bases: LiOH, NaOH, KOH, RbOH, $Ca(OH)_2$, $Sr(OH)_2$ and $Ba(OH)_2$
c. All weak acids have a K_a value less than 1 but greater than K_w. Some weak acids are in Table 14.2 of the text. All weak bases have a K_b value less than 1 but greater than K_w. Some weak bases are in Table 14.3 of the text.
d. All conjugate bases of weak acids are weak bases, i.e., all have a K_b value less than 1 but greater than K_w. Some examples of these are the conjugate bases of the weak acids in Table 14.2 of the text.

e. All conjugate acids of weak bases are weak acids, i.e., all have a K_a value less than 1 but greater than K_w. Some examples of these are the conjugate acids of the weak bases in Table 14.3 of the text.

f. Alkali metal ions (Li^+, Na^+, K^+, Rb^+, Cs^+) and heavier alkaline earth metal ions (Ca^{2+}, Sr^{2+}, Ba^{2+}) have no acidic or basic properties in water.

g. All conjugate bases of strong acids (Cl^-, Br^-, I^-, NO_3^-, ClO_4^-, HSO_4^-) have no basic properties in water ($K_b << K_w$) and only HSO_4^- has any acidic properties in water.

Lets apply these ideas to this problem to see what type of species are present. The letters in parenthesis is/are the generalization(s) above which identifies that species.

KOH: strong base (b)
KCl: neutral; K^+ and Cl^- have no acidic/basic properties (f and g).
KCN: CN^- is a weak base, $K_b = 1.0 \times 10^{-14}/6.2 \times 10^{-10} = 1.6 \times 10^{-5}$ (c and d). Ignore K^+(f).
NH_4Cl: NH_4^+ is a weak acid, $K_a = 5.6 \times 10^{-10}$ (c and e). Ignore Cl^-(g).
HCl: strong acid (a)

The most acidic solution will be the strong acid followed by the weak acid. The most basic solution will be the strong base followed by the weak base. The KCl solution will be between the acidic and basic solutions at pH = 7.00.

Most acidic → most basic; HCl > NH_4Cl > KCl > KCN > KOH

105. From the K_a values, acetic acid is a stronger acid than hypochlorous acid. Conversely, the conjugate base of acetic acid, $C_2H_3O_2^-$, will be a weaker base than the conjugate base of hypochlorous acid, OCl^-. Thus, the hypochlorite ion, OCl^-, is a stronger base than the acetate ion, $C_2H_3O_2^-$. In general, the stronger the acid, the weaker the conjugate base. This statement comes from the relationship $K_w = K_a \times K_b$ which holds for all conjugate acid-base pairs.

107. $NaN_3 \rightarrow Na^+ + N_3^-$; Azide, N_3^-, is a weak base since it is the conjugate base of a weak acid. All conjugate bases of weak acids are weak bases ($K_w < K_b < 1$). Ignore Na^+.

$$N_3^- + H_2O \rightleftharpoons HN_3 + OH^- \quad K_b = \frac{K_w}{K_a} = \frac{1.0 \times 10^{-14}}{1.9 \times 10^{-5}} = 5.3 \times 10^{-10}$$

	N_3^-		HN_3	OH^-
Initial	0.010 *M*		0	~0
	x mol/L of N_3^- reacts with H_2O to reach equilibrium			
Change	$-x$	→	$+x$	$+x$
Equil.	$0.010 - x$		x	x

$$K_b = \frac{[HN_3][OH^-]}{[N_3^-]} = 5.3 \times 10^{-10} = \frac{x^2}{0.010 - x} \approx \frac{x^2}{0.010} \quad \text{(assuming } x << 0.010)$$

$$x = [OH^-] = 2.3 \times 10^{-6}\,M;\ [H^+] = \frac{1.0 \times 10^{-14}}{2.3 \times 10^{-6}} = 4.3 \times 10^{-9}\,M \quad \text{Assumptions good.}$$

$[HN_3] = [OH^-] = 2.3 \times 10^{-6}\,M$; $[Na^+] = 0.010\,M$; $[N_3^-] = 0.010 - 2.3 \times 10^{-6} = 0.010\,M$

109. a. $CH_3NH_3Cl \rightarrow CH_3NH_3^+ + Cl^-$: $CH_3NH_3^+$ is a weak acid. Cl^- is the conjugate base of a strong acid. Cl^- has no basic (or acidic) properties.

$$CH_3NH_3^+ \rightleftharpoons CH_3NH_2 + H^+ \quad K_a = \frac{[CH_3NH_2][H^+]}{[CH_3NH_3^+]} = \frac{K_w}{K_b} = \frac{1.00 \times 10^{-14}}{4.38 \times 10^{-4}} = 2.28 \times 10^{-11}$$

	$CH_3NH_3^+$	$\rightleftharpoons$	CH_3NH_2	+	H^+
Initial	0.10 *M*		0		~0
	x mol/L $CH_3NH_3^+$ dissociates to reach equilibrium				
Change	$-x$	$\rightarrow$	$+x$		$+x$
Equil.	$0.10 - x$		x		x

$$K_a = 2.28 \times 10^{-11} = \frac{x^2}{0.10 - x} \approx \frac{x^2}{0.10} \quad \text{(assuming } x << 0.10)$$

$x = [H^+] = 1.5 \times 10^{-6}$ *M*; pH = 5.82 Assumptions good.

b. $NaCN \rightarrow Na^+ + CN^-$: CN^- is a weak base. Na^+ has no acidic (or basic) properties.

$$CN^- + H_2O \rightleftharpoons HCN + OH^- \quad K_b = \frac{K_w}{K_a} = \frac{1.0 \times 10^{-14}}{6.2 \times 10^{-10}} = 1.6 \times 10^{-5}$$

	CN^- + H_2O	$\rightleftharpoons$	HCN	+	OH^-
Initial	0.050 *M*		0		~0
	x mol/L CN^- reacts with H_2O to reach equilibrium				
Change	$-x$	$\rightarrow$	$+x$		$+x$
Equil.	$0.050 - x$		x		x

$$K_b = 1.6 \times 10^{-5} = \frac{[HCN][OH^-]}{[CN^-]} = \frac{x^2}{0.050 - x} \approx \frac{x^2}{0.050}$$

$x = [OH^-] = 8.9 \times 10^{-4}$ *M*; pOH = 3.05; pH = 10.95 Assumptions good.

111. All these salts contain Na^+ which has no acidic/basic properties and a conjugate base of a weak acid (except for NaCl where Cl^- is a neutral species.). All conjugate bases of weak acids are weak bases since K_b for these species are between K_w and 1. To identify the species, we will use the data given to determine the K_b value for the weak conjugate base. From the K_b value and data in Table 14.2 of the text, we can identify the conjugate base present by calculating the K_a value for the weak acid. We will use A^- as an abbreviation for the weak conjugate base.

	A^- + H_2O	$\rightleftharpoons$	HA	+	OH^-
Initial	0.100 mol/1.00 L		0		~0
	x mol/L A^- reacts with H_2O to reach equilibrium				
Change	$-x$	$\rightarrow$	$+x$		$+x$
Equil.	$0.100 - x$		x		x

$$K_b = \frac{[HA][OH^-]}{[A^-]} = \frac{x^2}{0.100 - x}$$; From the problem, pH = 8.07:

pOH = 14.00 - 8.07 = 5.93; $[OH^-] = x = 10^{-5.93} = 1.2 \times 10^{-6}$ *M*

$$K_b = \frac{(1.2 \times 10^{-6})^2}{0.100 - 1.2 \times 10^{-6}} = 1.4 \times 10^{-11} = K_b \text{ value for the conjugate base of a weak acid.}$$

The K_a value for the weak acid equals K_w/K_b: $K_a = \dfrac{1.0 \times 10^{-14}}{1.4 \times 10^{-11}} = 7.1 \times 10^{-4}$

From Table 14.2 of the text, this K_a value is closest to HF. Therefore, the unknown salt is NaF.

113. Major species present: $Al(H_2O)_6^{3+}$ ($K_a = 1.4 \times 10^{-5}$), NO_3^- (neutral) and H_2O ($K_w = 1.0 \times 10^{-14}$); $Al(H_2O)_6^{3+}$ is a stronger acid than water so it will be the dominant H^+ producer.

	$Al(H_2O)_6^{3+}$	$\rightleftharpoons$	$Al(H_2O)_5(OH)^{2+}$	+	H^+
Initial	0.050 M		0		~0
	x mol/L $Al(H_2O)_6^{3+}$ dissociates to reach equilibrium				
Change	$-x$	$\rightarrow$	$+x$		$+x$
Equil.	$0.050 - x$		x		x

$$K_a = 1.4 \times 10^{-5} = \frac{[Al(H_2O)_5(OH)^{2+}]\,[H^+]}{[Al(H_2O)_6]^{3+}} = \frac{x^2}{0.050 - x} \approx \frac{x^2}{0.050}$$

$x = 8.4 \times 10^{-4}\ M = [H^+]$; pH = -log ($8.4 \times 10^{-4}$) = 3.08; Assumptions good.

115. Reference Table 14.6 of the text and the solution to Exercise 14.103 for some generalizations on acid-base properties of salts.

a. $NaNO_3 \rightarrow Na^+ + NO_3^-$ neutral; Neither species has any acidic/basic properties.

b. $NaNO_2 \rightarrow Na^+ + NO_2^-$ basic; NO_2^- is a weak base and Na^+ has no effect on pH.

$NO_2^- + H_2O \rightleftharpoons HNO_2 + OH^-$ $\quad K_b = \dfrac{K_w}{K_{a,\,HNO_2}} = \dfrac{1.0 \times 10^{-14}}{4.0 \times 10^{-4}} = 2.5 \times 10^{-11}$

c. $NH_4NO_3 \rightarrow NH_4^+ + NO_3^-$ acidic; NH_4^+ is a weak acid and NO_3^- has no effect on pH.

$NH_4^+ \rightleftharpoons H^+ + NH_3$ $\quad K_a = \dfrac{K_w}{K_{b,\,NH_3}} = \dfrac{1.0 \times 10^{-14}}{1.8 \times 10^{-5}} = 5.6 \times 10^{-10}$

d. $NH_4NO_2 \rightarrow NH_4^+ + NO_2^-$ acidic; NH_4^+ is a weak acid ($K_a = 5.6 \times 10^{-10}$) and NO_2^- is a weak base ($K_b = 2.5 \times 10^{-11}$). Since $K_{a,\,NH_4^+} > K_{b,\,NO_2^-}$, then the solution is acidic.

$NH_4^+ \rightleftharpoons H^+ + NH_3$ $\quad K_a = 5.6 \times 10^{-10}$; $\quad NO_2^- + H_2O \rightleftharpoons HNO_2 + OH^-$ $\quad K_b = 2.5 \times 10^{-11}$

e. $NaOCl \rightarrow Na^+ + OCl^-$ basic; OCl^- is a weak base and Na^+ has no effect on pH.

$OCl^- + H_2O \rightleftharpoons HOCl + OH^-$ $\quad K_b = \dfrac{K_w}{K_{a,\,HOCl}} = \dfrac{1.0 \times 10^{-14}}{3.5 \times 10^{-8}} = 2.9 \times 10^{-7}$

f. $NH_4OCl \rightarrow NH_4^+ + OCl^-$ basic; NH_4^+ is a weak acid and OCl^- is a weak base. Since $K_{b,\,OCl^-} > K_{a,\,NH_4^+}$, then the solution is basic.

$NH_4^+ \rightleftharpoons NH_3 + H^+$ $K_a = 5.6 \times 10^{-10}$; $OCl^- + H_2O \rightleftharpoons HOCl + OH^-$ $K_b = 2.9 \times 10^{-7}$

Relationships Between Structure and Strengths of Acids and Bases

117. a. $HBrO < HBrO_2 < HBrO_3$; As the number of oxygen atoms attached to the bromine atom increases, acid strength increases.

b. $HIO_2 < HBrO_2 < HClO_2$; As the electronegativity of the central atom increases, acid strength increases.

c. $HBrO_3 < HClO_3$; Same reasoning as in b.

d. $H_2SO_3 < H_2SO_4$; Same reasoning as in a.

119. a. $H_2O < H_2S < H_2Se$; As the strength of the H – X bond decreases, acid strength increases.

b. $CH_3CO_2H < FCH_2CO_2H < F_2CHCO_2H < F_3CCO_2H$; As the electronegativity of neighboring atoms increase, acid strength increases.

c. $NH_4^+ < HONH_3^+$; Same reason as in b.

d. $NH_4^+ < PH_4^+$; Same reason as in a.

121. In general, metal oxides form basic solutions in water and nonmetal oxides form acidic solutions in water.

a. basic; $CaO(s) + H_2O(l) \rightarrow Ca(OH)_2(aq)$, $Ca(OH)_2$ is a strong base.

b. acidic; $SO_2(g) + H_2O(l) \rightarrow H_2SO_3(aq)$, H_2SO_3 is a weak diprotic acid.

c. acidic; $Cl_2O(g) + H_2O(l) \rightarrow 2\ HOCl(aq)$, HOCl is a weak acid.

Lewis Acids and Bases

123. A Lewis base is an electron pair donor and a Lewis acid is an electron pair acceptor.

a. $B(OH)_3$, acid; H_2O, base   b. Ag^+, acid; NH_3, base   c. BF_3, acid; NH_3, base

125. $Al(OH)_3(s) + 3\ H^+(aq) \rightarrow Al^{3+}(aq) + 3\ H_2O(l)$ (Bronsted-Lowry base, H^+ acceptor)

$Al(OH)_3(s) + OH^-(aq) \rightarrow Al(OH)_4^-(aq)$ (Lewis acid, electron pair acceptor)

127. Fe^{3+} should be the stronger Lewis acid. Fe^{3+} is smaller and has a greater positive charge. Because of this, Fe^{3+} will be more strongly attracted to lone pairs of electrons as compared to Fe^{2+}.

Additional Exercises

129. $2.48 \text{ g TlOH} \times \dfrac{1 \text{ mol}}{221.4 \text{ g}} = 1.12 \times 10^{-2}$ mol; TlOH is a strong base, so $[OH^-] = 1.12 \times 10^{-2}$ *M*.

$pOH = -\log[OH^-] = 1.951$; $pH = 14.000 - pOH = 12.049$

131. Let HBz = benzoic acid. A saturated solution is one where the maximum amount of solute (benzoic acid) has dissolved. Let C = equilibrium concentration of HBz in the saturated solution. Since benzoic acid is a weak acid ($K_a = 6.4 \times 10^{-5}$), then at equilibrium we also have H^+ and the conjugate base of benzoic acid (Bz^-) present. H^+ and Bz^- are both produced from the dissociation of HBz so their concentrations must be equal. Let $x = [H^+] = [Bz^-]$ at equilibrium. The general set-up for this problem is:

	HBz	⇌	H^+	+	Bz^-
Equil.	C		x		x

Solve for C, the equilibrium solubility of HBz, using the K_a expression.

$K_a = \dfrac{[H^+][Bz^-]}{[HBz]} = 6.4 \times 10^{-5} = \dfrac{x^2}{C}$; Since pH = 2.80: $x = [H^+] = [Bz^-] = 10^{-2.80} = 1.6 \times 10^{-3}$ *M*

$$C = \frac{(1.6 \times 10^{-3})^2}{6.4 \times 10^{-5}} = 4.0 \times 10^{-2}\,M$$

The equilibrium molar solubility of benzoic acid is 4.0×10^{-2} mol/L. The initial amount of benzoic acid that will dissolve in water is $4.0 \times 10^{-2}\,M + 1.6 \times 10^{-3}\,M = 4.2 \times 10^{-2}$ mol/L (this is the equilibrium concentration of benzoic acid plus the amount of benzoic acid that dissociated, x).

133. For this problem we will abbreviate $CH_2{=}CHCO_2H$ as Hacr and $CH_2{=}CHCO_2^-$ as acr^-.

a. Solving the weak acid problem:

	Hacr	⇌	H^+	+	acr^-	$K_a = 5.6 \times 10^{-5}$
Initial	0.10 *M*		~0		0	
Equil.	$0.10 - x$		x		x	

$\dfrac{x^2}{0.10 - x} = 5.6 \times 10^{-5} \approx \dfrac{x^2}{0.10}$, $x = [H^+] = 2.4 \times 10^{-3}$ *M*; pH = 2.62; Assumptions good.

b. $\% \text{ dissociation} = \dfrac{[H^+]}{[Hacr]_o} \times 100 = \dfrac{2.4 \times 10^{-3}}{0.10} \times 100 = 2.4\%$

c. acr^- is a weak base and the major source of OH^- in this solution.

acr^- + H_2O	$\rightleftharpoons$	Hacr	+ OH^-	$K_b = \frac{K_w}{K_a} = \frac{1.0 \times 10^{-14}}{5.6 \times 10^{-5}}$
Initial 0.050 *M*		0	~0	$K_b = 1.8 \times 10^{-10}$
Equil. 0.050 - x		x	x	

$$K_b = \frac{[Hacr][OH^-]}{[acr^-]} = 1.8 \times 10^{-10} = \frac{x^2}{0.050 - x} \approx \frac{x^2}{0.050}$$

$x = [OH^-] = 3.0 \times 10^{-6}$ *M*; pOH = 5.52; pH = 8.48 Assumptions good.

135. a. $Fe(H_2O)_6^{3+} + H_2O \rightleftharpoons Fe(H_2O)_5(OH)^{2+} + H_3O^+$

	$Fe(H_2O)_6^{3+}$	$Fe(H_2O)_5(OH)^{2+}$	H_3O^+
Initial	0.10 *M*	0	~0
Equil.	0.10 - x	x	x

$$K_a = \frac{[Fe(H_2O)_5(OH)^{2+}][H_3O^+]}{[Fe(H_2O)_6^{3+}]} = 6.0 \times 10^{-3} = \frac{x^2}{0.10 - x} \approx \frac{x^2}{0.10}$$

$x = 2.4 \times 10^{-2}$; Assumption is poor (x is 24% of 0.10). Using successive approximations:

$$\frac{x^2}{0.10 - 0.024} = 6.0 \times 10^{-3},\ x = 0.021$$

$$\frac{x^2}{0.10 - 0.021} = 6.0 \times 10^{-3},\ x = 0.022; \quad \frac{x^2}{0.10 - 0.022} = 6.0 \times 10^{-3},\ x = 0.022$$

$x = [H^+] = 0.022$ *M*; pH = 1.66

b. Because of the lower charge, $Fe^{2+}(aq)$ will not be as strong an acid as $Fe^{3+}(aq)$. A solution of iron(II) nitrate will be less acidic (have a higher pH) than a solution with the same concentration of iron(III) nitrate.

137. Solution is acidic from $HSO_4^- \rightleftharpoons H^+ + SO_4^{2-}$. Solving the weak acid problem:

	HSO_4^-	$\rightleftharpoons$	H^+	+	SO_4^{2-}	$K_a = 1.2 \times 10^{-2}$
Initial	0.10 *M*		~0		0	
Equil.	0.10 - x		x		x	

$$1.2 \times 10^{-2} = \frac{[H^+][SO_4^{2-}]}{[HSO_4^-]} = \frac{x^2}{0.10 - x} \approx \frac{x^2}{0.10},\ x = 0.035$$

Assumption is not good (x is 35% of 0.10). Using successive approximations:

$$\frac{x^2}{0.10 - x} \approx \frac{x^2}{0.10 - 0.035} = 1.2 \times 10^{-2},\ x = 0.028$$

$$\frac{x^2}{0.10 - 0.028} = 1.2 \times 10^{-2},\ x = 0.029; \quad \frac{x^2}{0.10 - 0.029} = 1.2 \times 10^{-2},\ x = 0.029$$

$x = [H^+] = 0.029\ M$; pH = 1.54

139. a. In the lungs, there is a lot of O_2 and the equilibrium favors $Hb(O_2)_4$. In the cells there is a deficiency of O_2 and the equilibrium favors HbH_4^{4+}.

b. CO_2 is a weak acid, $CO_2 + H_2O \rightleftharpoons HCO_3^- + H^+$. Removing CO_2 essentially decreases H^+. $Hb(O_2)_4$ is then favored and O_2 is not released by hemoglobin in the cells. Breathing into a paper bag increases $[CO_2]$ in the blood, thus increasing $[H^+]$ which shifts the reaction left.

c. CO_2 builds up in the blood and it becomes too acidic, driving the equilibrium to the left. Hemoglobin can't bind O_2 as strongly in the lungs. Bicarbonate ion acts as a base in water and neutralizes the excess acidity.

141. a. H_2SO_3 b. $HClO_3$ c. H_3PO_3

NaOH and KOH are soluble ionic compounds composed of Na^+ and K^+ cations and OH^- anions. All soluble ionic compounds dissolve to form the ions from which they are formed. In oxyacids, the compounds are all covalent compounds in which electrons are shared to form bonds (unlike ionic compounds). When these compounds are dissolved in water, the covalent bond between oxygen and hydrogen breaks to form H^+ ions.

Challenge Problems

143. Since this is a very dilute solution of NaOH, then we must worry about the amount of OH^- donated from the autoionization of water.

$$NaOH \rightarrow Na^+ + OH^-$$

$$H_2O \rightleftharpoons H^+ + OH^- \quad K_w = [H^+][OH^-] = 1.0 \times 10^{-14}$$

This solution, like all solutions, must be charged balance, that is [positive charge] = [negative charge]. For this problem, the charge balance equation is:

$$[Na^+] + [H^+] = [OH^-], \text{ where } [Na^+] = 1.0 \times 10^{-7}\ M \text{ and } [H^+] = \frac{K_w}{[OH^-]}$$

Substituting into the charge balance equation:

$$1.0 \times 10^{-7} + \frac{1.0 \times 10^{-14}}{[OH^-]} = [OH^-], \quad [OH^-]^2 - 1.0 \times 10^{-7}[OH^-] - 1.0 \times 10^{-14} = 0$$

Using the quadratic formula to solve:

$$[OH^-] = \frac{-(-1.0 \times 10^{-7}) \pm [(-1.0 \times 10^{-7})^2 - 4(1)(-1.0 \times 10^{-14})]^{1/2}}{2(1)}$$

$[OH^-] = 1.6 \times 10^{-7}\ M$; pOH = $-\log(1.6 \times 10^{-7}) = 6.80$; pH = 7.20

145. $HBrO \rightleftharpoons H^+ + BrO^-$ $K_a = 2 \times 10^{-9}$

	$HBrO$		H^+	BrO^-
Initial	1.0×10^{-6} *M*		~0	0
	x mol/L HBrO dissociates to reach equilibrium			
Change	$-x$	$\rightarrow$	$+x$	$+x$
Equil.	$1.0 \times 10^{-6} - x$		x	x

$$K_a = 2 \times 10^{-9} = \frac{x^2}{1.0 \times 10^{-6} - x} \approx \frac{x^2}{1.0 \times 10^{-6}};\ x = [H^+] = 4 \times 10^{-8}\ M;\ pH = 7.4$$

Lets check the assumptions. This answer is impossible! We can't add a small amount of an acid to a neutral solution and get a basic solution. The highest pH possible for an acid in water is 7.0. In the correct solution, we would have to take into account the autoionization of water.

147. Since NH_3 is so concentrated, we need to calculate the OH^- contribution from the weak base NH_3.

$NH_3 + H_2O \rightleftharpoons NH_4^+ + OH^-$ $K_b = 1.8 \times 10^{-5}$

	NH_3	NH_4^+	OH^-	
Initial	15.0 *M*	0	0.0100 *M*	(Assume no volume change.)
Equil.	$15.0 - x$	x	$0.0100 + x$	

$$K_b = 1.8 \times 10^{-5} = \frac{x(0.0100 + x)}{15.0 - x} \approx \frac{x(0.0100)}{15.0},\ x = 0.027;$$ Assumption is horrible (x is 270% of 0.0100).

Using the quadratic formula:

$$1.8 \times 10^{-5}(15.0 - x) = 0.0100\,x + x^2,\ x^2 + 0.0100\,x - 2.7 \times 10^{-4} = 0$$

$$x = 1.2 \times 10^{-2},\ [OH^-] = 1.2 \times 10^{-2} + 0.0100 = 0.022\ M$$

149. PO_4^{3-} is the conjugate base of HPO_4^{2-}. The K_a value for HPO_4^{2-} is $K_{a_3} = 4.8 \times 10^{-13}$.

$$PO_4^{3-}(aq) + H_2O(l) \rightleftharpoons HPO_4^{2-}(aq) + OH^-(aq) \quad K_b = \frac{K_w}{K_{a_3}} = \frac{1.0 \times 10^{-14}}{4.8 \times 10^{-13}} = 0.021$$

HPO_4^{2-} is the conjugate base of $H_2PO_4^-$ ($K_{a_2} = 6.2 \times 10^{-8}$).

$$HPO_4^{2-} + H_2O \rightleftharpoons H_2PO_4^- + OH^- \quad K_b = \frac{K_w}{K_{a_2}} = \frac{1.0 \times 10^{-14}}{6.2 \times 10^{-8}} = 1.6 \times 10^{-7}$$

$H_2PO_4^-$ is the conjugate base of H_3PO_4 ($K_{a_1} = 7.5 \times 10^{-3}$).

$$H_2PO_4^- + H_2O \rightleftharpoons H_3PO_4 + OH^- \quad K_b = \frac{K_w}{K_{a_1}} = \frac{1.0 \times 10^{-14}}{7.5 \times 10^{-3}} = 1.3 \times 10^{-12}$$

From the K_b values, PO_4^{3-} is the strongest base. This is expected since PO_4^{3-} is the conjugate base of the weakest acid (HPO_4^{2-}).

CHAPTER FIFTEEN

APPLICATIONS OF AQUEOUS EQUILIBRIA

Questions

13. A buffered solution must contain both a weak acid and a weak base. Most buffered solutions are prepared using a weak acid plus the conjugate base of the weak acid (which is a weak base). Buffered solutions are useful for controlling the pH of a solution since they resist pH change.

15. No, as long as there is both a weak acid and a weak base present, the solution will be buffered. If the concentrations are the same, the buffer will have the same capacity towards both added H^+ and OH^-. In addition, buffers with equal concentrations of weak acid and conjugate base have pH = pK_a.

17. No, since there are three colored forms, there must be two proton transfer reactions. Thus, there must be at least two acidic protons in the acid (orange) form of thymol blue.

19. The two forms of an indicator are different colors. The HIn form has one color and the In^- form has another color. To see only one color, that form must be in an approximately ten fold excess or greater over the other form. When the ratio of the two forms is less than 10, both colors are present. To go from $[HIn]/[In^-] = 10$ to $[HIn]/[In^-] = 0.1$ requires a change of 2 pH units (a 100 fold decrease in $[H^+]$) as the indicator changes from the HIn color to the In^- color.

Exercises

Buffers

21. When strong acid or strong base is added to an acetic acid/sodium acetate mixture, the strong acid/base is neutralized. The reaction goes to completion resulting in the strong acid/base being replaced with a weak acid/base which results in a new buffer solution. The reactions are:

$H^+(aq) + CH_3CO_2^-(aq) \rightarrow CH_3CO_2H(aq)$; $OH^- + CH_3CO_2H(aq) \rightarrow CH_3CO_2^-(aq) + H_2O(l)$

23. a. This is a weak acid problem. Let $HC_3H_5O_2$ = HOPr and $C_3H_5O_2^-$ = OPr^-.

$$HOPr \rightleftharpoons H^+ + OPr^- \qquad K_a = 1.3 \times 10^{-5}$$

	HOPr		H^+	OPr^-
Initial	0.100 *M*		~0	0
	x mol/L HOPr dissociates to reach equilibrium			
Change	$-x$	→	$+x$	$+x$
Equil.	$0.100 - x$		x	x

$$K_a = 1.3 \times 10^{-5} = \frac{[H^+][OPr^-]}{[HOPr]} = \frac{x^2}{0.100 - x} \approx \frac{x^2}{0.100}$$

$x = [H^+] = 1.1 \times 10^{-3}$ *M*; pH = 2.96 Assumptions good by the 5% rule.

b. This is a weak base problem.

$$OPr^- + H_2O \rightleftharpoons HOPr + OH^- \qquad K_b = \frac{K_w}{K_a} = 7.7 \times 10^{-10}$$

	OPr^-		HOPr	OH^-
Initial	0.100 *M*		0	~0
	x mol/L OPr^- reacts with H_2O to reach equilibrium			
Change	$-x$	→	$+x$	$+x$
Equil.	$0.100 - x$		x	x

$$K_b = 7.7 \times 10^{-10} = \frac{[HOPr][OH^-]}{[OPr^-]} = \frac{x^2}{0.100 - x} \approx \frac{x^2}{0.100}$$

$x = [OH^-] = 8.8 \times 10^{-6}$ *M*; pOH = 5.06; pH = 8.94 Assumptions good.

c. pure H_2O, $[H^+] = [OH^-] = 1.0 \times 10^{-7}$ *M*; pH = 7.00

d. This solution contains a weak acid and its conjugate base. This is a buffer solution. We will solve for the pH through the weak acid equilibrium reaction.

$$HOPr \rightleftharpoons H^+ + OPr^- \qquad K_a = 1.3 \times 10^{-5}$$

	HOPr		H^+	OPr^-
Initial	0.100 *M*		~0	0.100 *M*
	x mol/L HOPr dissociates to reach equilibrium			
Change	$-x$	→	$+x$	$+x$
Equil.	$0.100 - x$		x	$0.100 + x$

$$1.3 \times 10^{-5} = \frac{(0.100 + x)(x)}{0.100 - x} \approx \frac{(0.100)(x)}{0.100} = x = [H^+]$$

$[H^+] = 1.3 \times 10^{-5}$ *M*; pH = 4.89 Assumptions good.

Alternately, we can use the Henderson-Hasselbalch equation to calculate the pH of buffer solutions.

$$pH = pK_a + \log \frac{[Base]}{[Acid]} = pK_a + \log \frac{(0.100)}{(0.100)} = pK_a = -\log (1.3 \times 10^{-5}) = 4.89$$

The Henderson-Hasselbalch equation will be valid when an assumption of the type, $0.1 + x \approx 0.1$, that we just made in this problem is valid. From a practical standpoint, this will almost always be true for useful buffer solutions. If the assumption is not valid, the solution will have such a low buffering capacity it will not be of any use to control the pH. Note: The Henderson-Hasselbalch equation can only be used to solve for the pH of buffer solutions.

25. $0.100\, M\ HC_3H_5O_2$: percent dissociation $= \dfrac{[H^+]}{[HC_3H_5O_2]_o} \times 100 = \dfrac{1.1 \times 10^{-3}\, M}{0.100\, M} \times 100 = 1.1\%$

$0.100\, M\ HC_3H_5O_2 + 0.100\, M\ NaC_3H_5O_2$: % dissociation $= \dfrac{1.3 \times 10^{-5}}{0.100} \times 100 = 1.3 \times 10^{-2}\%$

The percent dissociation of the acid decreases from 1.1% to 1.3×10^{-2} % when $C_3H_5O_2^-$ is present. This is known as the common ion effect. The presence of the conjugate base of the weak acid inhibits the acid dissociation reaction.

27. a. We have a weak acid ($HOPr = HC_3H_5O_2$) and a strong acid (HCl) present. The amount of H^+ donated by the weak acid will be negligible. To prove it lets consider the weak acid equilibrium reaction:

$$HOPr \rightleftharpoons H^+ + OPr^- \qquad K_a = 1.3 \times 10^{-5}$$

	HOPr		H^+	OPr^-
Initial	$0.100\, M$		$0.020\, M$	0
	x mol/L HOPr dissociates to reach equilibrium			
Change	$-x$	$\rightarrow$	$+x$	$+x$
Equil.	$0.100 - x$		$0.020 + x$	x

$[H^+] = 0.020 + x \approx 0.020\, M$; pH = 1.70 Assumption good ($x = 6.5 \times 10^{-5}$ which is << 0.020).

Note: The H^+ contribution from the weak acid HOPr was negligible. The pH of the solution can be determined by only considering the amount of strong acid present.

b. Added H^+ reacts completely with the best base present, OPr^-.

$$OPr^- + H^+ \rightarrow HOPr$$

	OPr^-	H^+		HOPr	
Before	$0.100\, M$	$0.020\, M$		0	
Change	-0.020	-0.020	$\rightarrow$	$+0.020$	Reacts completely
After	0.080	0		$0.020\, M$	

After reaction, a weak acid, HOPr, and its conjugate base, OPr^-, are present. This is a buffer solution. Using the Henderson-Hasselbalch equation where $pK_a = -\log (1.3 \times 10^{-5}) = 4.89$:

$$pH = pK_a + \log \frac{[Base]}{[Acid]} = 4.89 + \log \frac{(0.080)}{(0.020)} = 5.49 \qquad \text{Assumptions good.}$$

c. This is a strong acid problem. $[H^+] = 0.020\ M$; pH = 1.70

d. Added H^+ reacts completely with the best base present, OPr^-.

	OPr^-	+	H^+	$\rightarrow$	HOPr	
Before	$0.100\ M$		$0.020\ M$		$0.100\ M$	
Change	-0.020		-0.020	$\rightarrow$	+0.020	Reacts completely
After	0.080		0		0.120	

A buffer solution results (weak acid + conjugate base). Using the Henderson-Hasselbalch equation:

$$pH = pK_a + \log \frac{[\text{Base}]}{[\text{Acid}]} = 4.89 + \log \frac{(0.080)}{(0.120)} = 4.71$$

29. a. OH^- will react completely with the best acid present, HOPr.

	HOPr	+	OH^-	$\rightarrow$	OPr^-	+	H_2O	
Before	$0.100\ M$		$0.020\ M$		0			
Change	-0.020		-0.020	$\rightarrow$	+0.020			Reacts completely
After	0.080		0		0.020			

A buffer solution results after the reaction. Using the Henderson-Hasselbalch equation:

$$pH = pK_a + \log \frac{[\text{Base}]}{[\text{Acid}]} = 4.89 + \log \frac{(0.020)}{(0.080)} = 4.29$$

b. We have a weak base and a strong base present at the same time. The amount of OH^- added by the weak base will be negligible. To prove it, lets consider the weak base equilibrium:

$$OPr^- + H_2O \rightleftharpoons HOPr + OH^- \quad K_b = 7.7 \times 10^{-10}$$

	OPr^-		HOPr	OH^-
Initial	$0.100\ M$		0	$0.020\ M$
	x mol/L OPr^- reacts with H_2O to reach equilibrium			
Change	$-x$	$\rightarrow$	$+x$	$+x$
Equil.	$0.100 - x$		x	$0.020 + x$

$[OH^-] = 0.020 + x \approx 0.020\ M$; pOH = 1.70; pH = 12.30 Assumption good.

Note: The OH^- contribution from the weak base OPr^- was negligible ($x = 3.9 \times 10^{-9}\ M$ as compared to $0.020\ M\ OH^-$ from the strong base). The pH can be determined by only considering the amount of strong base present.

c. This is a strong base in water. $[OH^-] = 0.020\ M$; pOH = 1.70; pH = 12.30

d. OH^- will react completely with HOPr, the best acid present.

	HOPr	+	OH^-	$\rightarrow$	OPr^-	+	H_2O	
Before	0.100 M		0.020 M		0.100 M			
Change	-0.020		-0.020	$\rightarrow$	+0.020			Reacts completely
After	0.080		0		0.120			

Using the Henderson-Hasselbalch equation to solve for the pH of the resulting buffer solution:

$$pH = pK_a + \log \frac{[Base]}{[Acid]} = 4.89 + \log \frac{(0.120)}{(0.080)} = 5.07$$

31. Consider all of the results to Exercises 15.23, 15.27, and 15.29:

Solution	Initial pH	after added acid	after added base
a	2.96	1.70	4.29
b	8.94	5.49	12.30
c	7.00	1.70	12.30
d	4.89	4.71	5.07

The solution in Exercise 15.23d is a buffer; it contains both a weak acid ($HC_3H_5O_2$) and a weak base ($C_3H_5O_2^-$). Solution d shows the greatest resistance to changes in pH when either strong acid or strong base is added, which is the primary property of buffers.

33. Major species: HNO_2, NO_2^- and Na^+. Na^+ has no acidic or basic properties. The appropriate equilibrium reaction to use is the K_a reaction of HNO_2 which contains both HNO_2 and NO_2^-. Solving the equilibrium problem (called a buffer problem):

	HNO_2	$\rightleftharpoons$	NO_2^-	+	H^+
Initial	1.00 M		1.00 M		~0
	x mol/L HNO_2 dissociates to reach equilibrium				
Change	$-x$	$\rightarrow$	$+x$		$+x$
Equil.	$1.00 - x$		$1.00 + x$		x

$$K_a = 4.0 \times 10^{-4} = \frac{[NO_2^-][H^+]}{[HNO_2]} = \frac{(1.00 + x)(x)}{(1.00 - x)} \approx \frac{1.00(x)}{1.00} \quad \text{(assuming } x << 1.00\text{)}$$

$x = 4.0 \times 10^{-4}\ M = [H^+]$; Assumptions good ($x$ is 4.0×10^{-2}% of 1.00).

$pH = -\log(4.0 \times 10^{-4}) = 3.40$

Note: We would get the same answer using the Henderson-Hasselbalch equation. Use whichever method is easiest for you.

35. Major species: HNO_2, NO_2^-, Na^+ and OH^-. The OH^- from the strong base will react with the best acid present (HNO_2). Any reaction involving a strong base is assumed to go to completion. The stoichiometry problem is:

	OH^-	+	HNO_2	$\rightarrow$	NO_2^- + H_2O	
Before	0.10 mol/1.00 L		1.00 *M*		1.00 *M*	
Change	-0.10 *M*		-0.10 *M*	$\rightarrow$	+0.10 *M*	Reacts completely
After	0		0.90		1.10	

After all the OH^- reacts, we are left with a solution containing a weak acid (HNO_2) and its conjugate base (NO_2^-). This is what we call a buffer problem. We will solve this buffer problem using the K_a equilibrium reaction.

	HNO_2	$\rightleftharpoons$	NO_2^-	+	H^+
Initial	0.90 *M*		1.10 *M*		~0
	x mol/L HNO_2 dissociates to reach equilibrium				
Change	$-x$	$\rightarrow$	$+x$		$+x$
Equil.	$0.90 - x$		$1.10 + x$		x

$$K_a = 4.0 \times 10^{-4} = \frac{(1.10 + x)(x)}{(0.90 - x)} \approx \frac{1.10(x)}{0.90}, \quad x = [H^+] = 3.3 \times 10^{-4}\ M;\ \ pH = 3.48;\ \ \text{Assumptions good.}$$

Note: The added NaOH to this buffer solution only changes the pH from 3.40 to 3.48. If the NaOH were added to 1.0 L of pure water, the pH would change from 7.00 to 13.00.

37. Major species: HNO_2, NO_2^-, H^+, Na^+, Cl^-; The added H^+ from the strong acid will react completely with the best base present (NO_2^-).

	H^+	+	NO_2^-	$\rightarrow$	HNO_2	
Before	$\frac{0.20\text{ mol}}{1.00\text{ L}}$		1.00 *M*		1.00 *M*	
Change	-0.20 *M*		-0.20 *M*	$\rightarrow$	+0.20 *M*	Reacts completely
After	0		0.80		1.20	

After all the H^+ has reacted, we have a buffer solution (a solution containing a weak acid and its conjugate base). Solving the buffer problem:

	HNO_2	$\rightleftharpoons$	NO_2^-	+	H^+
Initial	1.20 *M*		0.80 *M*		0
Equil.	$1.20 - x$		$0.80 + x$		$+x$

$$K_a = 4.0 \times 10^{-4} = \frac{(0.80 + x)(x)}{1.20 - x} \approx \frac{0.80(x)}{1.20}, \quad x = [H^+] = 6.0 \times 10^{-4}\ M;\ \ pH = 3.22;\ \ \text{Assumptions good.}$$

Note: The added HCl to this buffer solution only changes the pH from 3.40 to 3.22. If the HCl were added to 1.0 L of pure water, the pH would change from 7.00 to 0.70.

39. $75.0 \text{ g } NaC_2H_3O_2 \times \frac{1 \text{ mol}}{82.03 \text{ g}} = 0.914 \text{ mol}; \ [C_2H_3O_2^-] = \frac{0.914 \text{ mol}}{0.5000 \text{ L}} = 1.83\ M$

We will solve this buffer problem using the Henderson-Hasselbalch equation.

$$pH = pK_a + \log \frac{[C_2H_3O_2^-]}{[HC_2H_3O_2]} = -\log(1.8 \times 10^{-5}) + \log\left(\frac{1.83}{0.64}\right) = 4.74 + 0.46 = 5.20$$

41. a. $HC_2H_3O_2 \rightleftharpoons H^+ + C_2H_3O_2^- \qquad K_a = 1.8 \times 10^{-5}$

	$HC_2H_3O_2$		H^+	$C_2H_3O_2^-$
Initial	$0.10\ M$		~0	$0.25\ M$
	x mol/L $HC_2H_3O_2$ dissociates to reach equilibrium			
Change	$-x$	→	$+x$	$+x$
Equil.	$0.10 - x$		x	$0.25 + x$

$$1.8 \times 10^{-5} = \frac{x(0.25 + x)}{(0.10 - x)} \approx \frac{x(0.25)}{0.10} \quad \text{(assuming } 0.25 + x \approx 0.25 \text{ and } 0.10 - x \approx 0.10\text{)}$$

$x = [H^+] = 7.2 \times 10^{-6}\ M$; pH = 5.14   Assumptions good by the 5% rule.

Alternatively, we can use the Henderson-Hasselbalch equation:

$$pH = pK_a + \log \frac{[\text{Base}]}{[\text{Acid}]} \quad \text{where } pK_a = -\log(1.8 \times 10^{-5}) = 4.74$$

$$pH = 4.74 + \log \frac{(0.25)}{(0.10)} = 4.74 + 0.40 = 5.14$$

The Henderson-Hasselbalch equation will be valid when assumptions of the type, $0.10 - x \approx 0.10$, that we just made are valid. From a practical standpoint, this will almost always be true for useful buffer solutions. Note: The Henderson-Hasselbalch equation can only be used to solve for the pH of buffer solutions.

b. $pH = 4.74 + \log \frac{(0.10)}{(0.25)} = 4.74 + (-0.40) = 4.34$

c. $pH = 4.74 + \log \frac{(0.20)}{(0.080)} = 4.74 + 0.40 = 5.14$

d. $pH = 4.74 + \log \frac{(0.080)}{(0.20)} = 4.74 + (-0.40) = 4.34$

43. This is a buffer solution since a weak base (CH_3NH_2) and its conjugate acid ($CH_3NH_3^+$) are present at the same time. Using the Henderson Hasselbalch equation:

$$pH = pK_a + \log \frac{[\text{base}]}{[\text{acid}]}; \ K_a = \frac{K_w}{K_b} = \frac{1.00 \times 10^{-14}}{4.38 \times 10^{-4}} = 2.28 \times 10^{-11}$$

$$pH = -\log(2.28 \times 10^{-11}) + \log\left(\frac{0.50\ M}{0.70\ M}\right) = 10.642 + (-0.15) = 10.49$$

45. $[H^+]$ added = $\frac{0.010 \text{ mol}}{0.25 \text{ L}} = 0.040\ M$; The added H^+ reacts completely with NH_3 to form NH_4^+.

a.

	NH_3	+ H^+	$\rightarrow$	NH_4^+	
Before	$0.050\ M$	$0.040\ M$		$0.15\ M$	
Change	-0.040	-0.040	$\rightarrow$	+0.040	Reacts completely
After	0.010	0		0.19	

A buffer solution still exists after H^+ reacts completely. Using the Henderson-Hasselbalch equation:

$$pH = pK_a + \log \frac{[NH_3]}{[NH_4^+]} = -\log (5.6 \times 10^{-10}) + \log \left(\frac{0.010}{0.19}\right) = 9.25 + (-1.28) = 7.97$$

b.

	NH_3	+ H^+	$\rightarrow$	NH_4^+	
Before	$0.50\ M$	$0.040\ M$		$1.50\ M$	
Change	-0.040	-0.040	$\rightarrow$	+0.040	Reacts completely
After	0.46	0		1.54	

A buffer solution still exists. $pH = pK_a + \log \frac{[NH_3]}{[NH_4^+]} = 9.25 + \log \left(\frac{0.46}{1.54}\right) = 8.73$

The two buffers differ in their capacity and not the pH (both buffers had an initial pH = 8.77). Solution b has the greatest capacity since it has the largest concentrations of weak acid and conjugate base. Buffers with greater capacities will be able to absorb more H^+ or OH^- added.

47. $NH_4^+ \rightleftharpoons H^+ + NH_3 \quad K_a = \frac{K_w}{K_b} = 5.6 \times 10^{-10}$; $pK_a = -\log (5.6 \times 10^{-10}) = 9.25$; We will use the Henderson-Hasselbalch equation to calculate the concentration ratio necessary for each buffer.

$$pH = pK_a + \log \frac{[\text{base}]}{[\text{acid}]}, \quad pH = 9.25 + \log \frac{[NH_3]}{[NH_4^+]}$$

a. $9.00 = 9.25 + \log \frac{[NH_3]}{[NH_4^+]}$

$\log \frac{[NH_3]}{[NH_4^+]} = -0.25$

$\frac{[NH_3]}{[NH_4^+]} = 10^{-0.25} = 0.56$

b. $8.80 = 9.25 + \log \frac{[NH_3]}{[NH_4^+]}$

$\log \frac{[NH_3]}{[NH_4^+]} = -0.45$

$\frac{[NH_3]}{[NH_4^+]} = 10^{-0.45} = 0.35$

c. $10.00 = 9.25 + \log \frac{[NH_3]}{[NH_4^+]}$

$\frac{[NH_3]}{[NH_4^+]} = 10^{0.75} = 5.6$

d. $9.60 = 9.25 + \log \frac{[NH_3]}{[NH_4^+]}$

$\frac{[NH_3]}{[NH_4^+]} = 10^{0.35} = 2.2$

49. A best buffer has large and equal quantities of weak acid and conjugate base. Since [acid] = [base] for a best buffer, then $pH = pK_a + \log\frac{[base]}{[acid]} = pK_a + 0 = pK_a$.

The best acid choice for a pH = 7.00 buffer would be the weak acid with a pK_a close to 7 or $K_a \approx 1 \times 10^{-7}$. HOCl is the best choice in Table 14.2 ($K_a = 3.5 \times 10^{-8}$; $pK_a = 7.46$). To make this buffer, we need to calculate the [base]/[acid] ratio.

$$7.00 = 7.46 + \log\frac{[base]}{[acid]}, \quad \frac{[OCl^-]}{[HOCl]} = 10^{-0.46} = 0.35$$

Any $OCl^-/HOCl$ buffer in a concentration ratio of 0.35:1 will have a pH = 7.00. One possibility is [NaOCl] = 0.35 *M* and [HOCl] = 1.0 *M*.

51. The reaction $H^+ + NH_3 \rightarrow NH_4^+$ goes to completion for solutions b-d. After this reaction occurs, there must be both NH_3 and NH_4^+ in solution for it to be a buffer. The important components of each solution (after any reaction) are:

a. 0.05 *M* NH_4^+ and 0.05 *M* H^+ (two acids present, no buffer)

b. 0.05 *M* NH_4^+ (no NH_3 remains, no buffer)

c. 0.05 *M* NH_4^+ and 0.05 *M* H^+ (too much H^+ added)

d. 0.05 *M* NH_4^+ and 0.05 *M* NH_3 (a buffer solution results)

Only the combination in mixture d results in a buffer. Note that the concentrations are halved from the initial values. This is because equal volumes of two solutions were added together which halves the initial concentrations.

53. When OH^- is added, it converts $HC_2H_3O_2$ into $C_2H_3O_2^-$: $HC_2H_3O_2 + OH^- \rightarrow C_2H_3O_2^- + H_2O$

From this reaction, the moles of $C_2H_3O_2^-$ produced <u>equals</u> the moles of OH^- added. Also, the total concentration of acetic acid plus acetate ion must equal 2.0 *M* (assuming no volume change on addition of NaOH). Summarizing for each solution:

$$[C_2H_3O_2^-] + [HC_2H_3O] = 2.0\,M \text{ and } [C_2H_3O_2^-] \text{ produced} = [OH^-] \text{ added}$$

a. $pH = pK_a + \log\frac{[C_2H_3O_2^-]}{[HC_2H_3O_2]}$; For $pH = pK_a$, $\log\frac{[C_2H_3O_2^-]}{[HC_2H_3O_2]} = 0$

Therefore, $\frac{[C_2H_3O_2^-]}{[HC_2H_3O_2]} = 1.0$ and $[C_2H_3O_2^-] = [HC_2H_3O_2]$

Since $[C_2H_3O_2^-] + [HC_2H_3O_2] = 2.0\,M$, then $[C_2H_3O_2^-] = [HC_2H_3O_2] = 1.0\,M = [OH^-]$ added

To produce a 1.0 *M* $C_2H_3O_2^-$ solution we need to add 1.0 mol of NaOH to 1.0 L of the 2.0 *M* $HC_2H_3O_2$ solution. The resultant solution will have $pH = pK_a = 4.74$.

b. $4.00 = 4.74 + \log \frac{[C_2H_3O_2^-]}{[HC_2H_3O_2]}, \ \frac{[C_2H_3O_2^-]}{[HC_2H_3O_2]} = 10^{-0.74} = 0.18$

$[C_2H_3O_2^-] = 0.18\ [HC_2H_3O_2]$ or $[HC_2H_3O_2] = 5.6\ [C_2H_3O_2^-]$; Since $[C_2H_3O_2^-] + [HC_2H_3O_2] =$ 2.0 M, then:

$$[C_2H_3O_2^-] + 5.6\ [C_2H_3O_2^-] = 2.0\ M,\ [C_2H_3O_2^-] = \frac{2.0}{6.6} = 0.30\ M = [OH^-] \text{ added}$$

We need to add 0.30 mol of NaOH to 1.0 L of 2.0 M $HC_2H_3O_2$ solution to produce 0.30 M $C_2H_3O_2^-$. The resultant solution will have pH = 4.00.

c. $5.00 = 4.74 + \log \frac{[C_2H_3O_2^-]}{[HC_2H_3O_2]}, \ \frac{[C_2H_3O_2^-]}{[HC_2H_3O_2]} = 10^{0.26} = 1.8$

$1.8\ [HC_2H_3O_2] = [C_2H_3O_2^-]$ or $[HC_2H_3O_2] = 0.56\ [C_2H_3O_2^-]$; Since $[HC_2H_3O_2] + [C_2H_3O_2^-]$ = 2.0 M, then:

$$1.56\ [C_2H_3O_2^-] = 2.0\ M,\ [C_2H_3O_2^-] = 1.3\ M = [OH^-] \text{ added}$$

We need to add 1.3 mol of NaOH to 1.0 L of 2.0 M $HC_2H_3O_2$ to produce a solution with pH = 5.00.

Acid-Base Titrations

55.

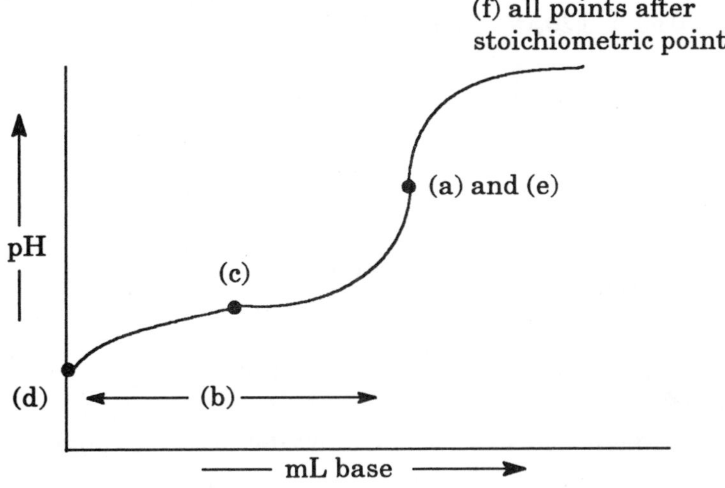

$HA + OH^- \rightarrow A^- + H_2O$; Added OH^- from the strong base converts the weak acid, HA, into its conjugate base, A^-. Initially before any OH^- is added (point d), HA is the dominant species present. After OH^- is added, both HA and A^- are present and a buffer solution results (region b). At the equivalence point (points a and e), exactly enough OH^- has been added to convert all of the weak acid, HA, into its conjugate base, A^-. Past the equivalence point (region f), excess OH^- is present. For the answer to part b, we included almost the entire buffer region. The maximum buffer region (or the region which is the best buffer solution) is around the halfway point to equivalence (point c). At this point, enough OH^- has been added to convert exactly one-half of the weak acid present initially into its conjugate base so $[HA] = [A^-]$ and $pH = pK_a$. A best buffer has about equal concentrations of weak acid and conjugate base present.

57. This is a strong acid (HCl) titrated by a strong base (NaOH). Added OH^- from the strong base will react completely with the H^+ present from the strong acid.

a. Only strong acid present. $[H^+] = 0.100\ M$; pH = 1.000

b. $\text{mmol OH}^- \text{ added} = 20.0 \text{ mL} \times \dfrac{0.20 \text{ mmol OH}^-}{\text{mL}} = 4.00 \text{ mmol OH}^-$

$\text{mmol H}^+ \text{ present} = 100.0 \text{ mL} \times \dfrac{0.100 \text{ mmol H}^+}{\text{mL}} = 10.0 \text{ mmol H}^+$

Note: The units mmoles are usually easier numbers to work with. The units for molarity are moles/L but are also equal to mmoles/mL.

	H^+	+	OH^-	$\rightarrow$	H_2O
Before	10.0 mmol		4.00 mmol		
Change	-4.00 mmol		-4.00 mmol		Reacts completely
After	6.0 mmol		0		

The excess H^+ determines pH. $[H^+]_{excess} = \dfrac{6.0 \text{ mmol H}^+}{100.0 \text{ mL} + 20.0 \text{ mL}} = 0.050\ M$; pH = 1.30

c. $\text{mmol OH}^- \text{ added} = 30.0 \text{ mL} \times 0.20\ M = 6.00 \text{ mmol OH}^-$

	H^+	+	OH^-	$\rightarrow$	H_2O
Before	10.0 mmol		6.00 mmol		
After	4.0 mmol		0		

$[H^+]_{excess} = \dfrac{4.0 \text{ mmol}}{(100.0 + 30.0) \text{ mL}} = 0.031\ M$; pH = 1.51

d. $\text{mmol OH}^- \text{ added} = 50.0 \text{ mL} \times 0.20\ M = 10.0 \text{ mmol OH}^-$; This is the equivalence point since we have added just enough OH^- to react with all the acid present. For a strong acid-strong base titration, pH = 7.00 at the equivalence point since only neutral species are present (Na^+, Cl^-, H_2O).

e. $\text{mmol OH}^- \text{ added} = 75.0 \text{ mL} \times 0.200\ M = 15.0 \text{ mmol OH}^-$

	H^+	+	OH^-	$\rightarrow$	H_2O
Before	10.0 mmol		15.0 mmol		
After	0		5.0 mmol		

Past the equivalence point, the pH is determined by the excess OH^- present.

$$[OH^-]_{excess} = \frac{5.0 \text{ mmol}}{(100.0 + 75.0) \text{ mL}} = 0.029\,M;\ \text{pOH} = 1.54;\ \text{pH} = 12.46$$

59. This is a weak acid ($HC_2H_3O_2$) titrated by a strong base (NaOH).

a. Only weak acid is present. Solving the weak acid problem:

	$HC_2H_3O_2$	$\rightleftharpoons$	H^+	+	$C_2H_3O_2^-$
Initial	$0.200\,M$		~0		0
	x mol/L $HC_2H_3O_2$ dissociates to reach equilibrium				
Change	$-x$	→	$+x$		$+x$
Equil.	$0.200 - x$		x		x

$$K_a = 1.8 \times 10^{-5} = \frac{x^2}{0.200 - x} = \frac{x^2}{0.200},\ x = [H^+] = 1.9 \times 10^{-3}\,M$$

pH = 2.72; Assumptions good.

b. The added OH^- will react completely with the best acid present, $HC_2H_3O_2$.

$$\text{mmol } HC_2H_3O_2 \text{ present} = 100.0 \text{ mL} \times \frac{0.200 \text{ mmol } HC_2H_3O_2}{\text{mL}} = 20.0 \text{ mmol } HC_2H_3O_2$$

$$\text{mmol } OH^- \text{ added} = 50.0 \text{ mL} \times \frac{0.100 \text{ mmol } OH^-}{\text{mL}} = 5.00 \text{ mmol } OH^-$$

	$HC_2H_3O_2$	+	OH^-	→	$C_2H_3O_2^-$ + H_2O	
Before	20.0 mmol		5.00 mmol		0	
Change	-5.00 mmol		-5.00 mmol	→	+5.00 mmol	Reacts completely
After	15.0 mmol		0		5.00 mmol	

After reaction of all the strong base, we have a buffer solution containing a weak acid ($HC_2H_3O_2$) and its conjugate base ($C_2H_3O_2^-$). We will use the Henderson-Hasselbalch equation to solve for the pH.

$$\text{pH} = \text{p}K_a + \log\frac{[C_2H_3O_2^-]}{[HC_2H_3O_2]} = -\log(1.8 \times 10^{-5}) + \log\left(\frac{5.00 \text{ mmol}/V_T}{15.0 \text{ mmol}/V_T}\right)$$ where V_T = total volume

$$\text{pH} = 4.74 + \log\left(\frac{5.00}{15.0}\right) = 4.74 + (-0.477) = 4.26$$

Note that the total volume cancels in the Henderson-Hasselbalch equation. For the [base]/[acid] term, the mole ratio equals the concentration ratio since the components of the buffer are always in the same volume of solution.

c. mmol OH^- added = 100.0 mL × 0.100 mmol OH^-/mL = 10.0 mmol OH^-; The same amount (20.0 mmol) of $HC_2H_3O_2$ is present as before (it never changes). As before, let the OH^- react to completion, then see what is remaining in solution after this reaction.

	$HC_2H_3O_2$	+	OH^-	→	$C_2H_3O_2^-$ + H_2O
Before	20.0 mmol		10.0 mmol		0
After	10.0 mmol		0		10.0 mmol

A buffer solution results after reaction. Since $[C_2H_3O_2^-] = [HC_2H_3O_2]$ = 10.0 mmol/total volume, then pH = pK_a. This is always true at the halfway point to equivalence for a weak acid/strong base titration, pH = pK_a.

$pH = -\log(1.8 \times 10^{-5}) = 4.74$

d. mmol OH^- added = 150.0 mL × 0.100 M = 15.0 mmol OH^-. Added OH^- reacts completely with the weak acid.

	$HC_2H_3O_2$	+	OH^-	→	$C_2H_3O_2^-$ + H_2O
Before	20.0 mmol		15.0 mmol		0
After	5.0 mmol		0		15.0 mmol

We have a buffer solution after all the OH^- reacts to completion. Using the Henderson-Hasselbalch equation:

$$pH = 4.74 + \log\frac{[C_2H_3O_2^-]}{[HC_2H_3O_2]} = 4.74 + \log\left(\frac{15.0 \text{ mmol}}{5.0 \text{ mmol}}\right)$$ (Total volume cancels, so we can use mol ratios.)

$pH = 4.74 + 0.48 = 5.22$

e. mmol OH^- added = 200.00 mL × 0.100 M = 20.0 mmol OH^-; As before, let the added OH^- react to completion with the weak acid, then see what is in solution after this reaction.

	$HC_2H_3O_2$	+	OH^-	→	$C_2H_3O_2^-$ + H_2O
Before	20.0 mmol		20.0 mmol		0
After	0		0		20.0 mmol

This is the equivalence point. Enough OH^- has been added to exactly neutralize all the weak acid present initially. All that remains that affects the pH at the equivalence point is the conjugate base of the weak acid, $C_2H_3O_2^-$. This is a weak base equilibrium problem.

$$C_2H_3O_2^- + H_2O \rightleftharpoons HC_2H_3O_2 + OH^- \quad K_b = \frac{K_w}{K_a} = \frac{1.0 \times 10^{-14}}{1.8 \times 10^{-5}} = 5.6 \times 10^{-10}$$

	$C_2H_3O_2^-$	$HC_2H_3O_2$	OH^-
Initial	20.0 mmol/300.0 mL	0	0
	x mol/L $C_2H_3O_2^-$ reacts with H_2O to reach equilibrium		
Change	$-x$ →	$+x$	$+x$
Equil.	$0.0667 - x$	x	x

$$K_b = 5.6 \times 10^{-10} = \frac{x^2}{0.0667 - x} \approx \frac{x^2}{0.0667}, \quad x = [OH^-] = 6.1 \times 10^{-6}\,M$$

pOH = 5.21; pH = 8.79; Assumptions good.

f. mmol OH^- added = 250.0 mL × 0.100 M = 25.0 mmol OH^-

	$HC_2H_3O_2$	+ OH^-	→ $C_2H_3O_2^-$ + H_2O
Before	20.0 mmol	25.0 mmol	0
After	0	5.0 mmol	20.0 mmol

After the titration reaction, we have a solution containing excess OH^- and a weak base, $C_2H_3O_2^-$. When a strong base and a weak base are both present, assume the amount of OH^- added from the weak base will be minimal, i.e., the pH past the equivalence point is determined by the amount of excess base.

$$[OH^-]_{excess} = \frac{5.0 \text{ mmol}}{100.0 \text{ mL} + 250.0 \text{ mL}} = 0.014\,M; \text{ pOH} = 1.85; \text{ pH} = 12.15$$

61. We will do sample calculations for the various parts of the titration. All results are summarized in Table 15.1 at the end of Exercise 15.63.

At the beginning of the titration, only the weak acid $HC_3H_5O_3$ is present.

$$HLac \rightleftharpoons H^+ + Lac^- \quad K_a = 10^{-3.86} = 1.4 \times 10^{-4} \quad HLac = HC_3H_5O_3, \; Lac^- = C_3H_5O_3^-$$

	HLac	H^+	Lac^-
Initial	0.100 M	~0	0
	x mol/L HLac dissociates to reach equilibrium		
Change	$-x$ →	$+x$	$+x$
Equil.	$0.100 - x$	x	x

$$1.4 \times 10^{-4} = \frac{x^2}{0.100 - x} = \frac{x^2}{0.100}, \quad x = [H^+] = 3.7 \times 10^{-3}\,M; \text{ pH} = 2.43 \quad \text{Assumptions good.}$$

Up to the stoichiometric point, we calculate the pH using the Henderson-Hasselbalch equation. This is the buffer region. For example, at 4.0 mL of NaOH added:

$$\text{initial mmol HLac present} = 25.0 \text{ mL} \times \frac{0.100 \text{ mmol}}{\text{mL}} = 2.50 \text{ mmol HLac}$$

$$\text{mmol } OH^- \text{ added} = 4.0 \text{ mL} \times \frac{0.100 \text{ mmol}}{\text{mL}} = 0.40 \text{ mmol } OH^-$$

Note: The units mmol are usually easier numbers to work with. The units for molarity are moles/L but are also equal to mmoles/mL.

The 0.40 mmol added OH^- converts 0.40 mmoles HLac to 0.40 mmoles Lac^- according to the equation:

$$HLac + OH^- \rightarrow Lac^- + H_2O \qquad \text{Reacts completely}$$

mmol HLac remaining = 2.50 - 0.40 = 2.10 mmol; mmol Lac^- produced = 0.40 mmol

We have a buffer solution. Using the Henderson-Hasselbalch equation where pK_a = 3.86:

$$pH = pK_a + \log\frac{[Lac^-]}{[HLac]} = 3.86 + \log\frac{(0.40)}{(2.10)}$$

(Total volume cancels, so we can use use the ratio of moles or mmoles.)

$$pH = 3.86 - 0.72 = 3.14$$

Other points in the buffer region are calculated in a similar fashion. Perform a stoichiometry problem first, followed by a buffer problem. The buffer region includes all points up to 24.9 mL OH^- added.

At the stoichiometric point (25.0 mL OH^- added), we have added enough OH^- to convert all of the HLac (2.50 mmol) into its conjugate base, Lac^-. All that is present is a weak base. To determine the pH, we perform a weak base calculation.

$$[Lac^-]_o = \frac{2.50\ \text{mmol}}{25.0\ \text{mL} + 25.0\ \text{mL}} = 0.0500\ M$$

	Lac^- + H_2O	$\rightleftharpoons$	HLac	+	OH^-
Initial	0.0500 M		0		0
	x mol/L Lac^- reacts with H_2O to reach equilibrium				
Change	$-x$	$\rightarrow$	$+x$		$+x$
Equil.	$0.0500 - x$		x		x

$$K_b = \frac{1.0 \times 10^{-14}}{1.4 \times 10^{-4}} = 7.1 \times 10^{-11}$$

$$K_b = \frac{x^2}{0.0500 - x} = \frac{x^2}{0.0500} = 7.1 \times 10^{-11}$$

$x = [OH^-] = 1.9 \times 10^{-6}\ M$; pOH = 5.72; pH = 8.28 Assumptions good.

Past the stoichiometric point, we have added more than 2.50 mmol of NaOH. The pH will be determined by the excess OH^- ion present. An example of this calculation follows.

At 25.1 mL: OH^- added = 25.1 mL × $\frac{0.100\ \text{mmol}}{\text{mL}}$ = 2.51 mmol OH^-; 2.50 mmol OH^- neutralizes all the weak acid present. The remainder is excess OH^-. excess OH^- = 2.51 - 2.50 = 0.01 mmol

$$[OH^-]_{excess} = \frac{0.01\ \text{mmol}}{(25.0 + 25.1)\ \text{mL}} = 2 \times 10^{-4}\ M;\ pOH = 3.7;\ pH = 10.3$$

All results are listed in Table 15.1 at the end of the solution to Exercise 15.63.

63. At beginning of the titration, only the weak base NH_3 is present. As always, solve for the pH using the K_b reaction for NH_3.

	NH_3 + H_2O	$\rightleftharpoons$	NH_4^+	+	OH^-	$K_b = 1.8 \times 10^{-5}$
Initial	$0.100\,M$		0		~0	
Equil.	$0.100 - x$		x		x	

$$K_b = \frac{x^2}{0.100 - x} = \frac{x^2}{0.100} = 1.8 \times 10^{-5}$$

$x = [OH^-] = 1.3 \times 10^{-3}\,M$; pOH = 2.89; pH = 11.11 Assumptions good.

In the buffer region (4.0 - 24.9 mL), we can use the Henderson-Hasselbalch equation:

$$K_a = \frac{1.0 \times 10^{-14}}{1.8 \times 10^{-5}} = 5.6 \times 10^{-10};\ pK_a = 9.25;\ pH = 9.25 + \log \frac{[NH_3]}{[NH_4^+]}$$

We must determine the amounts of NH_3 and NH_4^+ present after the added H^+ reacts completely with the NH_3. For example, after 8.0 mL HCl added:

$$\text{initial mmol } NH_3 \text{ present} = 25.0\text{ mL} \times \frac{0.100\text{ mmol}}{\text{mL}} = 2.50\text{ mmol } NH_3$$

$$\text{mmol } H^+ \text{ added} = 8.0\text{ mL} \times \frac{0.100\text{ mmol}}{\text{mL}} = 0.80\text{ mmol } H^+$$

Added H^+ reacts with NH_3 to completion: $NH_3 + H^+ \rightarrow NH_4^+$

mmol NH_3 remaining = 2.50 - 0.80 = 1.70 mmol; mmol NH_4^+ produced = 0.80 mmol

$$pH = 9.25 + \log \frac{1.70}{0.80} = 9.58$$ (Mole ratios can be used since the total volume cancels.)

Other points in the buffer region are calculated in similar fashion. Results are summarized in Table 15.1 at the end of Exercise 15.63.

At the stoichiometric point (25.0 mL H^+ added), just enough HCl has been added to convert all of the weak base (NH_3) into its conjugate acid (NH_4^+). Perform a weak acid calculation. $[NH_4^+]_o$ = 2.50 mmol/50.0 mL = 0.0500 M

	NH_4^+	$\rightleftharpoons$	H^+	+	NH_3	$K_a = 5.6 \times 10^{-10}$
Initial	$0.0500\,M$		0		0	
Equil.	$0.0500 - x$		x		x	

$$5.6 \times 10^{-10} = \frac{x^2}{0.0500 - x} = \frac{x^2}{0.0500},$$ $x = [H^+] = 5.3 \times 10^{-6}\,M$; pH = 5.28 Assumptions good.

Beyond the stoichiometric point, the pH is determined by the excess H^+. For example, at 28.0 mL of H^+ added:

$$H^+ \text{ added} = 28.0 \text{ mL} \times \frac{0.100 \text{ mmol}}{\text{mL}} = 2.80 \text{ mmol } H^+$$

$$\text{Excess } H^+ = 2.80 \text{ mmol} - 2.50 \text{ mmol} = 0.30 \text{ mmol excess } H^+$$

$$[H^+]_{excess} = \frac{0.30 \text{ mmol}}{(25.0 + 28.0) \text{ mL}} = 5.7 \times 10^{-3}\ M;\ \text{pH} = 2.24$$

All results are summarized below in Table 15.1.

Table 15.1: Summary of pH Results for Exercises 15.61 and 15.63 (Titration curves follow)

titrant mL	Exercise 15.61	Exercise 15.63
0.0	2.43	11.11
4.0	3.14	9.97
8.0	3.53	9.58
12.5	3.86	9.25
20.0	4.46	8.65
24.0	5.24	7.87
24.5	5.6	7.6
24.9	6.3	6.9
25.0	8.28	5.28
25.1	10.3	3.7
26.0	11.29	2.71
28.0	11.75	2.24
30.0	11.96	2.04

Note: The following plot contains the titration curves. The titration curves for Exercises 15.62 and 15.64 are included for your information.

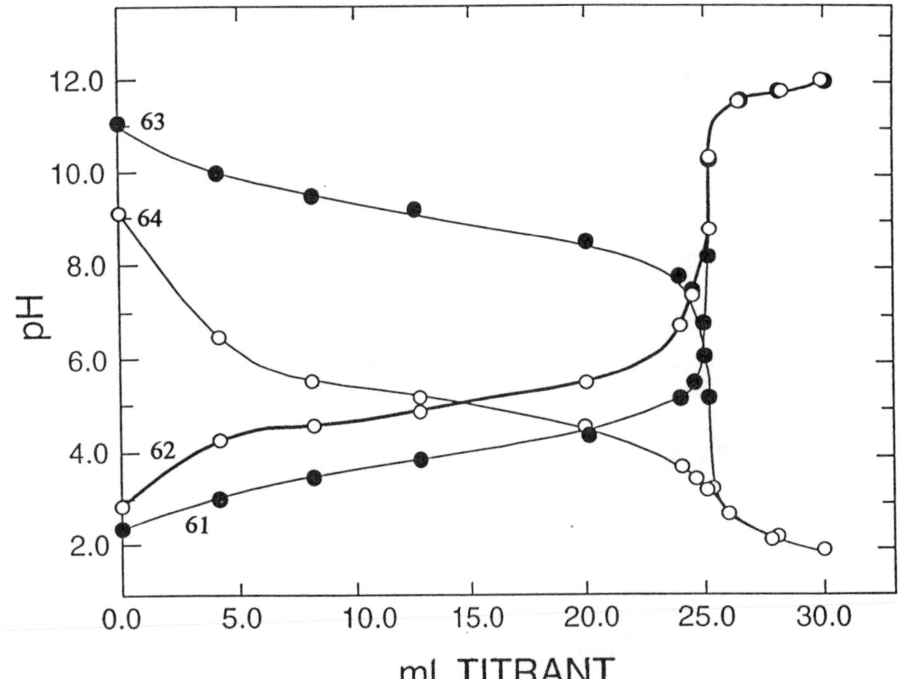

65. a. This is a weak acid/strong base titration. At the halfway point to equivalence, [weak acid] = [conjugate base], so pH = pK_a (always for a weak acid/strong base titration).

$pH = -\log(6.4 \times 10^{-5}) = 4.19$

mmol $HC_7H_5O_2$ present = 100. mL × 0.10 *M* = 10. mmol $HC_7H_5O_2$. For the equivalence point, 10. mmol of OH^- must be added. The volume of OH^- added to reach the equivalence point is:

$$10.\ \text{mmol OH}^- \times \frac{1\ \text{mL}}{0.10\ \text{mmol OH}^-} = 1.0 \times 10^2\ \text{mL OH}^-$$

At the equivalence point, 10. mmol of $HC_7H_5O_2$ is neutralized by 10. mmol of OH^- to produce 10. mmol of $C_7H_5O_2^-$. This is a weak base. The total volume of the solution is 100.0 mL + 1.0 × 10^2 mL = 2.0 × 10^2 mL. Solving the weak base equilibrium problem:

$$C_7H_5O_2^- + H_2O \rightleftharpoons HC_7H_5O_2 + OH^- \quad K_b = \frac{K_w}{K_a} = \frac{1.0 \times 10^{-14}}{6.4 \times 10^{-5}} = 1.6 \times 10^{-10}$$

	$C_7H_5O_2^-$	$HC_7H_5O_2$	OH^-
Initial	10. mmol/2.0 × 10^2 mL	0	0
Equil.	0.050 - x	x	x

$$K_b = 1.6 \times 10^{-10} = \frac{x^2}{0.050 - x} \approx \frac{x^2}{0.050},\ x = [OH^-] = 2.8 \times 10^{-6}\ M$$

pOH = 5.55; pH = 8.45 Assumptions good.

b. At the halfway point to equivalence for a weak base/strong acid titration, pH = pK_a since [weak base] = [conjugate acid].

$$K_a = \frac{K_w}{K_b} = \frac{1.0 \times 10^{-14}}{5.6 \times 10^{-4}} = 1.8 \times 10^{-11};\ pH = pK_a = -\log(1.8 \times 10^{-11}) = 10.74$$

For the equivalence point (mmol acid added = mmol base present):

mmol $C_2H_5NH_2$ present = 100.0 mL × 0.10 *M* = 10. mmol $C_2H_5NH_2$

$$\text{mL HNO}_3\ \text{added} = 10.\ \text{mmol H}^+ \times \frac{1\ \text{mL}}{0.20\ \text{mmol H}^+} = 50.\ \text{mL H}^+$$

The strong acid added completely converts the weak base into its conjugate acid. Therefore, at the equivalence point, $[C_2H_5NH_3^+]_o$ = 10. mmol/(100.0 + 50.) mL = 0.067 *M*. Solving the weak acid equilibrium problem:

$$C_2H_5NH_3^+ \rightleftharpoons H^+ + C_2H_5NH_2$$

	$C_2H_5NH_3^+$	H^+	$C_2H_5NH_2$
Initial	0.067 *M*	0	0
Equil.	0.067 - x	x	x

$$K_a = 1.8 \times 10^{-11} = \frac{x^2}{0.067 - x} \approx \frac{x^2}{0.067},\ x = [H^+] = 1.1 \times 10^{-6}\ M$$

pH = 5.96; Assumptions good.

c. In a strong acid/strong base titration, the halfway point has no special significance other than exactly one-half of the original amount of acid present has been neutralized.

mmol H^+ present = 100.0 mL × 0.50 *M* = 50. mmol H^+

$$\text{mL OH}^- \text{ added} = 25.\text{ mmol OH}^- \times \frac{1\text{ mL}}{0.25\text{ mmol}} = 1.0 \times 10^2 \text{ mL OH}^-$$

	H^+	+	OH^-	→	H_2O
Before	50. mmol		25 mmol		
After	25 mmol		0		

$$[H^+]_{excess} = \frac{25\text{ mmol}}{(100.0 + 1.0 \times 10^2)\text{ mL}} = 0.13\ M;\ \text{pH} = 0.89$$

At the equivalence point of a strong acid/strong base titration, only neutral speices are present (Na^+,Cl^-, H_2O) so the pH = 7.00.

67. $$\text{mmol OH}^- \text{ added} = 40.0\text{ mL} \times \frac{0.100\text{ mmol OH}^-}{\text{mL}} = 4.00\text{ mmol OH}^-$$

The added OH^- will react completely with the weak acid, HA.

	HA	+	OH^-	→	A^-	+	H_2O
Before	10.0 mmol		4.00 mmol		0		
After	6.0 mmol		0		4.00 mmol		

After reaction, a buffer solution exists.

$$\text{pH} = \text{p}K_a + \log\frac{[\text{base}]}{[\text{acid}]},\ 4.00 = \text{p}K_a + \log\left(\frac{4.00\text{ mmol}/V_T}{6.0\text{ mmol}/V_T}\right)$$

$$\text{p}K_a = 4.00 - \log(4.00/6.0) = 4.00 - (-0.18) = 4.18;\ K_a = 10^{-4.18} = 6.6 \times 10^{-5}$$

Indicators

69. $$\text{HIN} \rightleftharpoons \text{In}^- + \text{H}^+ \quad K_a = \frac{[\text{In}^-][\text{H}^+]}{[\text{HIn}]} = 1.0 \times 10^{-9}$$

a. In a very acid solution, the HIn form dominates so the solution will be yellow.

b. The color change occurs when the concentration of the more dominate form is approximately ten times as great as the less dominant form of the indicator.

$$\frac{[HIn]}{[In^-]} = \frac{10}{1};\ K_a = 1.0 \times 10^{-9} = \left(\frac{1}{10}\right)[H^+],\ [H^+] = 1 \times 10^{-8}\ M;\ pH = 8.0 \text{ at color change}$$

c. This is way past the equivalence point (100.0 mL OH^- added) so the solution is very basic and the In^- form of the indicator dominates. The solution will be blue.

71. When choosing an indicator, we want the color change of the indicator to occur approximately at the pH of the equivalence point. Since the pH generally changes very rapidly at the equivalence point, we don't have to be exact. This is especially true for strong acid/strong base titrations. Some choices where color change occurs at about the pH of the equivalence point are:

Exercise	pH at eq. pt.	Indicator
15.57	7.00	bromthymol blue or phenol red
15.59	8.79	o-cresolphthalein or phenolphthalein

73.

Exercise	pH at eq. pt.	Indicator
15.61	8.28	phenolphthalein
15.63	5.28	bromcresol green

75. The pH will be less than about 0.5 since crystal violet is yellow at a pH less than about 0.5. The methyl orange result only tells us that the pH is less than about 3.5.

77. a. yellow b. orange c. blue d. bluish-green

Solubility Equilibria

79. a. $AgC_2H_3O_2(s) \rightleftharpoons Ag^+(aq) + C_2H_3O_2^-(aq)$ $K_{sp} = [Ag^+]\,[C_2H_3O_2^-]$

b. $Al(OH)_3(s) \rightleftharpoons Al^{3+}(aq) + 3\ OH^-(aq)$ $K_{sp} = [Al^{3+}]\,[OH^-]^3$

c. $Ca_3(PO_4)_2(s) \rightleftharpoons 3\ Ca^{2+}(aq) + 2\ PO_4^{3-}(aq)$ $K_{sp} = [Ca^{2+}]^3\,[PO_4^{3-}]^2$

81. In our set-up, s = solubility of the ionic solid in mol/L. This is defined as the maximum amount of a salt which can dissolve. Since solids do not appear in the K_{sp} expression, we do not need to worry about their initial and equilibrium amounts.

a.

	$CaC_2O_4(s)$	$\rightleftharpoons$	$Ca^{2+}(aq)$	+	$C_2O_4^{2-}(aq)$
Initial			0		0
	s mol/L of $CaC_2O_4(s)$ dissolves to reach equilibrium				
Change	-s	$\rightarrow$	+s		+s
Equil.			s		s

From the problem, s = 4.8×10^{-5} mol/L.

$$K_{sp} = [Ca^{2+}]\,[C_2O_4^{2-}] = (s)(s) = s^2,\ K_{sp} = (4.8 \times 10^{-5})^2 = 2.3 \times 10^{-9}$$

b. $BiI_3(s) \rightleftharpoons Bi^{3+}(aq) + 3\ I^-(aq)$

			Bi^{3+}	I^-
Initial			0	0
	s mol/L of $BiI_3(s)$ dissolves to reach equilibrium			
Change	-s	→	+s	+3s
Equil.			s	3s

$K_{sp} = [Bi^{3+}]\ [I^-]^3 = (s)(3s)^3 = 27s^4$, $K_{sp} = 27(1.32 \times 10^{-5})^4 = 8.20 \times 10^{-19}$

83. $PbBr_2(s) \rightleftharpoons Pb^{2+}(aq) + 2\ Br^-(aq)$

			Pb^{2+}	Br^-
Initial			0	0
	s mol/L of $PbBr_2(s)$ dissolves to reach equilibrium			
Change	-s	→	+s	+2s
Equil.			s	2s

From the problem, $s = [Pb^{2+}] = 2.14 \times 10^{-2}\ M$. So:

$$K_{sp} = [Pb^{2+}]\ [Br^-]^2 = s(2s)^2 = 4s^3,\ K_{sp} = 4(2.14 \times 10^{-2})^3 = 3.92 \times 10^{-5}$$

85. In our set-ups, s = solubility in mol/L. Since solids do not appear in the K_{sp} expression, we do not need to worry about their initial or equilibrium amounts.

a. $Ag_3PO_4(s) \rightleftharpoons 3\ Ag^+(aq) + PO_4^{3-}(aq)$

			Ag^+	PO_4^{3-}
Initial			0	0
	s mol/L of $Ag_3PO_4(s)$ dissolves to reach equilibrium			
Change	-s	→	+3s	+s
Equil.			3s	s

$K_{sp} = 1.8 \times 10^{-18} = [Ag^+]^3\ [PO_4^{3-}] = (3s)^3(s) = 27s^4$

$27s^4 = 1.8 \times 10^{-18}$, $s = (6.7 \times 10^{-20})^{1/4} = 1.6 \times 10^{-5}$ mol/L = molar solubility

b. $CaCO_3(s) \rightleftharpoons Ca^{2+} + CO_3^{2-}$

		Ca^{2+}	CO_3^{2-}
Initial	s = solubility (mol/L)	0	0
Equil.		s	s

$K_{sp} = 8.7 \times 10^{-9} = [Ca^{2+}]\ [CO_3^{2-}] = s^2$, $s = 9.3 \times 10^{-5}$ mol/L

c. $Hg_2Cl_2(s) \rightleftharpoons Hg_2^{2+} + 2\ Cl^-$

		Hg_2^{2+}	Cl^-
Initial	s = solubility (mol/L)	0	0
Equil.		s	2s

$K_{sp} = 1.1 \times 10^{-18} = [Hg_2^{2+}]\ [Cl^-]^2 = (s)(2s)^2 = 4s^3$, $s = 6.5 \times 10^{-7}$ mol/L

87. Let s = solubility of $Al(OH)_3$ in mol/L. Note: Since solids do not appear in the K_{sp} expression, we do not need to worry about their initial or equilibrium amounts.

	$Al(OH)_3(s)$	$\rightleftharpoons$	$Al^{3+}(aq)$	+	$3\ OH^-(aq)$
Initial			0		1.0×10^{-7} *M* from water
	s mol/L of $Al(OH)_3(s)$ dissolves to reach equilibrium = molar solubility				
Change	-s	$\rightarrow$	+s		+3s
Equil.			s		$1.0 \times 10^{-7} + 3s$

$K_{sp} = 2 \times 10^{-32} = [Al^{3+}]\,[OH^-]^3 = (s)(1.0 \times 10^{-7} + 3s)^3 \approx s(1.0 \times 10^{-7})^3$

$s = \dfrac{2 \times 10^{-32}}{1.0 \times 10^{-21}} = 2 \times 10^{-11}$ mol/L Assumption good ($1.0 \times 10^{-7} + 3s \approx 1.0 \times 10^{-7}$).

89. a. Since both solids dissolve to produce 3 ions in solution, then we can compare values of K_{sp} to determine relative solubility. Since the K_{sp} for CaF_2 is the smallest, then $CaF_2(s)$ is the least soluble (in mol/L).

b. We must calculate solubilities since each salt yields a different number of ions when it dissolves.

	$Ca_3(PO_4)_2(s)$	$\rightleftharpoons$	$3\ Ca^{2+}(aq)$	+	$2\ PO_4^{3-}(aq)$	$K_{sp} = 1.3 \times 10^{-32}$
Initial	s = solubility (mol/L)		0		0	
Equil.			3s		2s	

$K_{sp} = [Ca^{2+}]^3\,[PO_4^{3-}]^2 = (3s)^3(2s)^2 = 108s^5,\ \ s = (1.3 \times 10^{-32}/108)^{1/5} = 1.6 \times 10^{-7}$ mol/L

	$FePO_4(s)$	$\rightleftharpoons$	$Fe^{3+}(aq)$	+	$PO_4^{3-}(aq)$	$K_{sp} = 1.0 \times 10^{-22}$
Initial	s = solubility (mol/L)		0		0	
Equil.			s		s	

$K_{sp} = [Fe^{3+}]\,[PO_4^{3-}] = s^2,\ \ s = \sqrt{1.0 \times 10^{-22}} = 1.0 \times 10^{-11}$ mol/L

Since the molar solubility of $FePO_4$ is the smallest, then $FePO_4$ is least soluble (in mol/L).

91. a. $Fe(OH)_3(s) \rightleftharpoons Fe^{3+} + 3\ OH^-$ pH = 7.0 so $[OH^-] = 1 \times 10^{-7}$ *M*

	$Fe(OH)_3(s)$	$\rightleftharpoons$	Fe^{3+}	+	$3\ OH^-$
Initial			0		1×10^{-7} *M*
	s mol/L of $Fe(OH)_3(s)$ dissolves to reach equilibrium = molar solubility				
Change	-s	$\rightarrow$	+s		+3s
Equil.			s		$3s + 1 \times 10^{-7}$

$K_{sp} = 4 \times 10^{-38} = [Fe^{3+}]\,[OH^-]^3 = (s)(3s + 1 \times 10^{-7})^3 \approx s(1 \times 10^{-7})^3$

$s = 4 \times 10^{-17}$ mol/L Assumption good ($3s << 1 \times 10^{-7}$).

b. $Fe(OH)_3(s) \rightleftharpoons Fe^{3+} + 3\ OH^-$ pH = 5.0 so $[OH^-] = 1 \times 10^{-9}\ M$

Initial			0	$1 \times 10^{-9}\ M$	(buffered)
	s mol/L dissolves to reach equilibrium				
Change	-s	→	+s	-----	(assume no pH change in buffer)
Equil.			s	1×10^{-9}	

$K_{sp} = 4 \times 10^{-38} = [Fe^{3+}]\ [OH^-]^3 = (s)(1 \times 10^{-9})^3$, $s = 4 \times 10^{-11}$ mol/L = molar solubility

c. $Fe(OH)_3(s) \rightleftharpoons Fe^{3+} + 3\ OH^-$ pH = 11.0 so $[OH^-] = 1 \times 10^{-3}\ M$

Initial			0	$0.001\ M$	(buffered)
	s mol/L dissolves to reach equilibrium				
Change	-s	→	+s	-----	(assume no pH change)
Equil.			s	0.001	

$K_{sp} = 4 \times 10^{-38} = [Fe^{3+}]\ [OH^-]^3 = (s)(0.001)^3$, $s = 4 \times 10^{-29}$ mol/L = molar solubility

Note: As $[OH^-]$ increases, solubility decreases. This is the common ion effect.

93. $Ca_3(PO_4)_2(s) \rightleftharpoons 3\ Ca^{2+}(aq) + 2\ PO_4^{3-}(aq)$

Initial			0	$0.20\ M$
	s mol/L of $Ca_3(PO_4)_2(s)$ dissolves to reach equilibrium			
Change	-s	→	+3s	+2s
Equil.			3s	0.20 + 2s

$K_{sp} = 1.3 \times 10^{-32} = [Ca^{2+}]^3\ [PO_4^{3-}]^2 = (3s)^3(0.20 + 2s)^2$

Assuming $0.20 + 2s \approx 0.20$: $1.3 \times 10^{-32} = (3s)^2(0.20)^2 = 27s^3(0.040)$

s = molar solubility = 2.3×10^{-11} mol/L; Assumptions good.

95. If the anion in the salt can act as a base in water, then the solubility of the salt will increase as the solution becomes more acidic. Added H^+ will react with the base forming the conjugate acid. As the basic anion is removed, more of the salt will dissolve to replenish the basic anion. The salts with basic anions are Ag_3PO_4, $CaCO_3$, $Pb_3(PO_4)_2$, $SrSO_4$ and Ag_2CO_3. Only Hg_2Cl_2 does not have any pH dependence since Cl^- is a terrible base (the conjugate base of a strong acid).

$$Ag_3PO_4(s) + H^+(aq) \rightarrow 3\ Ag^+(aq) + HPO_4^{2-}(aq) \xrightarrow{\text{excess } H^+} 3\ Ag^+(aq) + H_3PO_4(aq)$$

$$CaCO_3(s) + H^+ \rightarrow Ca^{2+} + HCO_3^- \xrightarrow{\text{excess } H^+} Ca^{2+} + H_2CO_3\ [H_2O(l) + CO_2(g)]$$

$$Pb_3(PO_4)_2(s) + 2\ H^+ \rightarrow 3\ Pb^{2+} + 2\ HPO_4^{2-} \xrightarrow{\text{excess } H^+} 3\ Pb^{2+} + 2\ H_3PO_4$$

$$SrSO_4(s) + H^+ \rightarrow Sr^{2+} + HSO_4^-$$

$$Ag_2CO_3(s) + H^+ \rightarrow 2\,Ag^+ + HCO_3^- \xrightarrow{\text{excess } H^+} 2\,Ag^+ + H_2CO_3\ [H_2O(l) + CO_2(g)]$$

97. Potentially, $BaSO_4(s)$ could form if Q is greater than K_{sp}.

$$BaSO_4(s) \rightleftharpoons Ba^{2+}(aq) + SO_4^{2-}(aq) \quad K_{sp} = 1.5 \times 10^{-9}$$

To calculate Q, we need the initial concentrations of Ba^{2+} and SO_4^{2-}.

$$[Ba^{2+}]_o = \frac{\text{mmoles } Ba^{2+}}{\text{total mL solution}} = \frac{75.0\text{ mL} \times \dfrac{0.020\text{ mmol } Ba^{2+}}{\text{mL}}}{75.0\text{ mL} \times 125\text{ mL}} = 0.0075\,M$$

$$[SO_4^{2-}]_o = \frac{\text{mmoles } SO_4^{2-}}{\text{total mL solution}} = \frac{125\text{ mL} \times \dfrac{0.040\text{ mmol } SO_4^{2-}}{\text{mL}}}{200.\text{ mL}} = 0.025\,M$$

$$Q = [Ba^{2+}]_o[SO_4^{2-}]_o = (0.0075\,M)(0.025\,M) = 1.9 \times 10^{-4}$$

$Q > K_{sp}$ (1.5×10^{-9}) so $BaSO_4(s)$ will form.

99. The concentrations of ions is large so Q will be greater than K_{sp} and $BaC_2O_4(s)$ will form. To solve this problem, we will assume that the precipitation reaction goes to completion, then we will solve an equilibrium problem to get the actual ion concentrations.

$$100.\text{ mL} \times \frac{0.200\text{ mmol } K_2C_2O_4}{\text{mL}} = 20.0\text{ mmol } K_2C_2O_4$$

$$150.\text{ mL} \times \frac{0.250\text{ mmol } BaBr_2}{\text{mL}} = 37.5\text{ mmol } BaBr_2$$

	$Ba^{2+}(aq)$ +	$C_2O_4^{2-}(aq)$	$\rightarrow$	$BaC_2O_4(s)$	$K = 1/K_{sp} >> 1$
Before	37.5 mmol	20.0 mmol		0	
Change	-20.0	-20.0	$\rightarrow$	+20.0	Reacts completely (K is large)
After	17.5	0		20.0	

New initial concentrations (after complete precipitation) are: $[Ba^{2+}] = \dfrac{17.5\text{ mmol}}{250.\text{ mL}} = 7.00 \times 10^{-2}\,M$

$$[K^+] = \frac{2(20.0\text{ mmol})}{250.\text{ mL}} = 0.160\,M;\quad [Br^-] = \frac{2(37.5\text{ mmol})}{250.\text{ mL}} = 0.300\,M$$

For K^+ and Br^-, these are also the final concentrations. For Ba^{2+} and $C_2O_4^{2-}$, we need to perform an equilibrium calculation.

$$BaC_2O_4(s) \rightleftharpoons Ba^{2+}(aq) + C_2O_4^{2-}(aq) \qquad K_{sp} = 2.3 \times 10^{-8}$$

		Ba^{2+}	$C_2O_4^{2-}$
Initial		$0.0700\ M$	0
	s mol/L $BaC_2O_4(s)$ dissolves to reach equilibrium		
Equil.		$0.0700 + s$	s

$K_{sp} = 2.3 \times 10^{-8} = [Ba^{2+}]\,[C_2O_4^{2-}] = (0.0700 + s)(s) \approx 0.0700\ s$

$s = [C_2O_4^{2-}] = 3.3 \times 10^{-7}$ mol/L; $[Ba^{2+}] = 0.0700\ M$ Assumption good ($s << 0.0700$).

101. $Ag_3PO_4\ (s) \rightleftharpoons 3\ Ag^+(aq) + PO_4^{3-}(aq)$; When Q is greater than K_{sp}, then precipitation will occur. We will calculate the $[Ag^+]_o$ necessary for $Q = K_{sp}$. Any $[Ag^+]_o$ greater than this calculated number will cause precipitation of $Ag_3PO_4(s)$. In this problem, $[PO_4^{3-}]_o = [Na_3PO_4]_o = 1.0 \times 10^{-5}\ M$.

$K_{sp} = 1.8 \times 10^{-18}$; $Q = 1.8 \times 10^{-18} = [Ag^+]_o^3\,[PO_4^{3-}]_o = [Ag^+]_o^3\,(1.0 \times 10^{-5}\ M)$

$$[Ag^+]_o = \left(\frac{1.8 \times 10^{-18}}{1.0 \times 10^{-5}}\right)^{1/3}, \quad [Ag^+]_o = 5.6 \times 10^{-5}\ M$$

When $[Ag^+]_o = [AgNO_3]_o$ is greater than $5.6 \times 10^{-5}\ M$, then precipitation of $Ag_3PO_4(s)$ will occur.

Complex Ion Equilibria

103. a.

$$\begin{array}{ll} Co^{2+} + NH_3 \rightleftharpoons CoNH_3^{2+} & K_1 \\ CoNH_3^{2+} + NH_3 \rightleftharpoons Co(NH_3)_2^{2+} & K_2 \\ Co(NH_3)_2^{2+} + NH_3 \rightleftharpoons Co(NH_3)_3^{2+} & K_3 \\ Co(NH_3)_3^{2+} + NH_3 \rightleftharpoons Co(NH_3)_4^{2+} & K_4 \\ Co(NH_3)_4^{2+} + NH_3 \rightleftharpoons Co(NH_3)_5^{2+} & K_5 \\ Co(NH_3)_5^{2+} + NH_3 \rightleftharpoons Co(NH_3)_6^{2+} & K_6 \\ \hline Co^{2+} + 6\ NH_3 \rightleftharpoons Co(NH_3)_6^{2+} & K_f = K_1K_2K_3K_4K_5K_6 \end{array}$$

Note: The various K's included are for your information. Each NH_3 adds with a corresponding K value associated with that reaction. The overall formation constant, K_f, for the overall reaction is equal to the product of all the stepwise K values.

b.

$$\begin{array}{ll} Ag^+ + NH_3 \rightleftharpoons AgNH_3^+ & K_1 \\ AgNH_3^+ + NH_3 \rightleftharpoons Ag(NH_3)_2^+ & K_2 \\ \hline Ag^+ + 2\ NH_3 \rightleftharpoons Ag(NH_3)_2^+ & K_f = K_1K_2 \end{array}$$

105. An ammonia solution is basic. Initially the reaction that occurs is $Cu^{2+}(aq) + 2\ OH^-(aq) \rightarrow Cu(OH)_2(s)$. As the concentration of NH_3 increases, the soluble complex ion $Cu(NH_3)_4^{2+}$ forms: $Cu(OH)_2(s) + 4\ NH_3(aq) \rightleftharpoons Cu(NH_3)_4^{2+}(aq) + 2\ OH^-(aq)$

107. The formation constant for HgI_4^{2-} is an extremely large number. Because of this, we will let the Hg^{2+} and I^- ions present initially react to completion, and then solve an equilibrium problem to determine the Hg^{2+} concentration.

	Hg^{2+}	+ $4\ I^-$	$\rightleftharpoons$	HgI_4^{2-}	$K = 1.0 \times 10^{30}$
Before	$0.010\ M$	$0.78\ M$		0	
Change	-0.010	-0.040	$\rightarrow$	+0.010	Reacts completely (K large)
After	0	0.74		0.010	New initial
	x mol/L HgI_4^{2-} dissociates to reach equilibrium				
Change	$+x$	$+4x$	$\leftarrow$	$-x$	
Equil.	x	$0.74 + 4x$		$0.010 - x$	

$$K = 1.0 \times 10^{30} = \frac{[HgI_4^{2-}]}{[Hg^{2+}][I^-]^4} = \frac{(0.010 - x)}{(x)(0.74 + 4x)^4};$$ Making normal assumptions:

$$1.0 \times 10^{30} \approx \frac{(0.010)}{(x)(0.74)^4}, \quad x = [Hg^{2+}] = 3.3 \times 10^{-32}\ M$$ Assumptions good.

Note: 3.3×10^{-32} mol/L corresponds to one Hg^{2+} ion per 5×10^7 L. It is very reasonable to approach the equilibrium in two steps. The reaction does essentially go to completion.

109. a.

	CuCl(s)	$\rightleftharpoons$	Cu^+	+	Cl^-
Initial	s = solubility (mol/L)		0		0
Equil.			s		s

$K_{sp} = 1.2 \times 10^{-6} = [Cu^+][Cl^-] = s^2, \ s = 1.1 \times 10^{-3}$ mol/L

b. Cu^+ forms the complex ion $CuCl_2^-$ in the presence of Cl^-. We will consider both the K_{sp} reaction and the complex ion reaction at the same time.

$CuCl(s) \rightleftharpoons Cu^+(aq) + Cl^-(aq)$ $\qquad K_{sp} = 1.2 \times 10^{-6}$

$Cu^+(aq) + 2\ Cl^-(aq) \rightleftharpoons CuCl_2^-(aq)$ $\qquad K_f = 8.7 \times 10^4$

$CuCl(s) + Cl^-(aq) \rightleftharpoons CuCl_2^-(aq)$ $\qquad K = K_{sp} \times K_f = 0.10$

Initial	$0.10\ M$	0
Equil.	0.10 - s	s

where s = solubility of CuCl(s) in mol/L

$$K = 0.10 = \frac{[CuCl_2^-]}{[Cl^-]} = \frac{s}{0.10 - s}, \quad 1.0 \times 10^{-2} - 0.10\ s = s, \ 1.10\ s = 1.0 \times 10^{-2}, \ s = 9.1 \times 10^{-3}\ \text{mol/L}$$

111. In $2.0\ M$ NH_3, the soluble complex ion $Ag(NH_3)_2^+$ forms which increases the solubility of AgCl(s). The reaction is: $AgCl(s) + 2\ NH_3 \rightleftharpoons Ag(NH_3)_2^+ + Cl^-$. In $2.0\ M$ NH_4NO_3, NH_3 is only formed by the dissociation of the weak acid NH_4^+. There is not enough NH_3 produced by this reaction to dissolve AgCl(s) by the formation of the complex ion.

Additional Exercises

113. $NH_3 + H_2O \rightleftharpoons NH_4^+ + OH^-$ $K_b = \frac{[NH_4^+][OH^-]}{[NH_3]}$; Taking the -log of the K_b expression:

$$-\log K_b = -\log [OH^-] - \log \frac{[NH_4^+]}{[NH_3]}, \quad -\log [OH^-] = -\log K_b + \log \frac{[NH_4^+]}{[NH_3]}$$

$$pOH = pK_b + \log \frac{[NH_4^+]}{[NH_3]} \text{ or } pOH = pK_b + \log \frac{[Acid]}{[Base]}$$

115. A best buffer is when pH ≈ pK_a since these solutions have about equal concentrations of weak acid and conjugate base. Therefore, choose combinations that yield a buffer where pH ≈ pK_a, i.e., look at the acids available and choose the one whose pK_a is closest to the pH.

a. Potassium fluoride + HCl will yield a buffer consisting of HF (pK_a = 3.14) and F^-.

b. Benzoic acid + NaOH will yield a buffer consisting of benzoic acid (pK_a = 4.19) and benzoate anion.

c. Sodium acetate + acetic acid (pK_a = 4.74) is the best choice for pH = 5.0 buffer since acetic acid has a pK_a value closest to 5.0.

d. HOCl and NaOH: This is the best choice to produce a conjugate acid/base pair with pH = 7.0. This mixture would yield a buffer consisting of HOCl (pK_a = 7.46) and OCl^-. Actually the best choice for a pH = 7.0 buffer is an equimolar mixture of ammonium chloride and sodium acetate. NH_4^+ is a weak acid ($K_a = 5.6 \times 10^{-10}$) and $C_2H_3O_2^-$ is a weak base ($K_b = 5.6 \times 10^{-10}$). A mixture of the two will give a buffer at pH = 7 since the weak acid and weak base are the same strengths. $NH_4C_2H_3O_2$ is commercially available and its solutions are used for pH = 7.0 buffers.

e. Ammonium chloride + NaOH will yield a buffer consisting of NH_4^+ (pK_a = 9.26) and NH_3.

117. a. $NH_3 + H_2O \rightleftharpoons OH^- + NH_4^+$; pH = 8.95; pOH = 5.05

$$K_b = 1.8 \times 10^{-5} = \frac{[OH^-][NH_4^+]}{[NH_3]} = \frac{(10^{-5.05})[NH_4^+]}{(0.500)}, \quad [NH_4^+] = 1.0\,M$$

b. $4.00 \text{ g NaOH} \times \frac{1 \text{ mol}}{40.00 \text{ g}} = 0.100$ mol; OH^- converts 0.100 mol NH_4^+ into 0.100 mol NH_3.

$NH_4^+ + OH^- \rightarrow NH_3 + H_2O$; After this reaction goes to completion, a buffer solution still exists where mol NH_4^+ = 1.0 - 0.100 = 0.9 mol and mol NH_3 = 0.500 + 0.100 = 0.600 mol. The pH of this solution is:

$$pH = 9.26 + \log\left(\frac{0.600}{0.9}\right) = 9.26 + (-0.2) = 9.1$$

119. At a pH = 0.00, the $[H^+] = 10^{-0.00} = 1.0\ M$. Begin with 1.0 L × 2.0 mol/L NaOH = 2.0 mol OH^-. We will need 2.0 mol HCl to neutralize the OH^- plus an additional 1.0 mol excess H^+ to reduce the pH to 0.00. We need 3.0 mol HCl total assuming 1.0 L of solution.

121. $HA + OH^- \rightarrow A^- + H_2O$ where HA = acetylsalicylic acid (assuming a monoprotic acid)

$$\text{mmol HA present} = 27.36\ \text{mL OH}^- \times \frac{0.5106\ \text{mmol OH}^-}{\text{mL OH}^-} \times \frac{1\ \text{mmol HA}}{\text{mmol OH}^-} = 13.97\ \text{mmol HA}$$

$$\text{Molar mass of HA} = \frac{\text{grams}}{\text{moles}} = \frac{2.51\ \text{g HA}}{13.97 \times 10^{-3}\ \text{mol HA}} = 180.\ \text{g/mol}$$

To determine the K_a value, use the pH data. After complete neutralization of acetylsalicylic acid by OH^-, we have 13.97 mmol of A^- produced from the neutralization reaction. A^- will react completely with the added H^+ and reform acetylsalicylic acid, HA.

$$\text{mmol H}^+ \text{ added} = 13.68\ \text{mL} \times \frac{0.5106\ \text{mmol H}^+}{\text{mL}} = 6.985\ \text{mmol H}^+$$

	A^-	+	H^+	→	HA	
Before	13.97 mmol		6.985 mmol		0	
Change	-6.985		-6.985	→	+6.985	Reacts completely
After	6.985 mmol		0		6.985 mmol	

We have back titrated this solution to the halfway point to equivalence where pH = pK_a (assuming HA is a weak acid). We know this because after H^+ reacts completely, equal mmol of HA and A^- are present which only occurs at the halfway point to equivalence. Assuming acetylsalicylic acid is a weak acid, then pH = pK_a = 3.48. $K_a = 10^{-3.48} = 3.3 \times 10^{-4}$

123.

	$Ca_5(PO_4)_3OH(s)$	⇌	$5\ Ca^{2+}$	+	$3\ PO_4^{3-}$	+	OH^-
Initial	s = solubility (mol/L)		0		0		1.0×10^{-7}
Equil.			5s		3s		$s + 1.0 \times 10^{-7} \approx s$

$K_{sp} = 6.8 \times 10^{-37} = [Ca^{2+}]^5 [PO_4^{3-}]^3 [OH^-] = (5s)^5(3s)^3(s)$

$6.8 \times 10^{-37} = (3125)(27)s^9$, $s = 2.7 \times 10^{-5}$ mol/L Assumption good.

The solubility of hydroxyapatite will increase as the solution gets more acidic since both phosphate and hydroxide can react with H^+.

	$Ca_5(PO_4)_3F(s)$	⇌	$5\ Ca^{2+}$	+	$3\ PO_4^{3-}$	+	F^-
Initial	s = solubility (mol/L)		0		0		0
Equil.			5s		3s		s

$K_{sp} = 1 \times 10^{-60} = (5s)^5(3s)^3(s) = (3125)(27)s^9$, $s = 6 \times 10^{-8}$ mol/L

The hydroxyapatite in the tooth enamel is converted to the less soluble fluorapatite by fluoride treated water. The less soluble fluorapatite is more difficult to remove, making teeth less susceptible to decay.

125. a.

	$Pb(OH)_2(s)$	$\rightleftharpoons$	Pb^{2+}	+	$2\ OH^-$
Initial	s = solubility (mol/L)		0		1.0×10^{-7} *M* from water
Equil.			s		$1.0 \times 10^{-7} + 2s$

$K_{sp} = 1.2 \times 10^{-15} = [Pb^{2+}]\,[OH^-]^2 = s(1.0 \times 10^{-7} + 2s)^2 \approx s(2s^2) = 4s^3$

$s = [Pb^{2+}] = 6.7 \times 10^{-6}\ M$; Assumption to ignore OH^- from water is good by the 5% rule.

b.

	$Pb(OH)_2(s)$	$\rightleftharpoons$	Pb^{2+}	+	$2\ OH^-$	
Initial			0		0.10 *M*	pH = 13.00, $[OH^-] = 0.10\ M$
	s mol/L $Pb(OH)_2(s)$ dissolves to reach equilibrium					
Equil.			s		0.10	(buffered solution)

$1.2 \times 10^{-15} = (s)(0.10)^2$, $s = [Pb^{2+}] = 1.2 \times 10^{-13}\ M$

c. We need to calculate the Pb^{2+} concentration in equilibrium with $EDTA^{4-}$. Since K is large for the formation of $PbEDTA^{2-}$, let the reaction go to completion then solve an equilibrium problem to get the Pb^{2+} concentration.

	Pb^{2+}	+	$EDTA^{4-}$	$\rightleftharpoons$	$PbEDTA^{2-}$	$K = 1.1 \times 10^{18}$
Before	0.010 *M*		0.050 *M*		0	
	0.010 mol/L Pb^{2+} reacts completely (large K)					
Change	-0.010		-0.010	$\rightarrow$	+0.010	Reacts completely
After	0		0.040		0.010	New initial
	x mol/L $PbEDTA^{2-}$ dissociates to reach equilibrium					
Equil.	*x*		0.040 + *x*		0.010 - *x*	

$$1.1 \times 10^{18} = \frac{(0.010 - x)}{(x)(0.040 + x)} \approx \frac{(0.010)}{x(0.040)},\quad x = [Pb^{2+}] = 2.3 \times 10^{-19}\ M$$ Assumptions good.

Now calculate the solubility quotient for $Pb(OH)_2$ to see if precipitation occurs. The concentration of OH^- is 0.10 *M* since we have a solution buffered at pH = 13.00.

$Q = [Pb^{2+}]_o[OH^-]_o^2 = (2.3 \times 10^{-19})(0.10)^2 = 2.3 \times 10^{-21} < K_{sp}\ (1.2 \times 10^{-15})$

$Pb(OH)_2(s)$ will not form since Q is less than K_{sp}.

Challenge Problems

127. At equivalence point: 16.00 mL × 0.125 mmol/mL = 2.00 mmol OH^- added; There must be 2.00 mmol HX present initially.

2.00 mL NaOH added = 2.00 mL × 0.125 mmol/mL = 0.250 mmol OH^-; 0.250 mmol of OH^- added will convert 0.250 mmol HX into 0.250 mmol X^-. Remaining HX = 2.00 - 0.250 = 1.75 mmol HX; This is a buffer solution where $[H^+] = 10^{-6.912} = 1.22 \times 10^{-7}$ *M*. Since total volume cancels:

$$K_a = \frac{[H^+][X^-]}{[HX]} = \frac{1.22 \times 10^{-7}(0.250)}{1.75} = 1.74 \times 10^{-8}$$

Note: We could solve for K_a using the Henderson-Hasselbalch equation.

129. For HOCl, $K_a = 3.5 \times 10^{-8}$ and $pK_a = -\log(3.5 \times 10^{-8}) = 7.46$; This will be a buffer solution since the pH is close to the pK_a value.

$$pH = pK_a + \log\frac{[OCl^-]}{[HOCl]},\quad 8.00 = 7.46 + \log\frac{[OCl^-]}{[HOCl]},\quad \frac{[OCl^-]}{[HOCl]} = 10^{0.54} = 3.5$$

1.00 L × 0.0500 *M* = 0.0500 mol HOCl initially. Added OH^- converts HOCl into OCl^-. The total moles of OCl^- and HOCl must equal 0.0500 mol. Solving where n = moles:

$$n_{OCl^-} + n_{HOCl} = 0.0500 \text{ and } n_{OCl^-} = 3.5\, n_{HOCl}$$

$$4.5\, n_{HOCl} = 0.0500,\quad n_{HOCl} = 0.011 \text{ mol};\quad n_{OCl^-} = 0.039 \text{ mol}$$

We need to add 0.039 mol NaOH to produce 0.039 mol OCl^-.

0.039 mol OH^- = V × 0.0100 *M*, V = 3.9 L NaOH

131. Phenolphthalein will change color at pH ~ 9. Phenolphthalein will mark the second end point. Therefore, at the phenolphthalein end point we will have titrated both protons on malonic acid.

$H_2Mal + 2\,OH^- \rightarrow 2\,H_2O + Mal^{2-}$ where H_2Mal = malonic acid

$$31.50 \text{ mL} \times \frac{0.0984 \text{ mmol NaOH}}{\text{mL}} \times \frac{1 \text{ mmol } H_2Mal}{2 \text{ mmol NaOH}} = 1.55 \text{ mmol } H_2Mal$$

$$[H_2Mal] = \frac{1.55 \text{ mmol}}{25.00 \text{ mL}} = 0.0620\, M$$

133.

$$AgBr(s) \rightleftharpoons Ag^+ + Br^- \qquad K_{sp} = 5.0 \times 10^{-13}$$

$$Ag^+ + 2\,S_2O_3^{2-} \rightleftharpoons Ag(S_2O_3)_2^{3-} \qquad K_f = 2.9 \times 10^{13}$$

$$AgBr(s) + 2\,S_2O_3^{2-} \rightleftharpoons Ag(S_2O_3)_2^{3-} + Br^- \qquad K = K_{sp} \times K_f = 14.5 \quad \text{(Carry extra sig. figs.)}$$

	$AgBr(s)$	+	$2\ S_2O_3^{3-}$	$\rightleftharpoons$	$Ag(S_2O_3)_2^{3-}$	+	Br^-
Initial			0.500 *M*		0		0
	s mol/L AgBr(s) dissolves to reach equilibrium						
Change	-s		-2s	→	+s		+s
Equil.			0.500 - 2s		s		s

$K = \frac{s^2}{0.500 - 2s} = 14.5$; Using the quadratic equation since s is not small:

$$s^2 = 7.25 - 29.0\ s,\ \ s^2 + 29.0\ s - 7.25 = 0;\ \ s = \frac{-29.0 + \sqrt{(29.0)^2 + 4(7.25)}}{2} = 0.248\ \text{mol/L}$$

$$1.00\ \text{L} \times \frac{0.248\ \text{mol AgBr}}{\text{L}} \times \frac{187.8\ \text{g AgBr}}{\text{mol AgBr}} = 46.6\ \text{g AgBr} = 47\ \text{g AgBr}$$

CHAPTER SIXTEEN

SPONTANEITY, ENTROPY, AND FREE ENERGY

Questions

7. A spontaneous process is one that occurs without any outside intervention.

9. a. Entropy increases; there is a greater volume accessible to the randomly moving gas molecules which increases disorder.

   b. The positional entropy doesn't change. There is no change in volume and thus, no change in the numbers of positions of the molecules. The total entropy (ΔS_{univ}) increases because the increase in temperature increases the energy disorder (ΔS_{surr}).

   c. Entropy decreases because the volume decreases (P and V are inversely related).

11. Living organisms need an external source of energy to carry out these processes. Green plants use the energy from sunlight to produce glucose from carbon dioxide and water by photosynthesis. In the human body, the energy released from the metabolism of glucose helps drive the synthesis of proteins. For all processes combined, ΔS_{univ} must be greater than zero (2nd law).

13. Dispersion increases the entropy of the universe since the more widely something is dispersed, the greater the disorder. We must do work to overcome this disorder. In terms of the 2nd law, it would be more advantageous to prevent contamination of the environment rather than to clean it up later. As a substance disperses, we have a much larger area that must be decontaminated.

15. $w_{max} = \Delta G$; When ΔG is negative, the magnitude of ΔG is equal to the maximum possible useful work obtainable from the process (at constant T and P). When ΔG is positive, the magnitude of ΔG is equal to the minimum amount of work that must be expended to make the process spontaneous. Due to waste energy (heat) in any real process, the amount of useful work obtainable from a spontaneous process is always less than w_{max} and for a nonspontaneous reaction, an amount of work greater than w_{max} must be applied to make the process spontaneous.

Exercises

Spontaneity, Entropy, and the Second Law of Thermodynamics: Free Energy

17. a, b and c; From our own experiences, salt water, colored water and rust form without any outside intervention. A bedroom, however, spontaneously gets cluttered. It takes an outside energy source to clean a bedroom.

19. We draw all of the possible arrangements of the two particles in the three levels.

2 kJ	___	___	x	___	x	xx
1 kJ	___	x	___	xx	x	___
0 kJ	xx	x	x	___	___	___
Total E =	0 kJ	1 kJ	2 kJ	2 kJ	3 kJ	4 kJ

The most likely total energy is 2 kJ.

21. a. H_2 at 100°C and 0.5 atm; Higher temperature and lower pressure means greater volume and hence, greater positional entropy.

b. N_2 at STP has the greater volume.

c. $H_2O(l)$ is more disordered than $H_2O(s)$.

23. Of the three phases (solid, liquid, and gas), solids are most ordered and gases are most disordered. Thus, a and b (melting and sublimation) involve an increase in the entropy of the system since going from a solid to a liquid or a solid to a gas increases disorder. For freezing (process c), a substance goes from the more disordered liquid state to the more ordered solid state, hence, entropy decreases.

25. a. To boil a liquid requires heat. Hence, this is an endothermic process. All endothermic processes decrease the entropy of the surroundings (ΔS_{surr} is negative).

b. This is an exothermic process. Heat is released when gas molecules slow down enough to form the solid. In exothermic processes, the entropy of the surroundings increases (ΔS_{surr} is positive).

27. $\Delta G = \Delta H - T\Delta S$; When ΔG is negative, then the process will be spontaneous.

a. $\Delta G = \Delta H - T\Delta S = 25 \times 10^3$ J - (300. K)(5.0 J/K) = 24,000 J, Not spontaneous

b. ΔG = 25,000 J - (300. K)(100. J/K) = -5000 J, Spontaneous

c. Without calculating ΔG, we know this reaction will be spontaneous at all temperatures. ΔH is negative and ΔS is positive ($-T\Delta S < 0$). ΔG will always be less than zero with these sign combinations for ΔH and ΔS.

d. $\Delta G = -1.0 \times 10^4$ J - (200. K)(-40. J/K) = -2000 J, Spontaneous

29. At the boiling point, $\Delta G = 0$ so $\Delta H = T\Delta S$.

$$\Delta S = \frac{\Delta H}{T} = \frac{31.4 \text{ kJ/mol}}{(273.2 + 61.7)\text{K}} = 9.38 \times 10^{-2} \text{ kJ/K•mol} = 93.8 \text{ J/K•mol}$$

31. a. $NH_3(s) \rightarrow NH_3(l)$; $\Delta G = \Delta H - T\Delta S = 5650$ J/mol - 200. K (28.9 J/K•mol)

$\Delta G = 5650$ J/mol - 5780 J/mol = -130 J/mol

Yes, NH_3 will melt since $\Delta G < 0$ at this temperature.

b. At the melting point, $\Delta G = 0$ so $T = \dfrac{\Delta H}{\Delta S} = \dfrac{5650 \text{ J/mol}}{28.9 \text{ J/K•mol}} = 196$ K.

Chemical Reactions: Entropy Changes and Free Energy

33. a. Decrease in disorder; $\Delta S°(-)$

b. Increase in disorder; $\Delta S°(+)$

c. Decrease in disorder ($\Delta n < 0$); $\Delta S°(-)$

d. Increase in disorder ($\Delta n > 0$); $\Delta S°(+)$

For c and d, concentrate on the gaseous products and reactants. When there are more gaseous product molecules than gaseous reactant molecules ($\Delta n > 0$), then $\Delta S°$ will be positive (disorder increases). When Δn is negative then $\Delta S°$ is negative (disorder decreases).

35. a. $C_{12}H_{22}O_{11}$; Larger molecule, more parts, more disorder so larger S value.

b. H_2O (0°C); A substance at 0 K has S = 0. As temperature increases, S increases.

c. $H_2S(g)$; A gas has greater disorder than a liquid.

37. a. $H_2(g) + 1/2\ O_2(g) \rightarrow H_2O(g)$; Since there are more molecules of reactant gases as compared to product molecules of gas ($\Delta n < 0$), then $\Delta S°$ will be negative. $\Delta S° = \Sigma n_p S°_{products} - \Sigma n_r S°_{reactants}$

$\Delta S° = 1$ mol $H_2O(g)$(189 J/K•mol) - [1 mol $H_2(g)$(131 J/K•mol) + 1/2 mol $O_2(g)$(205 J/K•mol)]

$\Delta S° = 189$ J/K - 234 J/K = -45 J/K

b. $3\ O_2(g) \rightarrow 2\ O_3(g)$; Since Δn of gases is negative, then $\Delta S°$ will be negative.

$\Delta S° = 2$ mol(239 J/K•mol) - [3 mol(205 J/K•mol)] = -137 J/K

c. $N_2(g) + O_2(g) \rightarrow 2\ NO(g)$; Here Δn of gases = 2 - 2 = 0. We can't easily predict if $\Delta S°$ is positive or negative.

$\Delta S° = 2(211) - (192 + 205) = 25$ J/K

39. $CS_2(g) + 3\ O_2(g) \rightarrow CO_2(g) + 2\ SO_2(g)$; $\Delta S° = S°_{CO_2} + 2\ S°_{SO_2} - [3\ S°_{O_2} + S°_{CS_2}]$

-143 J/K = 214 J/K + 2(248 J/K) - 3(205 J/K) - (1 mol)$S°_{CS_2}$, $S°_{CS_2} = 238$ J/K•mol

41. $P_4(s,\alpha) \rightarrow P_4(s,\beta)$

a. At T < -76.9°C, this reaction is spontaneous and the sign of ΔG is (-). At 76.9°C, $\Delta G = 0$ and above -76.9 °C, the sign of ΔG is (+). This is consistent with ΔH (-) and ΔS (-).

b. Since the sign of ΔS is negative, then the β form has the more ordered structure.

43. a. A bond is broken which requires energy so ΔH is positive. Since there are more product molecules of gas than reactant molecules of gas ($\Delta n > 0$), then ΔS will be positive.

b. $\Delta G = \Delta H - T\Delta S$; For this reaction to be spontaneous, the favorable entropy term must dominate. The reaction will be spontaneous at higher temperatures where the ΔS term dominates.

45. a.

	$CH_4(g)$	+ $2\,O_2(g)$	$\rightarrow$ $CO_2(g)$	+ $2\,H_2O(g)$	
ΔH_f°	-75 kJ/mol	0	-393.5	-242	
ΔG_f°	-51 kJ/mol	0	-394	-229	Data from Appendix 4
S°	186 J/K•mol	205	214	189	

$\Delta H^\circ = \Sigma n_p \Delta H^\circ_{f,\ products} - \Sigma n_r \Delta H^\circ_{f,\ reactants}$; $\Delta S^\circ = \Sigma n_p S^\circ_{products} - \Sigma n_r S^\circ_{reactants}$

$\Delta H^\circ = 2\text{ mol}(-242\text{ kJ/mol}) + 1\text{ mol}(-393.5\text{ kJ/mol}) - [1\text{ mol}(-75\text{ kJ/mol})] = -803\text{ kJ}$

$\Delta S^\circ = 2\text{ mol}(189\text{ J/K•mol}) + 1\text{ mol}(214\text{ J/K•mol})$

$- [1\text{ mol}(186\text{ J/K•mol}) + 2\text{ mol}(205\text{ J/K•mol})] = -4\text{ J/K}$

There are two ways to get ΔG°. We can use $\Delta G^\circ = \Delta H^\circ - T\Delta S^\circ$ (be careful of units):

$\Delta G^\circ = \Delta H^\circ - T\Delta S^\circ = -803 \times 10^3\text{ J} - (298\text{ K})(-4\text{ J/K}) = -8.018 \times 10^5\text{ J} = -802\text{ kJ}$

or we can use ΔG_f° values where $\Delta G^\circ = \Sigma n_p \Delta G^\circ_{f,\ products} - \Sigma n_r \Delta G^\circ_{f,\ reactants}$:

$\Delta G^\circ = 2\text{ mol}(-229\text{ kJ/mol}) + 1\text{ mol}(-394\text{ kJ/mol}) - [1\text{ mol}(-51\text{ kJ/mol})]$

$\Delta G^\circ = -801\text{ kJ}$ (Answers are the same within round off error)

b.

	$6\,CO_2(g)$	+ $6\,H_2O(l)$	$\rightarrow$ $C_6H_{12}O_6(s)$	+ $6\,O_2(g)$
ΔH_f°	-393.5 kJ/mol	-286	-1275	0
S°	214 J/K•mol	70.	212	205

$\Delta H° = -1275 - [6(-286) + 6(-393.5)] = 2802$ kJ

$\Delta S° = 6(205) + 212 - [6(214) + 6(70.)] = -262$ J/K

$\Delta G° = 2802$ kJ - (298 K)(-0.262 kJ/K) = 2880. kJ

c. $P_4O_{10}(s) + 6\,H_2O(l) \rightarrow 4\,H_3PO_4(s)$

	$P_4O_{10}(s)$	$H_2O(l)$	$H_3PO_4(s)$
$\Delta H_f°$ (kJ/mol)	-2984	-286	-1279
S° (J/K•mol)	229	70.	110.

$\Delta H° = 4$ mol(-1279 kJ/mol) - [1 mol(-2984 kJ/mol) + 6 mol(-286 kJ/mol)] = -416 kJ

$\Delta S° = 4(110.) - [229 + 6(70.)] = -209$ J/K

$\Delta G° = \Delta H° - T\Delta S° = -416$ kJ - (298 K)(-0.209 kJ/K) = -354 kJ

d. $HCl(g) + NH_3(g) \rightarrow NH_4Cl(s)$

	$HCl(g)$	$NH_3(g)$	$NH_4Cl(s)$
$\Delta H_f°$ (kJ/mol)	-92	-46	-314
S° (J/K•mol)	187	193	96

$\Delta H° = -314 - [-92 - 46] = -176$ kJ; $\Delta S° = 96 - [187 + 193] = -284$ J/K

$\Delta G° = \Delta H° - T\Delta S° = -176$ kJ - (298 K)(-0.284 kJ/K) = -91 kJ

47. $CH_4(g) + CO_2(g) \rightarrow CH_3CO_2H(l)$

$\Delta H° = -484 - [-75 + (-393.5)] = -16$ kJ; $\Delta S° = 160 - [186 + 214] = -240.$ J/K

$\Delta G° = \Delta H° - T\Delta S° = -16$ kJ - (298 K)(-0.240 kJ/K) = 56 kJ

This reaction is spontaneous only at temperatures below $T = \Delta H°/\Delta S° = 67$ K (where the favorable $\Delta H°$ term will dominate). This is not practical. Substances will be in condensed phases and rates will be very slow at this extremely low temperature.

$CH_3OH(g) + CO(g) \rightarrow CH_3CO_2H(l)$

$\Delta H° = -484 - [-110.5 + (-201)] = -173$ kJ; $\Delta S° = 160 - [198 + 240.] = -278$ J/K

$\Delta G° = -173 \text{ kJ} - (298 \text{ K})(-0.278 \text{ kJ/K}) = -90. \text{ kJ}$

This reaction also has a favorable enthalpy and an unfavorable entropy term. This reaction is spontaneous at temperatures below $T = \Delta H°/\Delta S° = 622$ K. The reaction of CH_3OH and CO will be preferred. It is spontaneous at high enough temperatures that the rates of reaction should be reasonable.

49.

$SO_3(g) \rightarrow SO_2(g) + 1/2\ O_2(g)$	$\Delta G° = -1/2\ (-142 \text{ kJ})$
$S(s) + 3/2\ O_2(g) \rightarrow SO_3(g)$	$\Delta G° = -371 \text{ kJ}$
$S(s) + O_2(g) \rightarrow SO_2(g)$	$\Delta G° = 71 \text{ kJ} - 371 \text{ kJ} = -300. \text{ kJ}$

51. $\Delta G° = \Sigma n_p \Delta G°_{f,\ products} - \Sigma n_r \Delta G°_{f,\ reactants}$, $-374 \text{ kJ} = -1105 \text{ kJ} - \Delta G°_{f,\ SF_4}$, $\Delta G°_{f,\ SF_4} = -731$ kJ/mol

53. $\Delta G° = \Sigma n_p \Delta G°_{f,\ products} - \Sigma n_r \Delta G°_{f,\ reactants}$

$\Delta G° = 1 \text{ mol}(-2698 \text{ kJ/mol}) + 6 \text{ mol}(-237 \text{ kJ/mol}) - [4 \text{ mol}(13 \text{ kJ/mol}) + 8 \text{ mol}(0)] = -4172 \text{ kJ}$

Free Energy: Pressure Dependence and Equilibrium

55. $\Delta G = \Delta G° + RT \ln Q$; For this reaction: $\Delta G = \Delta G° + RT \ln \dfrac{P_{NO_2} \times P_{O_2}}{P_{NO} \times P_{O_3}}$

$\Delta G° = 1 \text{ mol}(52 \text{ kJ/mol}) + 1 \text{ mol}(0) - [1 \text{ mol}(87 \text{ kJ/mol}) + 1 \text{ mol}(163 \text{ kJ/mol})] = -198 \text{ kJ}$

$$\Delta G = -198 \text{ kJ} + \frac{8.3145 \text{ J/K•mol}}{1000 \text{ J/kJ}} (298 \text{ K}) \ln \frac{(1.00 \times 10^{-7} \text{ atm})(1.00 \times 10^{-3} \text{ atm})}{(1.00 \times 10^{-6} \text{ atm})(2.00 \times 10^{-6} \text{ atm})}$$

$\Delta G = -198 \text{ kJ} + 9.69 \text{ kJ} = -188 \text{ kJ}$

57. $\Delta G = \Delta G° + RT \ln Q = \Delta G° + RT \ln \left(\dfrac{P^2_{SO_2} \times P_{O_2}}{P^2_{SO_3}} \right)$

$\Delta G° = 2 \text{ mol}(-300. \text{ kJ/mol}) - 2 \text{ mol}(-371 \text{ kJ/mol}) = 142 \text{ kJ}$

a. These are standard conditions so $\Delta G = \Delta G°$ since $Q = 1$ and $\ln Q = 0$. Since $\Delta G°$ is positive, then the reverse reaction is spontaneous. The reaction shifts left to reach equilibrium.

b. $\Delta G = 142 \times 10^3 \text{ J} + 8.3145 \text{ J/K•mol } (298 \text{ K}) \ln \dfrac{(1.00 \times 10^{-9})^2 (1.00 \times 10^{-9})}{(1.00)^2}$

$\Delta G = 1.42 \times 10^5 \text{ J} - 1.54 \times 10^5 \text{ J} = -1.2 \times 10^4 \text{ J}$

Since ΔG is negative, then the forward reaction is spontaneous so the reaction shifts to the right to reach equilibrium.

59. $NO(g) + O_3(g) \rightleftharpoons NO_2(g) + O_2(g)$; $\Delta G° = \Sigma n_p \Delta G°_{f,\ products} - \Sigma n_r \Delta G°_{f,\ reactants}$

$\Delta G°$ = 1 mol(52 kJ/mol) - [1 mol(87 kJ/mol) + 1 mol(163 kJ/mol)] = -198 kJ

$$\Delta G° = -RT \ln K,\ K = \exp\frac{-\Delta G°}{RT} = \exp\left(\frac{-(-1.98 \times 10^5\ J)}{8.3145\ J/K\bullet mol\ (298\ K)}\right) = e^{79.912} = 5.07 \times 10^{34}$$

Note: When determining exponents, we will round off after the calculation is complete. This helps eliminate excessive round off error.

61. a. $\Delta H°$ = 2 mol(-92 kJ/mol) - [1 mol(0) + 1 mol(0)] = -184 kJ

$\Delta S°$ = 2 mol(187 J/K•mol) - [1 mol(131 J/K•mol) + 1 mol(223 J/K•mol)] = 20. J/K

$\Delta G° = \Delta H° - T\Delta S° = -184 \times 10^3$ J - 298 K (20. J/K) = -1.90×10^5 J = -190. kJ

$$\Delta G° = -RT \ln K,\ \ln K = \frac{-\Delta G°}{RT} = \frac{-(-1.90 \times 10^5\ J)}{8.3145\ J/K\bullet mol\ (298\ K)} = 76.683,\ K = e^{76.683} = 2.01 \times 10^{33}$$

b. These are standard conditions so $\Delta G = \Delta G°$ = -190. kJ. Since ΔG is negative, then the forward reaction is spontaneous so the reaction shifts right to reach equilibrium.

63. a. $$\Delta G° = -RT \ln K = -\frac{8.3145\ J}{K\ mol}(298\ K) \ln(1.00 \times 10^{-14}) = 7.99 \times 10^4\ J = 79.9\ kJ/mol$$

b. $$\Delta G°_{313} = -RT \ln K = -\frac{8.3145\ J}{K\ mol}(313\ K) \ln(2.92 \times 10^{-14}) = 8.11 \times 10^4\ J = 81.1\ kJ/mol$$

65. The equation $\ln K = \frac{-\Delta H}{R}\left(\frac{1}{T}\right) + \frac{\Delta S°}{R}$ is in the form of a straight line equation (y = mx + b).

Therefore, if we graph ln K vs 1/T we get a straight line with slope = m = $-\Delta H°/R$. For an endothermic process, the slope is negative since $\Delta H°$ is positive. A sketch of the plot would look like:

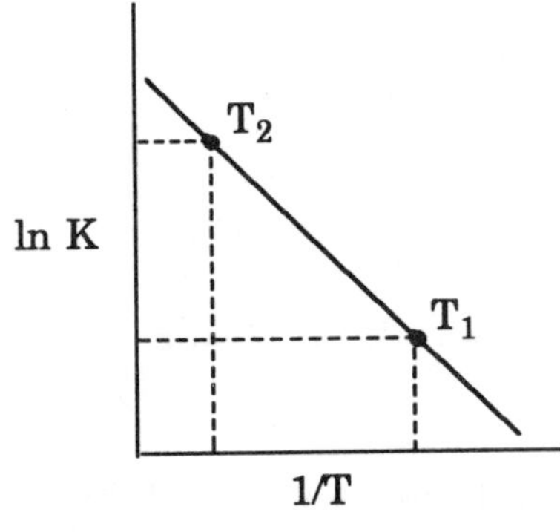

As we increase the temperature from T_1 to T_2 ($1/T_2 < 1/T_1$), we can see that ln K and, hence, K increases. A larger value of K means some more reactants can be converted to products (reaction shifts right).

Additional Exercises

67. It appears that the sum of the two processes has no net change. This is not so. By the second law of thermodynamics, ΔS_{univ} must have increased even though it looks as if we have gone through a cyclic process.

69. The introduction of mistakes is an effect of entropy. The purpose of redundant information is to provide a control to check the "correctness" of the transmitted information.

71. S (monoclinic) → S (rhombic); $\Delta H° = 0 - 0.30 = -0.30$ kJ; $\Delta S° = 31.73 - 32.55 = -0.82$ J/K

At the conversion temperature: $\Delta G° = 0$ so $\Delta H° = T\Delta S°$; $T = \frac{\Delta H°}{\Delta S°} = \frac{-3.0 \times 10^2 \text{ J}}{-0.82 \text{ J/K}} = 370$ K

73. $\Delta G°_{CO} = -RT \ln K_{CO}$ and $\Delta G°_{O_2} = -RT \ln K_{O_2}$

$\Delta G°_{CO} - \Delta G°_{O_2} = -RT \ln K_{CO} + RT \ln K_{O_2} = -RT \ln (K_{CO}/K_{O_2})$

$\Delta G°_{CO} - \Delta G°_{O_2} = -(8.3145 \text{ J/K•mol})(298 \text{ K}) \ln 210 = -13{,}000 \text{ J/mol} = -13 \text{ kJ/mol}$

75. From our answer to Exercise 16.63, $\Delta G° = 79.9$ kJ/mol at 25°C.

$$\Delta G = \Delta G° + RT \ln ([H^+][OH^-]) = 79.9 \text{ kJ/mol} + \frac{(8.3145 \text{ J/K•mol})(298 \text{ K})}{1000 \text{ J/kJ}} \ln ([H^+][OH^-])$$

a. $\Delta G = 79.9 + (8.3145 \times 298/1000) \ln [(1.00 \times 10^{-7}) \times (1.00 \times 10^{-7})] = 79.9 - 79.9 = 0$, At equilibrium.

b. $\Delta G = 0$, At equilibrium

c. $\Delta G = -34$ kJ, Shifts right

d. $\Delta G = 45.7$ kJ, Shifts left

e. $\Delta G = \Delta G° = 79.9$ kJ, Shifts left

Le Chatelier's principle gives the same results.

77. a. $\Delta G° = -RT \ln K$, $K = \exp(-\Delta G°/RT) = \exp\left(\frac{-(-30{,}500 \text{ J})}{8.3145 \text{ J/K•mol} \times 298 \text{ K}}\right) = 2.22 \times 10^5$

b. $C_6H_{12}O_6(s) + 6\ O_2(g) \rightarrow 6\ CO_2(g) + 6\ H_2O(l)$

$\Delta G° = 6 \text{ mol}(-394 \text{ kJ/mol}) + 6 \text{ mol}(-237 \text{ kJ/mol}) - 1 \text{ mol}(-911 \text{ kJ/mol}) = -2875 \text{ kJ}$

$\frac{2875 \text{ kJ}}{\text{mol glucose}} \times \frac{1 \text{ mol ATP}}{30.5 \text{ kJ}} = 94.3$ mol ATP; 94.3 molecules ATP/molecule glucose

This is an overstatement. The assumption that all of the free energy goes into this reaction is false. Actually only 38 moles of ATP are produced by metabolism of one mole of glucose.

79. The light source for the first reaction is necessary for kinetic reasons. The first reaction is just too slow to occur unless a light source is available. The kinetics of a reaction are independent of the thermodynamics of a reaction. Even though the first reaction is more favorable thermodynamically

(assuming standard conditions), it is unfavorable for kinetic reasons. The second reaction has a negative $\Delta G°$ value and is a fast reaction, so the second reaction occurs very quickly. When considering if a reaction will occur, thermodynamics and kinetics must both be considered.

Challenge Problems

81. a. From the plot, the activation energy of the reverse reaction is $E_a + (-\Delta G°) = E_a - \Delta G°$ ($\Delta G°$ is a negative number as drawn in the diagram).

$$k_f = A \exp\left(\frac{-E_a}{RT}\right) \text{ and } k_r = A \exp\left(\frac{-(E_a - \Delta G°)}{RT}\right), \quad \frac{k_f}{k_r} = \frac{A \exp\left(\frac{-E_a}{RT}\right)}{A \exp\left(\frac{-(E_a - \Delta G°)}{RT}\right)}$$

If the A factors are equal: $\frac{k_f}{k_r} = \exp\left(\frac{-E_a}{RT} + \frac{(E_a - \Delta G°)}{RT}\right) = \exp\left(\frac{-\Delta G°}{RT}\right)$

From $\Delta G° = -RT \ln K$, $K = \exp\left(\frac{-\Delta G°}{RT}\right)$; Since K and $\frac{k_f}{k_r}$ are both equal to the same expression, then $K = \frac{k_f}{k_r}$.

b. The catalyst will change the activation energy of both the forward and reverse reaction (but not change $\Delta G°$). Therefore, a catalyst must increase the rate of both the forward and reverse reactions.

83. a. $\Delta G° = G_B° - G_A° = 11{,}718 - 8996 = 2722$ J

$$K = \exp\left(\frac{-\Delta G°}{RT}\right) = \exp\left(\frac{-2722 \text{ J}}{(8.3145 \text{ J/K•mol})(298 \text{ K})}\right) = 0.333$$

b. Since Q = 1.00 > K, reaction shifts left. Let x = atm of B(g) which reacts to reach equilibrium.

	A(g)	⇌	B(g)
Initial	1.00 atm		1.00 atm
Equil.	$1.00 + x$		$1.00 - x$

$$\frac{1.00 - x}{1.00 + x} = 0.333, \quad 1.00 - x = 0.333 + 0.333\,x, \quad x = 0.50 \text{ atm}$$

$P_B = 1.00 - 0.50 = 0.50$ atm; $P_A = 1.00 + 0.50 = 1.50$ atm

c. $\Delta G = \Delta G° + RT \ln Q = \Delta G° + RT \ln (P_B/P_A)$

$\Delta G = 2722 \text{ J} + (8.3145)(298) \ln (0.50/1.50) = 2722 \text{ J} - 2722 \text{ J} = 0$ (carrying extra sig. figs.)

85. $K_p = P_{CO_2}$; To insure Ag_2CO_3 from decomposing, P_{CO_2} should be greater than K_p.

From Exercise 16.65, $\ln K = \frac{-\Delta H°}{RT} + \frac{\Delta S°}{R}$. For two conditions of K and T, the equation is:

$$\ln \frac{K_2}{K_1} = \frac{\Delta H°}{R}\left(\frac{1}{T_1} - \frac{1}{T_2}\right)$$

Let $T_1 = 25°C = 298$ K, $K_1 = 6.23 \times 10^{-3}$ torr; $T_2 = 110.°C = 383$ K, $K_2 = ?$

$$\ln \frac{K_2}{6.23 \times 10^{-3}\ \text{torr}} = \frac{79.14 \times 10^3\ \text{J/mol}}{8.3145\ \text{J/K•mol}}\left(\frac{1}{298\ \text{K}} - \frac{1}{383\ \text{K}}\right)$$

$$\ln \frac{K_2}{6.23 \times 10^{-3}} = 7.1,\quad \frac{K_2}{6.23 \times 10^{-3}} = e^{7.1} = 1.2 \times 10^3,\ K_2 = 7.5\ \text{torr}$$

To prevent decomposition of Ag_2CO_3, the partial pressure of CO_2 should be greater than 7.5 torr.

CHAPTER SEVENTEEN

ELECTROCHEMISTRY

Review of Oxidation - Reduction Reactions

13. Oxidation: increase in oxidation number; loss of electrons

    Reduction: decrease in oxidation number; gain of electrons

15. The species oxidized shows an increase in oxidation numbers and is called the reducing agent. The species reduced shows a decrease in oxidation numbers and is the oxidizing agent. The pertinent oxidation numbers are listed by the substance oxidized and the substance reduced.

	Redox?	Ox. Agent	Red. Agent	Substance Oxidized	Substance Reduced
a.	Yes	H_2O	CH_4	CH_4 (C, -4 → +2)	H_2O (H, +1 → 0)
b.	Yes	$AgNO_3$	Cu	Cu (0 → +2)	$AgNO_3$ (Ag, +1 → 0)
c.	Yes	HCl	Zn	Zn (0 → +2)	HCl (H, +1 → 0)
d.	No; There is no change in any oxidation numbers.				

Questions

17. In a galvanic cell, a spontaneous redox reaction occurs which produces an electric current. In an electrolytic cell, electricity is used to force a redox reaction to occur that is not spontaneous.

19. a. Cathode: The electrode at which reduction occurs.

    b. Anode: The electrode at which oxidation occurs.

    c. Oxidation half-reaction: The half-reaction in which electrons are products. In a galvanic cell, the oxidation half-reaction always occurs at the anode.

    d. Reduction half-reaction: The half-reaction in which electrons are reactants. In a galvanic cell, the reduction half-reaction always occurs at the cathode.

21. As a battery discharges, E_{cell} decreases, eventually reaching zero. A charged battery is not at equilibrium. At equilibrium $E_{cell} = 0$ and $\Delta G = 0$. We get no work out of an equilibrium system. A battery is useful to us because it can do work as it approaches equilibrium.

23. Both fuel cells and batteries are galvanic cells. However, fuel cells, unlike batteries, have the reactants continuously supplied and can produce a current indefinitely.

Galvanic Cells, Cell Potentials, Standard Reduction Potentials, and Free Energy

25. A typical galvanic cell diagram is:

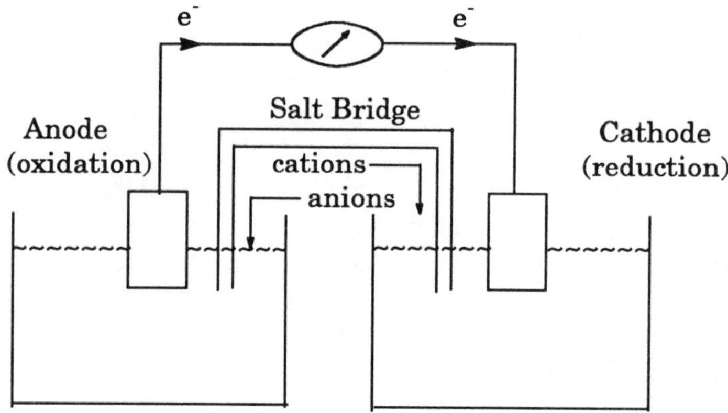

The diagram for all cells will look like this. The contents of each half-cell compartment will be identified for each reaction, with all solute concentrations at 1.0 *M* and all gases at 1.0 atm. For Exercise 17.25 and 17.29, the flow of ions through the salt bridge was not asked for in the questions. If asked, however, cations always flow into the cathode compartment and anions always flow into the anode compartment. This is required to keep each compartment electrically neutral.

a. Table 17.1 of the text lists balanced reduction half-reactions for many substances. For this overall reaction, we need the Cl_2 to Cl^- reduction half-reaction and the Cr^{3+} to $Cr_2O_7^{2-}$ oxidation half-reaction. Manipulating these two half-reactions gives the overall balanced equation.

$$(Cl_2 + 2\,e^- \rightarrow 2\,Cl^-) \times 3$$
$$7\,H_2O + 2\,Cr^{3+} \rightarrow Cr_2O_7^{2-} + 14\,H^+ + 6\,e^-$$

$$7\,H_2O(l) + 2\,Cr^{3+}(aq) + 3\,Cl_2(g) \rightarrow Cr_2O_7^{2-}(aq) + 6\,Cl^-(aq) + 14\,H^+(aq)$$

The contents of each compartment is:

Cathode: Pt electrode; Cl_2 bubbled into solution, Cl^- in solution

Anode: Pt electrode; Cr^{3+}, H^+, and $Cr_2O_7^{2-}$ in solution

We need a nonreactive metal to use as the electrode in each case, since all of the reactants and products are in solution. Pt is a common choice. Another possibility is graphite.

b. $Cu^{2+} + 2\,e^- \rightarrow Cu$
$Mg \rightarrow Mg^{2+} + 2e^-$

$Cu^{2+}(aq) + Mg(s) \rightarrow Cu(s) + Mg^{2+}(aq)$

Cathode: Cu electrode; Cu^{2+} in solution

Anode: Mg electrode; Mg^{2+} in solution

27. To determine E° for the overall cell reaction, we must add the standard reduction potential to the standard oxidation potential ($E^\circ_{cell} = E^\circ_{red} + E^\circ_{ox}$). Reference Table 17.1 for values of standard reduction potentials. Remember that $E^\circ_{ox} = -E^\circ_{red}$ and that standard potentials are <u>not</u> multiplied by the integer used to obtain the overall balanced equation.

25a. $E^\circ_{cell} = E^\circ_{Cl_2 \rightarrow Cl^-} + E^\circ_{Cr^{3+} \rightarrow Cr_2O_7^{2-}} = 1.36\ V + (-1.33\ V) = 0.03\ V$

25b. $E^\circ_{cell} = E^\circ_{Cu^{2+} \rightarrow Cu} + E^\circ_{Mg \rightarrow Mg^{2+}} = 0.34\ V + 2.37\ V = 2.71\ V$

29. Reference Exercise 17.25 for a typical galvanic cell design. The contents of each half-cell compartment is identified below with all solute concentrations at 1.0 *M* and all gases at 1.0 atm. For each pair of half-reactions, the half-reaction with the largest standard reduction potential will be the cathode reaction and the half-reaction with the smallest reduction potential will be reversed to become the anode reaction. Only this combination gives a spontaneous overall reaction, i.e., a reaction with a positive overall standard cell potential.

a. $Cl_2 + 2\,e^- \rightarrow 2\,Cl^-$ $E^\circ = 1.36\ V$
$2\,Br^- \rightarrow Br_2 + 2\,e^-$ $-E^\circ = -1.09\ V$

$Cl_2(g) + 2\,Br^-(aq) \rightarrow Br_2(aq) + 2\,Cl^-(aq)$ $E^\circ_{cell} = 0.27\ V$

The contents of each compartment is:

Cathode: Pt electrode; $Cl_2(g)$ bubbled in, Cl^- in solution

Anode: Pt electrode; Br_2 and Br^- in solution

b. $(2\,e^- + 2\,H^+ + IO_4^- \rightarrow IO_3^- + H_2O) \times 5$ $E^\circ = 1.60\ V$
$(4\,H_2O + Mn^{2+} \rightarrow MnO_4^- + 8\,H^+ + 5\,e^-) \times 2$ $-E^\circ = -1.51\ V$

$10\,H^+ + 5\,IO_4^- + 8\,H_2O + 2\,Mn^{2+} \rightarrow 5\,IO_3^- + 5\,H_2O + 2\,MnO_4^- + 16\,H^+$ $E^\circ_{cell} = 0.09\ V$

This simplifies to:

$3\,H_2O(l) + 5\,IO_4^-(aq) + 2\,Mn^{2+}(aq) \rightarrow 5\,IO_3^-(aq) + 2\,MnO_4^-(aq) + 6\,H^+(aq)$ $E^\circ_{cell} = 0.09\ V$

Cathode: Pt electrode; IO_4^-, IO_3^-, and H_2SO_4 (as a source of H^+) in solution

Anode: Pt electrode; Mn^{2+}, MnO_4^- and H_2SO_4 in solution

31. In standard line notation, the anode is listed first and the cathode is listed last. A double line separates the two compartments. By convention, the electrodes are on the ends with all solutes and gases towards the middle. A single line is used to indicate a phase change. We also included all concentrations.

25a. Pt | Cr^{3+} (1.0 *M*), $Cr_2O_7^{2-}$ (1.0 *M*), H^+ (1.0 *M*) || Cl_2 (1.0 atm)| Cl^- (1.0 *M*) | Pt

25b. Mg | Mg^{2+} (1.0 *M*) || Cu^{2+} (1.0 *M*) | Cu

29a. Pt | Br^- (1.0 *M*), Br_2 (1.0 *M*) || Cl_2 (1.0 atm) | Cl^- (1.0 *M*) | Pt

29b. Pt | Mn^{2+} (1.0 *M*), MnO_4^- (1.0 *M*), H^+ (1.0 *M*) || IO_4^- (1.0 *M*), IO_3^- (1.0 *M*), H^+ (1.0 *M*) | Pt

33. a.

$$Pb^{2+} + 2\,e^- \rightarrow Pb \qquad E° = -0.13\ V$$
$$Fe \rightarrow Fe^{2+} + 2\,e^- \qquad -E° = 0.44\ V$$
$$Pb^{2+}(aq) + Fe(s) \rightarrow Pb(s) + Fe^{2+}(aq) \qquad E°_{cell} = 0.31\ V$$

b.

$$(MnO_4^- + 4\,H^+ + 3\,e^- \rightarrow MnO_2 + 2\,H_2O) \times 2 \qquad E° = 1.68\ V$$
$$(Mn^{2+} + 2\,H_2O \rightarrow MnO_2 + 4\,H^+ + 2\,e^-) \times 3 \qquad -E° = -1.21\ V$$
$$2\,MnO_4^-(aq) + 3\,Mn^{2+}(aq) + 2\,H_2O(l) \rightarrow 5\,MnO_2(s) + 4\,H^+(aq) \qquad E°_{cell} = 0.47\ V$$

35. a.

$$2\,Ag^+ + 2\,e^- \rightarrow 2\,Ag \qquad E° = 0.80\ V$$
$$Cu \rightarrow Cu^{2+} + 2\,e^- \qquad -E° = -0.34\ V$$
$$2\,Ag^+ + Cu \rightarrow Cu^{2+} + 2\,Ag \qquad E°_{cell} = 0.46\ V$$

Spontaneous at standard conditions ($E°_{cell} > 0$).

b.

$$Zn^{2+} + 2\,e^- \rightarrow Zn \qquad E° = -0.76\ V$$
$$Ni \rightarrow Ni^{2+} + 2\,e^- \qquad -E° = 0.23\ V$$
$$Zn^{2+} + Ni \rightarrow Zn + Ni^{2+} \qquad E°_{cell} = -0.53\ V$$

Not spontaneous at standard conditions ($E°_{cell} < 0$).

37. $\Delta G° = -nFE°_{cell}$; Reference Exercises 17.25, 17.27 and 17.29 for balanced reactions and standard cell potentials. The balanced reactions are necessary to determine n, the moles of electrons transferred.

25a. $7\,H_2O + 2\,Cr^{3+} + 3\,Cl_2 \rightarrow Cr_2O_7^{2-} + 6\,Cl^- + 14\,H^+ \quad E°_{cell} = 0.03\ V = 0.03\ J/C$

$$\Delta G° = -nFE°_{cell} = -(6\ mol\ e^-)(96{,}485\ C/mol\ e^-)(0.03\ J/C) = -1.7 \times 10^4\ J = -20\ kJ$$

25b. $\Delta G° = -(2\ mol\ e^-)(96{,}485\ C/mol\ e^-)(2.71\ J/C) = -5.23 \times 10^5\ J = -523\ kJ$

29a. $\Delta G° = -(2\ mol\ e^-)(96{,}485\ C/mol\ e^-)(0.27\ J/C) = -5.21 \times 10^4\ J = -52\ kJ$

29b. $\Delta G° = -(10\ mol\ e^-)(96{,}485\ C/mol\ e^-)(0.09\ J/C) = -8.7 \times 10^4\ J = -90\ kJ$

39. $Cl_2 + 2\,e^- \rightarrow 2\,Cl^-$ $\qquad E° = 1.36$ V

$(ClO_2^- \rightarrow ClO_2 + e^-) \times 2$ $\qquad -E° = -0.954$ V

$2\,ClO_2^-(aq) + Cl_2(g) \rightarrow 2\,ClO_2(aq) + 2\,Cl^-(aq)$ $\qquad E°_{cell} = 0.41$ V = 0.41 J/C

$\Delta G° = -nFE°_{cell} = -(2 \text{ mol } e^-)(96{,}485 \text{ C/mol } e^-)(0.41 \text{ J/C}) = -7.91 \times 10^4 \text{ J} = -79 \text{ kJ}$

41. Since the cells are at standard conditions, then $w_{max} = \Delta G = \Delta G° = -nFE°_{cell}$. See Exercise 17.33 for the balanced overall equations and for $E°_{cell}$.

33a. $w_{max} = -(2 \text{ mol } e^-)(96{,}485 \text{ C/mol } e^-)(0.31 \text{ J/C}) = -6.0 \times 10^4 \text{ J} = -60. \text{ kJ}$

33b. $w_{max} = -(6 \text{ mol } e^-)(96{,}485 \text{ C/mol } e^-)(0.47 \text{ J/C}) = -2.7 \times 10^5 \text{ J} = -270 \text{ kJ}$

43. $2\,H_2O + 2\,e^- \rightarrow H_2 + 2\,OH^-$ $\quad \Delta G° = \Sigma n_p \Delta G°_{f,\text{ products}} - \Sigma n_r \Delta G°_{f,\text{ reactants}} = 2(-157) - [2(-237)] = 160. \text{ kJ}$

$$\Delta G° = -nFE°,\ E° = \frac{-\Delta G°}{nF} = \frac{-1.60 \times 10^5 \text{ J}}{(2 \text{ mol } e^-)(96{,}485 \text{ C/mol } e^-)} = -0.829 \text{ J/C} = -0.829 \text{ V}$$

The two values agree (-0.83 V in Table 17.1).

45. $\Delta G° = -nFE° = -(1 \text{ mol } e^-)(96{,}485 \text{ C/mol } e^-)(0.80 \text{ J/C})(1 \text{ kJ}/1000 \text{ J}) = -77 \text{ kJ}$

$-77 \text{ kJ} = \Delta G°_{f,\,Ag} - [\Delta G°_{f,\,Ag^+} + \Delta G°_{f,\,e^-}],\ -77 \text{ kJ} = 0 - [\Delta G°_{f,\,Ag^+} + 0],\ \Delta G°_{f,\,Ag^+} = 77 \text{ kJ/mol}$

47. Good oxidizing agents are easily reduced. Oxidizing agents are on the left side of the reduction half-reactions listed in Table 17.1. We look for the largest, most positive standard reduction potentials to correspond to the best oxidizing agents. The ordering from worst to best oxidizing agents is:

	Mg^{2+} <	Fe^{2+} <	Fe^{3+} <	$Cr_2O_7^{2-}$ <	Cl_2 <	MnO_4^-
E°(V)	-2.37	-0.44	0.77	1.33	1.36	1.68

49. a. $2\,H^+ + 2\,e^- \rightarrow H_2$ $\quad E° = 0.00$ V; $Cu \rightarrow Cu^{2+} + 2\,e^-$ $\quad -E° = -0.34$ V

$E°_{cell} = -0.34$ V; No, H^+ cannot oxidize Cu to Cu^{2+} at standard conditions ($E°_{cell} < 0$).

b. $2\,H^+ + 2\,e^- \rightarrow H_2$ $\quad E° = 0.00$ V; $Mg \rightarrow Mg^{2+} + 2\,e^-$ $\quad -E° = 2.37$ V

$E°_{cell} = 2.37$ V; Yes, H^+ can oxidize Mg to Mg^{2+} at standard conditions ($E°_{cell} > 0$).

c. $Fe^{3+} + e^- \rightarrow Fe^{2+}$ $\quad E° = 0.77$ V; $2\,I^- \rightarrow I_2 + 2\,e^-$ $\quad -E° = -0.54$ V

$E°_{cell} = 0.77 - 0.54 = 0.23$ V; Yes, Fe^{3+} can oxidize I^- to I_2.

d. $Fe^{3+} + e^- \rightarrow Fe^{2+}$ $\quad E° = 0.77$ V; $2\,Br^- \rightarrow Br_2 + 2\,e^-$ $\quad -E° = -1.09$ V

$E°_{cell} = 0.77 - 1.09 = -0.32$ V; No, Fe^{3+} cannot oxidize Br^- to Br_2.

51. $Br_2 + 2\,e^- \rightarrow 2\,Br^-$ $E° = 1.09$ V $La^{3+} + 3\,e^- \rightarrow La$ $E° = -2.37$ V
$2\,H^+ + 2\,e^- \rightarrow H_2$ $E° = 0.00$ V $Ca^{2+} + 2\,e^- \rightarrow Ca$ $E° = -2.76$ V
$Cd^{2+} + 2\,e^- \rightarrow Cd$ $E° = -0.40$ V

a. Oxidizing agents are on the left side of the above reduction half-reactions. Br_2 is the best oxidizing agent (largest E°).

b. Reducing agents are on the right side of the reduction half-reactions. Ca is the best reducing agent (largest -E°).

c. $MnO_4^- + 8\,H^+ + 5\,e^- \rightarrow Mn^{2+} + 4\,H_2O$ $E° = 1.51$ V; Permanganate can oxidize Br^-, H_2, Cd and Ca at standard conditions. When MnO_4^- is coupled with these reagents, $E°_{cell}$ is positive. Note: La is not one of the choices given in the question or it would have been included.

d. $Zn \rightarrow Zn^{2+} + 2\,e^-$ $-E° = 0.76$ V; Zinc can reduce Br_2 and H^+ since $E°_{cell} > 0$.

53. a. $2\,Br^- \rightarrow Br_2 + 2\,e^-$ $-E° = -1.09$ V; $2\,Cl^- \rightarrow Cl_2 + 2\,e^-$ $-E° = -1.36$ V; $E° > 1.09$ V to oxidize Br^-; $E° < 1.36$ V to not oxidize Cl^-; $Cr_2O_7^{2-}$, O_2, MnO_2, and IO_3^- are all possible since when all of these oxidizing agents are coupled with Br^- give $E°_{cell} > 0$ and when coupled with Cl^- give $E°_{cell} < 0$ (assuming standard conditions).

b. $Mn \rightarrow Mn^{2+} + 2\,e^-$ $-E° = 1.18$; $Ni \rightarrow Ni^{2+} + 2\,e^-$ $-E° = 0.23$ V; Any oxidizing agent with $-0.23\text{ V} > E° > -1.18$ V will work. $PbSO_4$, Cd^{2+}, Fe^{2+}, Cr^{3+}, Zn^{2+} and H_2O will be able to oxidize Mn but not oxidize Ni (assuming standard conditions).

The Nernst Equation

55.
$$H_2O_2 + 2\,H^+ + 2\,e^- \rightarrow 2\,H_2O \qquad E° = 1.78\text{ V}$$
$$(Ag \rightarrow Ag^+ + e^-) \times 2 \qquad -E° = -0.80\text{ V}$$
$$H_2O_2(aq) + 2\,H^+(aq) + 2\,Ag(s) \rightarrow 2\,H_2O(l) + 2\,Ag^+(aq) \qquad E°_{cell} = 0.98\text{ V}$$

a. A galvanic cell is based on spontaneous chemical reactions. At standard conditions, this reaction produces a voltage of 0.98 V. Any change in concentration that increases the tendency of the forward reaction to occur will increase the cell potential. Conversely, any change in concentration that decreases the tendency of the forward reaction to occur (increases the tendency of the reverse reaction to occur) will decrease the cell potential. Using Le Chatelier's principle, increasing the reactant concentrations of H_2O_2 and H^+ from 1.0 *M* to 2.0 *M* will drive the forward reaction further to right (will further increase the tendency of the forward reaction to occur). Therefore, E_{cell} will be greater than $E°_{cell}$.

b. Here, we decreased the reactant concentration of H^+ and increased the product concentration of Ag^+ from the standard conditions. This decreases the tendency of the forward reaction to occur which will decrease E_{cell} as compared to $E°_{cell}$ ($E_{cell} < E°_{cell}$).

57. At 25°C, $E_{cell} = E^\circ_{cell} - \frac{0.0592}{n} \log Q$ where $Q = [Ag^+]^2 / [H_2O_2][H^+]^2$; See Exercise 17.55 for the overall balanced equation and for E°_{cell}.

a. $E_{cell} = 0.98\text{ V} - \frac{0.0592}{2} \log \frac{(1.0)^2}{(2.0)(2.0)^2} = 0.98\text{ V} - (-0.027\text{ V}) = 1.01\text{ V}$

b. $E_{cell} = 0.98\text{ V} - \frac{0.0592}{2} \log \frac{(2.0)^2}{(1.0)(1.0 \times 10^{-7})^2} = 0.98\text{ V} - 0.43\text{ V} = 0.55\text{ V}$

59. For concentration cells, the driving force for the reaction is the difference in ion concentrations between the anode and cathode. In order to equalize the ion concentrations, the anode always has the smaller ion concentration. The general set-up for this concentration cell is:

Cathode:	$Ag^+(x\,M) + e^- \rightarrow Ag$	$E^\circ = 0.80\text{ V}$
Anode:	$Ag \rightarrow Ag^+(y\,M) + e^-$	$-E^\circ = -0.80\text{ V}$
	$Ag^+(\text{cathode}, x\,M) \rightarrow Ag^+(\text{anode}, y\,M)$	$E^\circ_{cell} = 0.00\text{ V}$

$$E_{cell} = E^\circ_{cell} - \frac{0.0592}{n} \log Q = \frac{-0.0592}{1} \log \frac{[Ag^+]_{anode}}{[Ag^+]_{cathode}}$$

For each concentration cell, we will calculate the cell potential using the above equation. Remember that the anode always has the smaller ion concentration.

a. Since both compartments are at standard conditions ($[Ag^+] = 1.0\,M$) then $E_{cell} = E^\circ_{cell} = 0$ V. No voltage is produced since no reaction occurs. Concentration cells only produce a voltage when the ion concentrations are not equal.

b. Cathode = 2.0 M Ag^+; Anode = 1.0 M Ag^+; Electron flow is always from the anode to the cathode so electrons flow to the right in the diagram.

$$E_{cell} = \frac{-0.0592}{n} \log \frac{[Ag^+]_{anode}}{[Ag^+]_{cathode}} = \frac{-0.0592}{1} \log \frac{1.0}{2.0} = 0.018\text{ V}$$

c. Cathode = 1.0 M Ag^+; Anode = 0.10 M Ag^+; Electrons flow to the left in the diagram.

$$E_{cell} = \frac{-0.0592}{n} \log \frac{[Ag^+]_{anode}}{[Ag^+]_{cathode}} = \frac{-0.0592}{1} \log \frac{0.10}{1.0} = 0.059\text{ V}$$

d. Cathode = 1.0 M Ag^+; Anode = 4.0×10^{-5} M Ag^+; Electrons flow to the left in the diagram.

$$E_{cell} = \frac{-0.0592}{1} \log \frac{4.0 \times 10^{-5}}{1.0} = 0.26\text{ V}$$

e. Since the ion concentrations are the same, then $\log([Ag^+]_{anode}/[Ag^+]_{cathode}) = \log(1.0) = 0$ and $E_{cell} = 0$. No electron flow occurs.

61.

$$5\ e^- + 8\ H^+ + MnO_4^- \rightarrow Mn^{2+} + 4\ H_2O \qquad E° = 1.51\ V$$
$$(Fe^{2+} \rightarrow Fe^{3+} + e^-) \times 5 \qquad -E° = -0.77\ V$$

$$8\ H^+(aq) + MnO_4^-(aq) + 5\ Fe^{2+}(aq) \rightarrow 5\ Fe^{3+}(aq) + Mn^{2+}(aq) + 4\ H_2O(l) \qquad E°_{cell} = 0.74\ V$$

$$E_{cell} = E°_{cell} - \frac{0.0592}{n} \log Q = 0.74\ V - \frac{0.0592}{5} \log \frac{[Fe^{3+}]^5\,[Mn^{2+}]}{[Fe^{2+}]^5\,[MnO_4^-]\,[H^+]^8}$$; pH = 4.0 so $H^+ = 1 \times 10^{-4}\ M$

$$E_{cell} = 0.74 - \frac{0.0592}{5} \log \frac{(1 \times 10^{-6})^5\,(1 \times 10^{-6})}{(1 \times 10^{-3})^5\,(1 \times 10^{-2})\,(1 \times 10^{-4})^8}$$

$$E_{cell} = 0.74 - \frac{0.0592}{5} \log (1 \times 10^{13}) = 0.74\ V - 0.15\ V = 0.59\ V = 0.6\ V$$ (1 sig. fig. due to concentrations)

Yes, $E_{cell} > 0$ so reaction will occur as written.

63.

$$Cu^{2+} + 2\ e^- \rightarrow Cu \qquad E° = 0.34\ V$$
$$Zn \rightarrow Zn^{2+} + 2\ e^- \qquad -E° = 0.76\ V$$

$$Cu^{2+}(aq) + Zn(s) \rightarrow Zn^{2+}(aq) + Cu(s) \qquad E°_{cell} = 1.10\ V$$

Since Zn^{2+} is a product in the reaction, then the Zn^{2+} concentration increases from 1.00 *M* to 1.20 *M*. This means that the reactant concentration of Cu^{2+} must decrease from 1.00 *M* to 0.80 *M* (from the 1:1 mol ratio in the balanced reaction).

$$E_{cell} = E°_{cell} - \frac{0.0592}{n} \log Q = 1.10\ V - \frac{0.0592}{2} \log \frac{[Zn^{2+}]}{[Cu^{2+}]}$$

$$E_{cell} = 1.10\ V - \frac{0.0592}{2} \log \frac{1.20}{0.80} = 1.10\ V - 0.0052\ V = 1.09\ V$$

65. $Cu^{2+}(aq) + H_2(g) \rightarrow 2\ H^+(aq) + Cu(s)$ $\quad E°_{cell} = 0.34\ V - 0.00V = 0.34\ V$ and n = 2

Since $P_{H_2} = 1.0$ atm and $[H^+] = 1.0\ M$: $E_{cell} = E°_{cell} - \frac{0.0592}{2} \log \frac{1}{[Cu^{2+}]}$

a. $E_{cell} = 0.34\ V - \frac{0.0592}{2} \log \frac{1}{2.5 \times 10^{-4}} = 0.34\ V - 0.11 = 0.23\ V$

b. $0.195\ V = 0.34\ V - \frac{0.0592}{2} \log \frac{1}{[Cu^{2+}]}$, $\log \frac{1}{[Cu^{2+}]} = 4.90$, $[Cu^{2+}] = 10^{-4.90} = 1.3 \times 10^{-5}\ M$

Note: When determining exponents, we will carry extra significant figures.

67. $Cu^{2+}(aq) + H_2(g) \rightarrow 2\ H^+(aq) + Cu(s)$ $\quad E°_{cell} = 0.34\ V - 0.00\ V = 0.34\ V$ and n = 2

Since $P_{H_2} = 1.0$ atm and $[H^+] = 1.0\ M$: $E_{cell} = E°_{cell} - \frac{0.0592}{2} \log \frac{1}{[Cu^{2+}]}$

Use the K_{sp} expression to calculate the Cu^{2+} concentration in the cell.

$Cu(OH)_2(s) \rightleftharpoons Cu^{2+}(aq) + 2\,OH^-(aq)$ $K_{sp} = 1.6 \times 10^{-19} = [Cu^{2+}][OH^-]^2$

From problem, $[OH^-] = 0.10\,M$, so: $[Cu^{2+}] = \dfrac{1.6 \times 10^{-19}}{(0.10)^2} = 1.6 \times 10^{-17}\,M$

$$E_{cell} = E^\circ_{cell} - \frac{0.0592}{2}\log\frac{1}{[Cu^{2+}]} = 0.34\text{ V} - \frac{0.0592}{2}\log\frac{1}{1.6 \times 10^{-17}} = 0.34 - 0.50 = -0.16\text{ V}$$

Since $E_{cell} < 0$, then the forward reaction is not spontaneous, but the reverse reaction is spontaneous. The Cu electrode becomes the anode and $E_{cell} = 0.16$ V for the reverse reaction. The cell reaction is: $2\,H^+(aq) + Cu(s) \rightarrow Cu^{2+}(aq) + H_2(g)$.

69. See Exercises 17.25, 17.27 and 17.29 for balanced reactions and standard cell potentials. Balanced reactions are necessary to determine n, the moles of electrons transferred.

25a. $7\,H_2O + 2\,Cr^{3+} + 3\,Cl_2 \rightarrow Cr_2O_7^{2-} + 6\,Cl^- + 14\,H^+$ $E^\circ_{cell} = 0.03\text{ V} = 0.03\text{ J/C}$

$E_{cell} = E^\circ_{cell} - \dfrac{0.0592}{n}\log Q$: At equilibrium, $E_{cell} = 0$ and $Q = K$, so $E^\circ_{cell} = \dfrac{0.0592}{n}\log K$

$$\log K = \frac{nE^\circ}{0.0592} = \frac{6(0.03)}{0.0592} = 3.04,\ K = 10^{3.04} = 1 \times 10^3$$

Note: When determining exponents, we will round off after the calculation is complete in order to help eliminate excessive round off error.

25b. $\log K = \dfrac{2(2.71)}{0.0592} = 91.554,\ K = 3.58 \times 10^{91}$

29a. $\log K = \dfrac{2(0.27)}{0.0592} = 9.12,\ K = 1.3 \times 10^9$

29b. $\log K = \dfrac{10(0.09)}{0.0592} = 15.20,\ K = 2 \times 10^{15}$

71. a.

$2\,e^- + 2\,H_2O \rightarrow 2\,OH^- + H_2$	$E^\circ = -0.83$ V
$(Na \rightarrow Na^+ + e^-) \times 2$	$-E^\circ = 2.71$ V
$2\,Na(s) + 2\,H_2O(l) \rightarrow 2\,NaOH(aq) + H_2(g)$	$E^\circ_{cell} = 1.88\text{ V} = 1.88\text{ J/C}$

$\Delta G^\circ = -nFE^\circ_{cell} = -(2\text{ mol }e^-)(96{,}485\text{ C/mol }e^-)(1.88\text{ J/C}) = -3.63 \times 10^5\text{ J} = -363\text{ kJ}$

From the Nernst equation: $E^\circ_{cell} = \dfrac{0.0592}{n}\log K,\ \log K = \dfrac{nE^\circ}{0.0592}$

$$\log K = \frac{2(1.88)}{0.0592} = 63.514,\ K = 10^{63.514} = 3.27 \times 10^{63}$$

Note: When determing exponents, we will round off after the calculation is complete.

b.

$$(2\ H^+ + 2\ e^- \rightarrow H_2) \times 3/2 \qquad E° = 0.00\ V$$
$$Al \rightarrow Al^{3+} + 3\ e^- \qquad -E° = 1.66\ V$$

$$3\ H^+(aq) + Al(s) \rightarrow Al^{3+}(aq) + 3/2\ H_2(g) \qquad E°_{cell} = 1.66\ V$$

$$\Delta G° = -(3\ mol\ e^-)(96{,}485\ C/mol\ e^-)(1.66\ J/C) = -4.80 \times 10^5\ J = -480.\ kJ$$

$$\log K = \frac{3(1.66)}{0.0592} = 84.122,\ K = 1.32 \times 10^{84}$$

c.

$$Br_2 + 2\ e^- \rightarrow 2\ Br^- \qquad E° = 1.09\ V$$
$$2\ I^- \rightarrow I_2 + 2\ e^- \qquad -E° = -0.54V$$

$$Br_2(aq) + 2\ I^-(aq) \rightarrow 2\ Br^-(aq) + I_2(s) \qquad E°_{cell} = 0.55\ V$$

$$\Delta G° = -(2\ mol\ e^-)(96{,}485\ C/mol\ e^-)(0.55\ J/C) = -1.1 \times 10^5\ J = -110\ kJ$$

$$\log K = \frac{2(0.55)}{0.0592} = 18.58,\ K = 3.8 \times 10^{18}$$

73. a.

$$Fe^{2+} + 2\ e^- \rightarrow Fe \qquad E° = -0.44\ V$$
$$Zn \rightarrow Zn^{2+} + 2\ e^- \qquad -E° = 0.76\ V$$

$$Fe^{2+}(aq) + Zn(s) \rightarrow Zn^{2+}(aq) + Fe(s) \qquad E°_{cell} = 0.32\ V = 0.32\ J/C$$

b. $\Delta G° = -nFE°_{cell} = -(2\ mol\ e^-)(96{,}485\ C/mol\ e^-)(0.32\ J/C) = -6.2 \times 10^4\ J = -62\ kJ$

$$E°_{cell} = \frac{0.0592}{n}\log K,\ \log K = \frac{nE°}{0.0592} = \frac{2(0.32)}{0.0592} = 10.81,\ K = 10^{10.81} = 6.5 \times 10^{10}$$

c. $E_{cell} = E°_{cell} - \frac{0.0592}{n}\log Q = 0.32\ V - \frac{0.0592}{2}\log\frac{[Zn^{2+}]}{[Fe^{2+}]}$

$$E_{cell} = 0.32 - \frac{0.0592}{2}\log\frac{0.10}{1.0 \times 10^{-5}} = 0.32 - 0.12 = 0.20\ V$$

75.

$$PbSO_4(s) + 2\ e^- \rightarrow Pb + SO_4^{2-} \qquad E° = -0.35\ V$$
$$Pb \rightarrow Pb^{2+} + 2\ e^- \qquad -E° = 0.13\ V$$

$$PbSO_4\ (s) \rightarrow Pb^{2+}(aq) + SO_4^{2-}(aq) \qquad E°_{cell} = -0.22\ V$$

For this overall reaction $K = K_{sp}$ so $E°_{cell} = \frac{0.0592}{n}\log K_{sp}$.

$$\log K_{sp} = \frac{nE°}{0.0592} = \frac{2(-0.22)}{0.0592} = -7.43,\quad K_{sp} = 10^{-7.43} = 3.7 \times 10^{-8}$$

77. $e^- + AgI \rightarrow Ag + I^-$ $\quad E^\circ_{AgI} = ?$

$Ag \rightarrow Ag^+ + e^-$ $\quad -E^\circ = -0.80\text{ V}$

$AgI(s) \rightarrow Ag^+(aq) + I^-(aq)$ $\quad E^\circ_{cell} = E^\circ_{AgI} - 0.80$ $\quad K = K_{sp} = 1.5 \times 10^{-16}$

For this overall reaction:

$$E^\circ_{cell} = \frac{0.0592}{n}\log K_{sp} = \frac{0.0592}{1}\log(1.5 \times 10^{-16}) = -0.94\text{ V}$$

$$E^\circ_{cell} = -0.94\text{ V} = E^\circ_{AgI} - 0.80\text{ V},\ E^\circ_{AgI} = -0.94 + 0.80 = -0.14\text{ V}$$

Electrolysis

79. a. $Al^{3+} + 3\,e^- \rightarrow Al$; 3 mol e^- are needed to produce 1 mol Al from Al^{3+}.

$$1.0 \times 10^3\text{ g Al} \times \frac{1\text{ mol Al}}{26.98\text{ g Al}} \times \frac{3\text{ mol e}^-}{\text{mol Al}} \times \frac{96{,}485\text{ C}}{\text{mol e}^-} \times \frac{1\text{ s}}{100.0\text{ C}} = 1.07 \times 10^5\text{ s} = 30.\text{ hours}$$

b. $$1.0\text{ g Ni} \times \frac{1\text{ mol Ni}}{58.69\text{ g Ni}} \times \frac{2\text{ mol e}^-}{\text{mol Ni}} \times \frac{96{,}485\text{ C}}{\text{mol e}^-} \times \frac{1\text{ s}}{100.0\text{ C}} = 33\text{ s}$$

c. $$5.0\text{ mol Ag} \times \frac{1\text{ mol e}^-}{\text{mol Ag}} \times \frac{96{,}485\text{ C}}{\text{mol e}^-} \times \frac{1\text{ s}}{100.0\text{ C}} = 4.8 \times 10^3\text{ s} = 1.3\text{ hours}$$

81. $$15\text{A} = \frac{15\text{ C}}{\text{s}} \times \frac{60\text{ s}}{\text{min}} \times \frac{60\text{ min}}{\text{h}} = 5.4 \times 10^4\text{ C of charge passed in 1 hour}$$

a. $$5.4 \times 10^4\text{ C} \times \frac{1\text{ mol e}^-}{96{,}485\text{ C}} \times \frac{1\text{ mol Co}}{2\text{ mol e}^-} \times \frac{58.93\text{ g Co}}{\text{mol Co}} = 16\text{ g Co}$$

b. $$5.4 \times 10^4\text{ C} \times \frac{1\text{ mol e}^-}{96{,}485\text{ C}} \times \frac{1\text{ mol Hf}}{4\text{ mol e}^-} \times \frac{178.5\text{ g Hf}}{\text{mol Hf}} = 25\text{ g Hf}$$

c. $2\,I^- \rightarrow I_2 + 2\,e^-$; $$5.4 \times 10^4\text{ C} \times \frac{1\text{ mol e}^-}{96{,}485\text{ C}} \times \frac{1\text{ mol I}_2}{2\text{ mol e}^-} \times \frac{253.8\text{ g I}_2}{\text{mol I}_2} = 71\text{ g I}_2$$

d. $CrO_3(l) \rightarrow Cr^{6+} + 3\,O^{2-}$; 6 mol e^- are needed to produce 1 mol Cr from molten CrO_3.

$$5.4 \times 10^4\text{ C} \times \frac{1\text{ mol e}^-}{96{,}485\text{ C}} \times \frac{1\text{ mol Cr}}{6\text{ mol e}^-} \times \frac{52.00\text{ g Cr}}{\text{mol Cr}} = 4.9\text{ g Cr}$$

83. $$600.\text{ s} \times \frac{5.00\text{ C}}{\text{s}} \times \frac{1\text{ mol e}^-}{96{,}485\text{ C}} \times \frac{1\text{ mol M}}{3\text{ mol e}^-} = 1.04 \times 10^{-2}\text{ mol M where M = unknown metal}$$

Atomic mass = $\dfrac{1.19\text{ g M}}{1.04 \times 10^{-2}\text{ mol M}} = \dfrac{114\text{ g}}{\text{mol}}$; The element is indium, In. Indium forms 3+ ions.

85. F_2 is produced at the anode: $2\ F^- \rightarrow F_2 + 2\ e^-$

$$2.00\ h \times \frac{60\ min}{h} \times \frac{60\ s}{min} \times \frac{10.0\ C}{s} \times \frac{1\ mol\ e^-}{96{,}485\ C} = 0.746\ mol\ e^-$$

$$0.746\ mol\ e^- \times \frac{1\ mol\ F_2}{2\ mol\ e^-} = 0.373\ mol\ F_2;\ \ PV = nRT,\ \ V = \frac{nRT}{P}$$

$$V = \frac{(0.373\ mol)\ (0.08206\ L{\bullet}atm/K{\bullet}mol)\ (298\ K)}{1.00\ atm} = 9.12\ L\ F_2$$

K is produced at the cathode: $K^+ + e^- \rightarrow K$

$$0.746\ mol\ e^- \times \frac{1\ mol\ K}{mol\ e^-} \times \frac{39.10\ g\ K}{mol\ K} = 29.2\ g\ K$$

87. $$\frac{150. \times 10^3\ g\ C_6H_8N_2}{h} \times \frac{1\ h}{60\ min} \times \frac{1\ min}{60\ s} \times \frac{1\ mol\ C_6H_8N_2}{108.14\ g\ C_6H_8N_2} \times \frac{2\ mol\ e^-}{mol\ C_6H_8N_2} \times \frac{96{,}485\ C}{mol\ e^-}$$

$$= 7.44 \times 10^4\ C/s \text{ or a current of } 7.44 \times 10^4\ A$$

89. $$2.30\ min \times \frac{60\ s}{min} = 138\ s;\ \ 138\ s \times \frac{2.00\ C}{s} \times \frac{1\ mol\ e^-}{96{,}485\ C} \times \frac{1\ mol\ Ag}{mol\ e^-} = 2.86 \times 10^{-3}\ mol\ Ag$$

$[Ag^+] = 2.86 \times 10^{-3}\ mol\ Ag^+/0.250\ L = 1.14 \times 10^{-2}\ M$

91. The metal ion with the most positive reduction potential will plate out first at the cathode. In this case, Pt will plate out first.

93. Reduction occurs at the cathode and oxidation occurs at the anode. First determine all the species present, then look up pertinent reduction and/or oxidation potentials in Table 17.1 for all these species. The cathode reaction will be the reaction with the most positive reduction potential and the anode reaction will be the reaction with the most positive oxidation potential.

a. Species present: K^+ and F^-; K^+ can be reduced to K and F^- can be oxidized to F_2 (from Table 17.1). The reactions are:

Cathode: $K^+ + e^- \rightarrow K$ $E° = -2.92\ V$
Anode: $2\ F^- \rightarrow F_2 + 2\ e^-$ $-E° = -2.87\ V$

b. Species present: Cu^{2+} and Cl^-; Cu^{2+} can be reduced and Cl^- can be oxidized. The reactions are:

Cathode: $Cu^{2+} + 2\ e^- \rightarrow Cu$ $E° = 0.34\ V$
Anode: $2\ Cl^- \rightarrow Cl_2 + 2\ e^-$ $-E° = -1.36\ V$

c. Species present: Mg^{2+} and I^-; Mg^{2+} can be reduced and I^- can be oxidized. The reactions are:

Cathode: $Mg^{2+} + 2\ e^- \rightarrow Mg$ $E° = -2.37\ V$
Anode: $2\ I^- \rightarrow I_2 + 2\ e^-$ $-E° = -0.54\ V$

95. Species present: Na^+, SO_4^{2-} and H_2O. From the potentials, H_2O is the more easily reduced than Na^+ and H_2O is more easily oxidized than SO_4^{2-}. The reactions are:

Cathode: $2\ H_2O + 2\ e^- \rightarrow H_2 + 2\ OH^-$; Anode: $2\ H_2O \rightarrow O_2 + 4\ H^+ + 4\ e^-$

97.
$$(Au(CN)_4^- + 3e^- \rightarrow Au + 4\ CN^-) \times 4$$
$$(4\ OH^- \rightarrow O_2 + 2\ H_2O + 4\ e^-) \times 3$$

$$4\ Au(CN)_4^-(aq) + 12\ OH^-(aq) \rightarrow 4\ Au(s) + 16\ CN^-(aq) + 3\ O_2(g) + 6\ H_2O(l)$$

$$1.00 \times 10^3\ g\ Au \times \frac{1\ mol\ Au}{197.0\ g} \times \frac{3\ mol\ O_2}{4\ mol\ Au} = 3.81\ mol\ O_2$$

$$V = \frac{nRT}{P} = \frac{3.81\ mol\left(\frac{0.08206\ L\ atm}{mol\ K}\right)(298\ K)}{740.\ torr \times \frac{1\ atm}{760\ torr}} = 95.7\ L\ O_2$$

Additional Exercises

99. The half-reaction for the SCE is:

$Hg_2Cl_2 + 2\ e^- \rightarrow 2\ Hg + 2\ Cl^-$ $\qquad E_{SCE} = 0.242\ V$

For a spontaneous reaction to occur, E_{cell} must be positive. Using the standard reduction potentials in Table 17.1 and the given SCE potential, deduce which combination will produce a positive overall cell potential.

a. $Cu^{2+} + 2\ e^- \rightarrow Cu$ $\qquad E° = 0.34\ V$

$E_{cell} = 0.34 - 0.242 = 0.10\ V$; SCE is the anode.

b. $Fe^{3+} + e^- \rightarrow Fe^{2+}$ $\qquad E° = 0.77\ V$

$E_{cell} = 0.77 - 0.242 = 0.53\ V$; SCE is the anode.

c. $AgCl + e^- \rightarrow Ag + Cl^-$ $\qquad E° = 0.22\ V$

$E_{cell} = 0.242 - 0.22 = 0.02\ V$; SCE is the cathode.

d. $Al^{3+} + 3\ e^- \rightarrow Al$ $\qquad E° = -1.66\ V$

$E_{cell} = 0.242 + 1.66 = 1.90\ V$; SCE is the cathode.

e. $Ni^{2+} + 2\ e^- \rightarrow Ni$ $\qquad E° = -0.23\ V$

$E_{cell} = 0.242 + 0.23 = 0.47\ V$; SCE is the cathode.

101. a. Possible reaction: $I_2(s) + 2\ Cl^-(aq) \rightarrow 2\ I^-(aq) + Cl_2(g)$ $E^\circ_{cell} = 0.54\ V - 1.36\ V = -0.82\ V$
This reaction is not spontaneous at standard conditions. No reaction occurs.

b. Possible reaction: $Cl_2(g) + 2\ I^-(aq) \rightarrow I_2(s) + 2\ Cl^-(aq)$ $E^\circ_{cell} = 0.82\ V$; This reaction is spontaneous at standard conditions. The reaction will occur.

c. Possible reaction: $2\ Ag(s) + Cu^{2+}(aq) \rightarrow Cu(s) + 2\ Ag^+(aq)$ $E^\circ_{cell} = -0.46\ V$; No reaction occurs.

d. Fe^{2+} can be oxidized or reduced, all of which have negative potentials. The other species present are H^+, SO_4^{2-}, H_2O, and O_2 from air. Only O_2 in the presence of H^+ has a large enough standard reduction potential to oxidize Fe^{2+} to Fe^{3+}. All other combinations, including the possible reduction of Fe^{2+}, give negative cell potentials. The reaction is: $4\ Fe^{2+}(aq) + 4\ H^+(aq) + O_2(g) \rightarrow 4\ Fe^{3+}(aq) + 2\ H_2O(l)$. $E^\circ_{cell} = 1.23 - 0.77 = 0.46\ V$; Spontaneous

103.

$$Al^{3+} + 3\ e^- \rightarrow Al \qquad E^\circ = -1.66\ V$$
$$Al + 6\ F^- \rightarrow AlF_6^{3-} + 3\ e^- \qquad -E^\circ = 2.07\ V$$

$$Al^{3+}(aq) + 6\ F^-(aq) \rightarrow AlF_6^{3-}(aq) \qquad E^\circ_{cell} = 0.41\ V \qquad K = ?$$

$$\log K = \frac{nE^\circ}{0.0592} = \frac{3(0.41)}{0.0592} = 20.78,\ K = 10^{20.78} = 6.0 \times 10^{20}$$

105. In a simplified view of the corrosion process, the half-reactions are:

Cathode: $O_2 + 2\ H_2O + 4\ e^- \rightarrow 4\ OH^-$
Anode: $(Fe \rightarrow Fe^{2+} + 2\ e^-) \times 2$

$$O_2(g) + 2\ H_2O(l) + 2\ Fe(s) \rightarrow 2\ Fe^{2+}(aq) + 4\ OH^-(aq)$$

Since OH^- is a product in this reaction, then added H^+ will react with OH^- to form H_2O. As OH^- is removed, this increases the tendency for the forward reaction to occur. Hence, corrosion is a greater problem under acidic conditions in our simplified view of this process.

107. Consider the strongest oxidizing agent combined with the strongest reducing agent from Table 17.1:

$$F_2 + 2\ e^- \rightarrow 2\ F^- \qquad E^\circ = 2.87\ V$$
$$(Li \rightarrow Li^+ + e^-) \times 2 \qquad -E^\circ = 3.05\ V$$

$$F_2(g) + 2\ Li(s) \rightarrow 2\ Li^+(aq) + 2\ F^-(aq) \qquad E^\circ_{cell} = 5.92\ V$$

The claim is impossible. The strongest oxidizing agent and reducing agent when combined only give E°_{cell} of about 6 V.

109. a.

$$O_2 + 2\ H_2O + 4\ e^- \rightarrow 4\ OH^- \qquad E^\circ = 0.40\ V$$
$$(H_2 + 2\ OH^- \rightarrow 2\ H_2O + 2\ e^-) \times 2 \qquad -E^\circ = 0.83\ V$$

$$2\ H_2(g) + O_2(g) \rightarrow 2\ H_2O(l) \qquad E^\circ_{cell} = 1.23\ V = 1.23\ J/C$$

Since standard conditions are assumed, then $w_{max} = \Delta G°$ for 2 mol H_2O produced.

$\Delta G° = -nFE°_{cell} = -(4 \text{ mol } e^-)(96{,}485 \text{ C/mol } e^-)(1.23 \text{ J/C}) = -475{,}000 \text{ J} = -475 \text{ kJ}$

For 1.00×10^3 g H_2O produced, w_{max} is:

$$1.00 \times 10^3 \text{ g } H_2O \times \frac{1 \text{ mol } H_2O}{18.02 \text{ g } H_2O} \times \frac{-475 \text{ kJ}}{2 \text{ mol } H_2O} = -13{,}200 \text{ kJ} = w_{max}$$

The work done can be no larger than the free energy change. The best that could happen is that all of the free energy released goes into doing work, but this does not occur in any real process since there is always waste energy in any real process. Fuel cells are more efficient in converting chemical energy into electrical energy; they are also less massive. The major disadvantage is that they are expensive. In addition, $H_2(g)$ and $O_2(g)$ are an explositve mixture, if ignited; much more so than fossil fuels.

111. If the metal, M, forms +1 ions, then the atomic mass of M would be:

$$\text{mol M} = 150. \text{ s} \times \frac{1.25 \text{ C}}{\text{s}} \times \frac{1 \text{ mol } e^-}{96{,}485 \text{ C}} \times \frac{1 \text{ mol M}}{1 \text{ mol } e^-} = 1.94 \times 10^{-3} \text{ mol M}$$

$$\text{Atomic mass of M} = \frac{0.109 \text{ g M}}{1.94 \times 10^{-3} \text{ mol M}} = 56.2 \text{ g/mol}$$

From the periodic table, the only metal with an atomic mass close to 56.2 g/mol is iron; but iron does not form stable +1 ions. If M forms +2 ions, then the atomic mass would be:

$$\text{mol M} = 150. \text{ s} \times \frac{1.25 \text{ C}}{\text{s}} \times \frac{1 \text{ mol } e^-}{96{,}485 \text{ C}} \times \frac{1 \text{ mol M}}{2 \text{ mol } e^-} = 9.72 \times 10^{-4} \text{ mol M}$$

$$\text{Atomic mass of M} = \frac{0.109 \text{ g M}}{9.72 \times 10^{-4} \text{ mol M}} = 112 \text{ g/mol}$$

Cadmium has an atomic mass of 112.4 g/mol and does form stable +2 ions. Cd^{2+} is a much more logical choice than Fe^+.

Challenge Problems

113. $\Delta G° = -nFE° = \Delta H° - T\Delta S°, \; E° = \dfrac{T\Delta S°}{nF} - \dfrac{\Delta H°}{nF}$

If we graph E° vs. T we should get a straight line (y = mx + b). The slope of the line is equal to $\Delta S°/nF$ and the y-intercept is equal to $-\Delta H°/nF$. From the equation above, E° will have a small temperature dependence when $\Delta S°$ is close to zero.

115. a. $\Delta G° = \Sigma n_p \Delta G°_{f,\text{ products}} - \Sigma n_r \Delta G°_{f,\text{ reactants}} = 2(-480.) + 3(86) - [3(-40.)] = -582 \text{ kJ}$

From oxidation numbers, n = 6. $\Delta G° = -nFE°, \; E° = \dfrac{-\Delta G°}{nF} = \dfrac{-(-582{,}000 \text{ J})}{6(96{,}485) \text{ C}} = 1.01 \text{ V}$

$$\log K = \frac{nE°}{0.0592} = \frac{6(1.01)}{0.0592} = 102.365,\ K = 10^{102.365} = 2.32 \times 10^{102}$$

b. $3 \times (2\ e^- + Ag_2S \rightarrow 2\ Ag + S^{2-})$ $\quad E°_{Ag_2S} = ?$

$2 \times (Al \rightarrow Al^{3+} + 3\ e^-)$ $\quad -E° = 1.66$ V

$3\ Ag_2S(s) + 2\ Al(s) \rightarrow 6\ Ag(s) + 3\ S^{2-}(aq) + 2\ Al^{3+}(aq)$ $\quad E°_{cell} = 1.01\ V = E°_{Ag_2S} + 1.66$

$E°_{Ag_2S} = 1.01 - 1.66 = -0.65$ V

117. $Pb^{2+} + 2\ e^- \rightarrow Pb$ $\quad E° = -0.13$ V

$Zn \rightarrow Zn^{2+} + 2\ e^-$ $\quad -E° = 0.76$ V

$Pb^{2+}(aq) + Zn(s) \rightarrow Zn^{2+}(aq) + Pb(s)$ $\quad E°_{cell} = 0.63$ V

$Zn(OH)_2(s) \rightleftharpoons Zn^{2+}(aq) + 2\ OH^-(aq)$ $\ K_{sp} = [Zn^{2+}]\,[OH^-]^2$; We must determine $[Zn^{2+}]$ in order to calculate K_{sp}. Using the Nernst equation for this cell:

$$E_{cell} = E°_{cell} - \frac{0.0592}{n}\log Q,\quad E_{cell} = 1.05\ V = 0.63\ V - \frac{0.0592}{2}\log\frac{[Zn^{2+}]}{[Pb^{2+}]}$$

$$1.05\ V = 0.63\ V - \frac{0.0592}{2}\log\frac{[Zn^{2+}]}{1.0},\ \log[Zn^{2+}] = -14.19,\ [Zn^{2+}] = 10^{-14.19} = 6.5 \times 10^{-15}\ M$$

$$K_{sp} = [Zn^{2+}]\,[OH^-]^2 = (6.5 \times 10^{-15})(0.10)^2 = 6.5 \times 10^{-17}$$

119. a. $E_{cell} = E_{ref} + 0.05916\ pH,\ 0.480\ V = 0.250\ V + 0.05916\ pH$

$$pH = \frac{0.480 - 0.250}{0.05916} = 3.888;\ \text{Uncertainty} = \pm 1\ mV = \pm 0.001\ V$$

$$pH_{max} = \frac{0.481 - 0.250}{0.05916} = 3.905;\quad pH_{min} = \frac{0.479 - 0.250}{0.05916} = 3.871$$

So if the uncertainty in potential is ± 0.001 V, then the uncertainty in pH is ± 0.017 or about ± 0.02 pH units. For this measurement, $[H^+] = 10^{-3.888} = 1.29 \times 10^{-4}\ M$. For an error of + 1 mV, $[H^+] = 10^{-3.905} = 1.24 \times 10^{-4}\ M$. For an error of -1 mV, $[H^+] = 10^{-3.871} = 1.35 \times 10^{-4}\ M$. So the uncertainty in $[H^+]$ is $\pm 0.06 \times 10^{-4}\ M = \pm 6 \times 10^{-6}\ M$.

b. From part a, we will be within ± 0.02 pH units if we measure the potential to the nearest ± 0.001 V (1 mV).

CHAPTER EIGHTEEN

THE REPRESENTATIVE ELEMENTS: GROUPS 1A THROUGH 4A

Questions

1. The gravity of the earth is not strong enough to keep H_2 in the atmosphere.

3. Ionic, covalent, and metallic (or interstitial); The ionic and covalent hydrides are true compounds obeying the law of definite proportions and differ from each other in the type of bonding. The interstitial hydrides are more like solid solutions of hydrogen with a transition metal, and do not obey the law of definite proportions.

5. Hydrogen forms many compounds in which the oxidation state is +1, as do the Group 1A elements. For example H_2SO_4 and HCl compared to Na_2SO_4 and NaCl. On the other hand, hydrogen forms diatomic H_2 molecules and is a nonmetal, while the Group 1A elements are metals. Hydrogen also forms compounds with a -1 oxidation state, which is not characteristic of Group 1A metals, e.g., NaH.

7. Group IA and IIA metals are all easily oxidized. They must be produced in the absence of materials (H_2O, O_2) that are capable of oxidizing them.

9. In graphite, planes of carbon atoms slide easily along each other. In addition, graphite is not volatile. The lubricant will not be lost when used in a high vacuum environment.

11. The bonds in SnX_4 compounds have a large covalent character. SnX_4 acts as discrete molecules held together by weak dispersion forces. SnX_2 compounds are ionic and are held in the solid state by strong ionic forces. Since the intermolecular forces are weaker for SnX_4 compounds, they are more volatile.

13. Size decreases from left to right and increases going down the periodic table. So going one element right and one element down would result in a similar size for the two elements diagonal to each other. The ionization energies will be similar for the diagonal elements since the periodic trends also oppose each other. Electron affinities are harder to predict, but atoms with similar size and ionization energy should also have similar electron affinities.

Exercises

Group 1A Elements

15. a. $\Delta H^\circ = -110.5 - [-75 + (-242)] = 207$ kJ; $\Delta S^\circ = 198 + 3(131) - [186 + 189] = 216$ J/K

b. $\Delta G° = \Delta H° - T\Delta S°$; $\Delta G° = 0$ when $T = \frac{\Delta H°}{\Delta S°} = \frac{207 \times 10^3 \text{ J}}{216 \text{ J/K}} = 958 \text{ K}$

At T > 958 K and standard pressures, the favorable $\Delta S°$ term dominates and the reaction is spontaneous ($\Delta G° < 0$).

17. sodium oxide: Na_2O; sodium superoxide: NaO_2; sodium peroxide: Na_2O_2

19. a. $Li_3N(s) + 3\ HCl(aq) \rightarrow 3\ LiCl(aq) + NH_3(aq)$

b. $Rb_2O(s) + H_2O(l) \rightarrow 2\ RbOH(aq)$

c. $Cs_2O_2(s) + 2\ H_2O(l) \rightarrow 2\ CsOH(aq) + H_2O_2(aq)$

d. $NaH(s) + H_2O(l) \rightarrow NaOH(aq) + H_2(g)$

21. $2\ Li(s) + 2\ C_2H_2(g) \rightarrow 2\ LiC_2H(s) + H_2(g)$; This is an oxidation-reduction reaction.

Group 2A Elements

23. barium oxide: BaO; barium peroxide: BaO_2

25. $Mg_3N_2(s) + 6\ H_2O(l) \rightarrow 2\ NH_3(g) + 3\ Mg^{2+}(aq) + 6\ OH^-(aq)$

$Mg_3P_2(s) + 6\ H_2O(l) \rightarrow 2\ PH_3(g) + 3\ Mg^{2+}(aq) + 6\ OH^-(aq)$

27.

H H H
N
Be
:Cl: :Cl:

Geometry is trigonal planar about Be.

Be uses sp^2 hybrid orbitals.

N uses sp^3 hybrid orbitals.

$BeCl_2$ is a Lewis acid.

29. $\frac{1 \text{ mg F}^-}{\text{L}} \times \frac{1 \text{ g}}{1000 \text{ mg}} \times \frac{1 \text{ mol F}^-}{19.00 \text{ g F}^-} = 5.3 \times 10^{-5}\ M\ F^- = 5 \times 10^{-5}\ M\ F^-$

$CaF_2(s) \rightleftharpoons Ca^{2+}(aq) + 2\ F^-(aq)$ $K_{sp} = [Ca^{2+}][F^-]^2 = 4.0 \times 10^{-11}$; Precipitation will occur when $Q > K_{sp}$. Lets calculate $[Ca^{2+}]$ so that $Q = K_{sp}$.

$Q = 4.0 \times 10^{-11} = [Ca^{2+}]_o[F^-]_o^2 = [Ca^{2+}]_o(5 \times 10^{-5})^2$, $[Ca^{2+}] = 2 \times 10^{-2}\ M$

$CaF_2(s)$ will precipitate when $[Ca^{2+}]_o > 2 \times 10^{-2}\ M$. Therefore, hard water should have a calcium ion concentration of less than $2 \times 10^{-2}\ M$ to avoid precipitate formation.

31. $Ca^{2+} + 2\ e^- \rightarrow Ca$; $1.00 \times 10^3\ kg \times \frac{1000\ g}{kg} \times \frac{1\ mol\ Ca}{40.08\ g} \times \frac{2\ mol\ e^-}{mol\ Ca} \times \frac{96{,}485\ C}{mol\ e^-} = 4.81 \times 10^9\ C$

$$\frac{4.81 \times 10^9\ C}{8.00\ h} \times \frac{1\ h}{60\ min} \times \frac{1\ min}{60\ s} = \frac{1.67 \times 10^5\ C}{s} = 1.67 \times 10^5\ A;\ Ca^{2+} + 2\ Cl^- \rightarrow Ca + Cl_2$$

$$1.00 \times 10^6\ g\ Ca \times \frac{1\ mol\ Ca}{40.08\ g} \times \frac{1\ mol\ Cl_2}{mol\ Ca} \times \frac{70.90\ g\ Cl_2}{mol\ Cl_2} = 1.77 \times 10^6\ g\ of\ Cl_2 = 1.77 \times 10^3\ kg\ Cl_2$$

Group 3A Elements

33. a. thallium(I) hydroxide b. indium(III) sulfide

c. Gallium(III) oxide, but is commonly called gallium oxide.

35. $B_2H_6 + 3\ O_2 \rightarrow 2\ B(OH)_3$

37. $Ga_2O_3(s) + 6\ H^+(aq) \rightarrow 2\ Ga^{3+}(aq) + 3\ H_2O(l)$

$Ga_2O_3(s) + 2\ OH^-(aq) + 3\ H_2O(l) \rightarrow 2\ Ga(OH)_4^-(aq)$

$In_2O_3(s) + 6\ H^+(aq) \rightarrow 2\ In^{3+}(aq) + 3\ H_2O(l)$; $In_2O_3(s) + OH^-(aq) \rightarrow$ no reaction

39. $2\ In(s) + 3\ F_2(g) \rightarrow 2\ InF_3(s)$; $2\ In(s) + 3\ Cl_2(g) \rightarrow 2\ InCl_3(s)$; $4\ In(s) + 3\ O_2(g) \rightarrow 2\ In_2O_3(s)$

$2\ In(s) + 6\ HCl(aq) \rightarrow 3\ H_2(g) + 2\ InCl_3(aq)$ or $2\ In(s) + 6\ HCl(g) \rightarrow 3\ H_2(g) + 2\ InCl_3(s)$

Group 4A Elements

41. a. Linear about all carbons b. sp

43. a. $K_2SiF_6(s) + 4\ K(l) \rightarrow 6\ KF(s) + Si(s)$

b. K_2SiF_6 is an ionic compound, composed of K^+ cations and SiF_6^{2-} anions. The SiF_6^{2-} anion is held together by covalent bonds. The structure is:

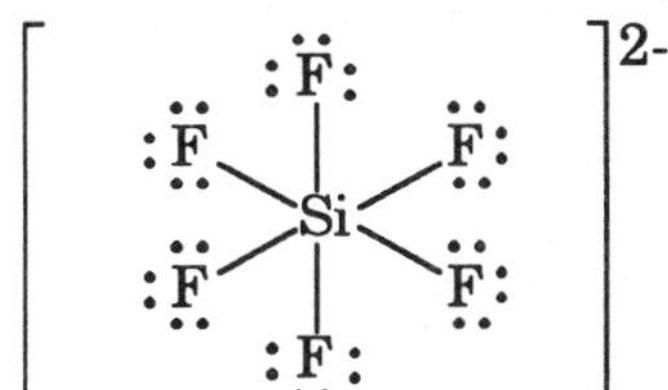

The anion is octahedral with Si utilizing d^2sp^3 hybrid orbitals to form the Si–F bonds.

45. Lead is very toxic. As the temperature of the water increases, the solubility of lead will increase. Drinking hot tap water from pipes containing lead solder could result in higher lead concentrations in the body.

47. tin(II) fluoride

49. The π electrons are free to move in graphite, thus giving it a greater conductivity (lower resistance). The electrons have the greatest mobility within sheets of carbon atoms, resulting in a lower resistance in the basal plane. Electrons in diamond are not mobile (high resistance). The structure of diamond is uniform in all directions; thus, there is no directional dependence of the resistivity.

Additional Exercises

51. $K^+(out) \rightarrow K^+(in)$; $E = E° - \frac{0.0592}{1} \log\left(\frac{[K^+]_{in}}{[K^+]_{out}}\right)$; E° = 0 for this reaction

$$E = -0.0592 \log\left(\frac{0.15}{5.0 \times 10^{-3}}\right) = -0.087 \text{ V}$$

ΔG = work = -nFE = -(1 mol e⁻)(96,485 C/mol e⁻)(-0.087 J/C) = 8400 J = 8.4 kJ = work

53. Strontium and calcium are both alkaline earth metals so both have similar chemical properties. Since milk is a good source of calcium, strontium could replace some calcium in milk without much difficulty.

55. Assuming $AlCl_3$ is covalent, then a single $AlCl_3$ has the Lewis structure:

```
 ..     ..
:Cl—Al—Cl:
 ''  |  ''
    :Cl:
     ''
```

The Al has room for a pair of electrons. It can act as a Lewis acid. The only lone pairs are on Cl, so the dimer structure is:

```
 ..      ..      ..
:Cl     :Cl:     Cl:
 '' \   /   \   / ''
     Al       Al
 .. /   \   /   \ ..
:Cl     :Cl:     Cl:
 ''      ''      ''
```

57. $LiAlH_4$: H, -1 (metal hydride); Li, +1; Al, $0 = +1 + x + 4(-1)$, $x = +3$ = oxidation state of Al

59. Ga(I): $[Ar]4s^2 3d^{10}$, no unpaired e^-; Ga(III): $[Ar]3d^{10}$, no unpaired e^-

Ga(II): $[Ar]4s^1 3d^{10}$, 1 unpaired e^-; Note: s electrons are lost before the d electrons.

If the compound contained Ga(II) it would be paramagnetic and if the compound contained Ga(I) and Ga(III) it would be diamagnetic. This can easily be determined by measuring the mass of a sample in the presence and in the absence of a magnetic field. Paramagnetic compounds will have an apparent greater mass in a magnetic field.

61.

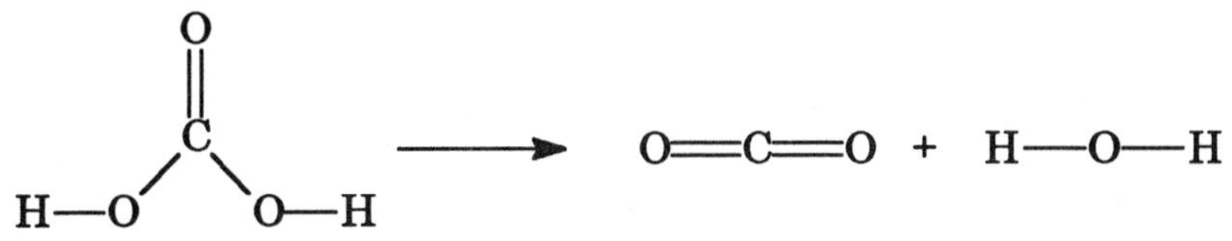

O
||
C
H—O O—H ⟶ O═C═O + H—O—H

Bonds broken: 2 C − O (358 kJ/mol)
1 O − H (467 kJ/mol)

Bonds formed: 1 C ═ O (799 kJ/mol)
1 O − H (467 kJ/mol)

$\Delta H = 2(358) + 467 - (799 + 467) = -83$ kJ; ΔH is favorable for the decomposition of H_2CO_3 to CO_2 and H_2O. ΔS is also favorable for the decomposition as there is an increase in disorder. Hence, H_2CO_3 will spontaneously decompose to CO_2 and H_2O.

63. From Table 18.14 in the text, Sn and Pb can reduce H^+ to H_2.

$Sn(s) + 2\ H^+(aq) \rightarrow Sn^{2+}(aq) + H_2(g)$; $Pb(s) + 2\ H^+(aq) \rightarrow Pb^{2+}(aq) + H_2(g)$

Challenge Problems

65. Na^+ can oxidize Na^- to Na. The purpose of the cryptand is to encapsulate the Na^+ ion so that it does not come in contact with the Na^- ion and oxidize it to sodium metal.

67. Carbon cannot form the fifth bond necessary for the transition state since carbon doesn't have low energy d orbitals available to expand the octet.

CHAPTER NINETEEN

THE REPRESENTATIVE ELEMENTS: GROUPS 5A THROUGH 8A

Questions

1. The reaction $N_2(g) + 3\ H_2(g) \rightarrow 2\ NH_3(g)$ is exothermic. Thus, the equilibrium constant K decreases as the temperature increases. Lower temperatures are favored for maximum yield of ammonia. However, at lower temperatures the rate is slow; without a catalyst the rate is too slow for the process to be feasible. The discovery of a catalyst increased the rate of reaction at a lower temperature favored by thermodynamics.

3. The pollution provides nitrogen and phosphorous nutrients so the algae can grow. The algae consume oxygen, causing fish to die.

5. Plastic sulfur consists of long S_n chains of sulfur atoms. As plastic sulfur becomes brittle, the long chains break down into S_8 rings.

7. Fluorine is the most reactive of the halogens because it is the most electronegative atom and the bond in the F_2 molecule is very weak.

9. Helium is unreactive and doesn't combine with any other elements. It is a very light gas and would easily escape the earth's gravitational pull as the planet was formed.

Exercises

Group 5A Elements

11. NO_4^{3-}

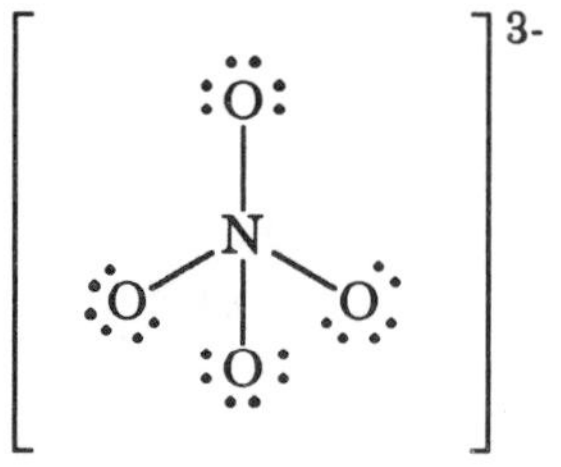

Both NO_4^{3-} and PO_4^{3-} have 32 valence electrons so both have similar Lewis structures. From the Lewis structure for NO_4^{3-}, the central N atom has a tetrahedral arrangement of electron pairs. N is small. There is probably not enough room for all 4 oxygen atoms around N. P is larger, thus, PO_4^{3-} is stable.

PO_3^-

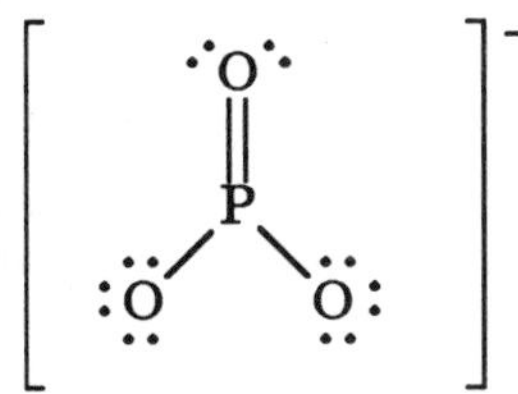

PO_3^- and NO_3^- both have 24 valence electrons so both have similar Lewis structures. From the Lewis structure for PO_3^-, PO_3^- has a trigonal arrangement of electron pairs about the central P atom (two single bonds and one double bond). P = O bonds are not particularly stable, while N = O bonds are stable. Thus, NO_3^- is stable.

13. a. $8\ H^+(aq) + 2\ NO_3^-(aq) + 3\ Cu(s) \rightarrow 3\ Cu^{2+}(aq) + 4\ H_2O(l) + 2\ NO(g)$

b. $NH_4NO_3(s) \xrightarrow{Heat} N_2O(g) + 2\ H_2O(g)$

c. $NO(g) + NO_2(g) + 2\ KOH(aq) \rightarrow 2\ KNO_2(aq) + H_2O(l)$

15. NH_3: sp^3; N_2H_4: both Ns are sp^3; NH_2OH: sp^3; N_2: sp; N_2O: central N, sp

NO: sp^2; N_2O_3: both Ns are sp^2; NO_2: sp^2; HNO_3: sp^2

17. $27.37 \times 10^9\ \text{lb}\ NH_3 \times \frac{1\ \text{kg}}{2.2046\ \text{lb}} \times \frac{1000\ \text{g}}{1\ \text{kg}} \times \frac{1\ \text{mol}\ NH_3}{17.03\ \text{g}\ NH_3} = 7.290 \times 10^{11}\ \text{mol}\ NH_3$

$\text{mol}\ N_2 = 7.290 \times 10^{11}\ \text{mol}\ NH_3 \times \frac{1\ \text{mol}\ N_2}{2\ \text{mol}\ NH_3} = 3.645 \times 10^{11}\ \text{mol}\ N_2$; STP = 273.15 K and 1.000 atm

$$V_{N_2} = \frac{nRT}{P} = \frac{3.645 \times 10^{11}\ \text{mol} \times \frac{0.08206\ \text{L atm}}{\text{K}\bullet\text{mol}} \times 273.15\ \text{K}}{1.000\ \text{atm}} = 8.170 \times 10^{12}\ \text{L}\ N_2$$

Since T and P are constant, then moles and volume are directly proportional to each other. From the balanced reaction, there is a 3:1 mol ratio between H_2 and N_2, so the volume of H_2 required is 3 times the volume of N_2 required.

$$V_{H_2} = V_{N_2} \times \frac{3\ \text{mol}\ H_2}{1\ \text{mol}\ N_2} = 8.170 \times 10^{12}\ \text{L} \times 3 = 2.451 \times 10^{13}\ \text{L}\ H_2$$

19.

$H_2N-NH_2\ (l) + 2\ F-F\ (g) \longrightarrow 4\ H-F(g) + N\equiv N\ (g)$

Bonds broken:	Bonds formed:
1 N – N (160. kJ/mol)	4 H – F (565 kJ/mol)
4 N – H (391 kJ/mol)	1 N≡ N (941 kJ/mol)
2 F – F (154 kJ/mol)	

$\Delta H = 160. + 4(391) + 2(154) - [4(565) + 941] = 2032 \text{ kJ} - 3201 \text{ kJ} = -1169 \text{ kJ}$

21. $1/2\ N_2(g) + 1/2\ O_2(g) \rightarrow NO(g)$ $\Delta G° = \Delta G°_{f,\ NO} = 87$ kJ/mol; By definition, $\Delta G_f°$ for a compound equals the free energy change that would accompany the formation of 1 mol of that compound from its elements in their standard states. NO (and some other oxides of nitrogen) have weaker bonds as compared to the triple bond of N_2 and the double bond of O_2. Because of this, NO (and some other oxides of nitrogen) have higher (positive) free energies as compared to the relatively stable N_2 and O_2 molecules.

23. M.O model:

NO^+: $(\sigma_{2s})^2(\sigma_{2s}*)^2(\pi_{2p})^4(\sigma_{2p})^2$, Bond order = (8 - 2)/2 = 3, 0 unpaired e^- (diamagnetic)

NO: $(\sigma_{2s})^2(\sigma_{2s}*)^2(\pi_{2p})^4(\sigma_{2p})^2(\pi_{2p}*)^1$, B.O. = 2.5, 1 unpaired e^- (paramagnetic)

NO^-: $(\sigma_{2s})^2(\sigma_{2s}*)^2(\pi_{2p})^4(\sigma_{2p})^2(\pi_{2p*})^2$, B.O. = 2, 2 unpaired e^- (paramagnetic)

Lewis structures: NO^+: [:N≡O:]$^+$

NO: ·N=O: ⟷ :N=O: ⟷ ·N=O:

NO^-: [:N=O:]$^-$

The two models only give the same results for NO^+ (a triple bond with no unpaired electrons). Lewis structure are not adequate for NO and NO^-. The M.O. model gives a better representation for all three species. For NO, Lewis structures are poor for odd electron species. For NO^-, both models predict a double bond but only the MO model correctly predicts that NO^- is paramagnetic.

25. a. $H_3PO_4 > H_3PO_3$; The strongest acid has the most oxygen atoms.

b. $H_3PO_4 > H_2PO_4^- > HPO_4^{2-}$; Acid strength decreases as protons are removed.

27. To complete the Lewis structures, just add lone pairs to the terminal atoms to complete the octets. See part d for the Lewis structures (32 valence electrons).

a. Yes, both have 4 sets of electrons about the P. We would predict a tetrahedral structure for both.

b. The hybridization is sp^3 for each P since both structures are tetrahedral.

c. P has to use one of its d orbitals to form the π bond.

d. The formal charges for the O and P atoms are next to the atoms in the following Lewis structures. In both structures all Cl atoms have a formal charge of zero.

$$:\ddot{O}:^{-1} \text{ bonded to } P^{+1} \text{ with three } :\ddot{Cl}: \quad \longleftrightarrow \quad \ddot{O}^{0} = P^{0} \text{ with three } :\ddot{Cl}:$$

The structure with the P = O bond is favored on the basis of formal charge since it has a zero formal charge for all atoms in the structure.

29. Production of antimony:

$$2\ Sb_2S_3(s) + 9\ O_2(g) \rightarrow 2\ Sb_2O_3(s) + 6\ SO_2(g);\quad 2\ Sb_2O_3(s) + 3\ C(s) \rightarrow 4\ Sb(s) + 3\ CO_2(g)$$

Production of bismuth:

$$2\ Bi_2S_3(s) + 9\ O_2(g) \rightarrow 2\ Bi_2O_3(s) + 6\ SO_2(g);\quad 2\ Bi_2O_3(s) + 3\ C(s) \rightarrow 4\ Bi(s) + 3\ CO_2(g)$$

Group 6A Elements

31. $O{=}O{-}O \rightarrow O{=}O + O$

Break O – O bond: $\Delta H = \frac{146\ kJ}{mol} \times \frac{1\ mol}{6.022 \times 10^{23}} = 2.42 \times 10^{-22}\ kJ = 2.42 \times 10^{-19}\ J$

A photon of light must contain at least 2.42×10^{-19} J to break one O – O bond.

$$E_{photon} = \frac{hc}{\lambda},\ \lambda = \frac{hc}{E} = \frac{(6.626 \times 10^{-34}\ J\ s)(2.998 \times 10^{8}\ m/s)}{2.42 \times 10^{-19}\ J} = 8.21 \times 10^{-7}\ m = 821\ nm$$

33. a. oxidation - reduction reaction

b. From Table 17.1 of the text, the reduction of NO_3^- to NO has a favorable (positive) reduction potential. Thus, NO(g) is the expected product.

c.

$$(3e^- + 3\ H^+ + HNO_3 \rightarrow NO + 2\ H_2O) \times 2$$
$$(S^{2-} \rightarrow S + 2e^-) \times 3$$

$$6\ H^+(aq) + 2\ HNO_3(aq) + 3\ S^{2-}(aq) \rightarrow 3\ S(s) + 2\ NO(g) + 4\ H_2O(l)$$

35. SF_5^- has 6 + 5(7) + 1 = 42 valence electrons.

square pyramid

Group 7A Elements

37. a. ClF_5, 7 + 5(7) = 42 e^-

b. IF_3, 7 + 3(7) = 28 e^-

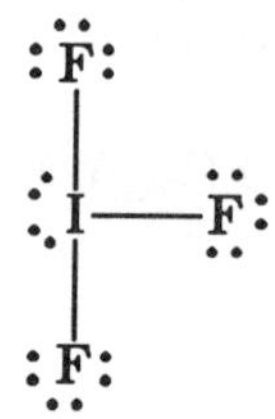

Square pyramid; d^2sp^3

T-shaped; dsp^3

c. $FBrO_2$, 7 + 7 + 2(6) = 26 e^-

Trigonal pyramid; sp^3

39. The balanced half-reactions and the overall balanced reaction for this oxidation-reduction reaction are:

$$2\,e^- + 2\,H^+ + XeF_2 \rightarrow Xe + 2\,HF$$
$$H_2O + BrO_3^- \rightarrow BrO_4^- + 2\,H^+ + 2\,e^-$$

$$H_2O(l) + BrO_3^-(aq) + XeF_2(aq) \rightarrow BrO_4^-(aq) + Xe(g) + 2\,HF(aq)$$

41. $ClO^- + H_2O + 2\,e^- \rightarrow 2\,OH^- + Cl^-$ $\quad E° = 0.90$ V

$2\,NH_3 + 2\,OH^- \rightarrow N_2H_4 + 2\,H_2O + 2\,e^-$ $\quad -E° = 0.10$ V

$ClO^-(aq) + 2\,NH_3(aq) \rightarrow Cl^-(aq) + N_2H_4(aq) + H_2O(l)$ $\quad E°_{cell} = 1.00$ V

Since E°_{cell} is positive for this reaction, then at standard conditions ClO^- can spontaneously oxidize NH_3 to the somwhat toxic N_2H_4.

Group 8A Elements

43. Xe has one more valence electron than I. Thus, the isoelectric species will have I plus one extra electron substituted for Xe, giving a species with a net minus one charge.

a. IO_4^- b. IO_3^- c. IF_2^- d. IF_4^- e. IF_6^- f. IOF_3^-

45. XeF_2 can react with oxygen to produce explosive xenon oxides and oxyfluorides.

Additional Exercises

47. As the halogen atoms get larger, it becomes more difficult to fit three halogen atoms around the small nitrogen atom, and the NX_3 molecule becomes less stable.

49. OCN^- has 6 + 4 + 5 + 1 = 16 valence electrons.

$[\,O{=}C{=}N\,]^-$ ⟷ $[\,O{-}C{\equiv}N\,]^-$ ⟷ $[\,O{\equiv}C{-}N\,]^-$

Formal charge	0	0	-1	-1	0	0	+1	0	-2

Only the first two resonance structures should be important. The third places a positive formal charge on the most electronegative atom in the ion and a -2 formal charge on N.

CNO^-:

$[\,C{=}N{=}O\,]^-$ ⟷ $[\,C{\equiv}N{-}O\,]^-$ ⟷ $[\,C{-}N{\equiv}O\,]^-$

Formal charge	-2	+1	0	-1	+1	-1	-3	+1	+1

All of the resonance structures for fulminate (CNO^-) involve greater formal charges than in cyanate (OCN^-), making fulminate more reactive (less stable).

51. Hypochlorite can act as an oxidizing agent. For example, it is capable of oxidizing I^- to I_2. If a solution containing I^- turns brown when BiOCl is added, then BiOCl is bismuth(I) hypochlorite. The brown color indicates production of I_2. If the solution doesn't change color, then it is bismuthylchloride.

53. TeF_5^- has 6 + 5(7) + 1 = 42 valence electrons.

$$\left[TeF_5 \right]^-$$

The lone pair of electrons around Te exerts a stronger repulsion than the bonding pairs, pushing the four square planar F's away from the lone pair and thus reducing the bond angles between the axial F atom and the square planar F atoms.

Challenge Problems

55. For the reaction:

$$O_2N{-}N{=}O \longrightarrow NO_2 + NO$$

the activation energy must in some way involve the breaking of a nitrogen-nitrogen single bond. For the reaction:

$$O_2N{-}N{=}O \longrightarrow O_2 + N_2O$$

at some point nitrogen-oxygen bonds must be broken. N – N single bonds (160. kJ/mol) are weaker than N – O single bonds (201 kJ/mol). In addition, resonance structures indicate that there is more double bond character in the N – O bonds than in the N – N bond. Thus, NO_2 and NO are preferred by kinetics because of the lower activation energy.

57. a. As we go down the family, K_a increases. This is consistent with the bond to hydrogen getting weaker.

b. Po is below Te, so K_a should be larger. The K_a value for H_2Po should be on the order of 10^{-2}.

CHAPTER TWENTY

TRANSITION METALS AND COORDINATION CHEMISTRY

Questions

5. a. Ligand: Species that donates a pair of electrons to form a covalent bond to a metal ion. Ligands act as Lewis bases (electron pair donors).

   b. Chelate: Ligand that can form more than one bond to a metal ion.

   c. Bidentate: Ligand that forms two bonds to a metal ion.

   d. Complex ion: Metal ion plus ligands.

7. Both electrons in the bond originally came from one of the atoms in the bond.

9. a. Isomers: Species with the same formulas but different properties. See text for examples of the following types of isomers.

   b. Structural isomers: Isomers that have one or more bonds that are different.

   c. Stereoisomers: Isomers that contain the same bonds but differ in how the atoms are arranged in space.

   d. Coordination isomers: Structural isomers that differ in the atoms that make up the complex ion.

   e. Linkage isomers: Structural isomers that differ in how one or more ligands are attached to the transition metal.

   f. Geometric isomers: (Cis - trans isomerism); Stereoisomers that differ in the positions of atoms with respect to a rigid ring, bond, or each other.

   g. Optical isomers: Stereoisomers that are nonsuperimposable mirror images of each other; that is, they are different in the same way that our left and right hands are different.

11. Cu^{2+}: $[Ar]3d^9$; Cu^+: $[Ar]3d^{10}$; Cu(II) is d^9 and Cu(I) is d^{10}. Color is a result of the electron transfer between split d orbitals. This cannot occur for the filled d orbitals in Cu(I). Cd^{2+}, like Cu^+, is also d^{10}. We would not expect $Cd(NH_3)_4Cl_2$ to be colored since the d orbitals are filled in this Cd^{2+} complex.

13. The d-orbital splitting in tetrahedral complexes is less than one-half the d-orbital splitting in octahedral complexes. There are no known ligands powerful enough to produce the strong-field case, hence all tetrahedral complexes are weak-field or high spin.

15. $Fe_2O_3(s) + 6\ H_2C_2O_4(aq) \rightarrow 2\ Fe(C_2O_4)_3^{3-}(aq) + 3\ H_2O(l) + 6\ H^+(aq)$; The oxalate anion forms a soluble complex ion with iron in rust (Fe_2O_3) which allows rust stains to be removed.

17. The lanthanide elements are located just before the 5d transition metals. The lanthanide contraction is the steady decrease in the atomic radii of the lanthanide elements when going from left to right across the periodic table. As a result of the lanthanide contraction, the sizes of the 4d and 5d elements are very similar (see Exercise 7.132). This leads to a greater similarity in the chemistry of the 4d and 5d elements in a given vertical group.

19. Advantages: cheap energy cost; less air pollution; Disadvantages: chemicals used in hydrometallurgy are expensive and sometimes toxic.

Exercises

Transition Metals

21. a. Ni: $[Ar]4s^23d^8$

 b. Cd: $[Kr]5s^24d^{10}$

 c. Zr: $[Kr]5s^24d^2$

 d. Os: $[Xe]6s^24f^{14}5d^6$

23. Transition metal ions lose the s electrons before the d electrons.

 a. Ni^{2+}: $[Ar]3d^8$

 b. Cd^{2+}: $[Kr]4d^{10}$

 c. Zr^{3+}: $[Kr]4d^1$ Zr^{4+}: [Kr]

 d. Os^{2+}: $[Xe]4f^{14}5d^6$; Os^{3+}: $[Xe]4f^{14}5d^5$

25. Transition metal ions lose the s electrons before the d electrons. Also, Pt is an exception to the normal filling order of electrons (see Figure 7.27).

 a. Co: $[Ar]4s^23d^7$

 Co^{2+}: $[Ar]3d^7$

 Co^{3+}: $[Ar]3d^6$

 b. Pt: $[Xe]6s^14f^{14}5d^9$

 Pt^{2+}: $[Xe]4f^{14}5d^8$

 Pt^{4+}: $[Xe]4f^{14}5d^6$

 c. Fe: $[Ar]4s^23d^6$

 Fe^{2+}: $[Ar]3d^6$

 Fe^{3+}: $[Ar]3d^5$

27. a. molybdenum(IV) sulfide; molybdenum(VI) oxide

 b. MoS_2, +4; MoO_3, +6 (Oxygen and sulfur are each in the -2 oxidation state.)

Coordination Compounds

29. Ammonia solutions are basic. The precipitation reaction is $Cu^{2+}(aq) + 2\ OH^-(aq) \rightarrow Cu(OH)_2(s)$. As the concentration of NH_3 increases, the soluble complex ion $Cu(NH_3)_4^{2+}$ forms: $Cu(OH)_2(s) + 4\ NH_3(aq) \rightarrow Cu(NH_3)_4^{2+}(aq) + 2\ OH^-(aq)$.

31. Only $[Cr(NH_3)_6]Cl_3$ will form a precipitate since only this compound will have Cl^- ions in solution. The Cl^- ions in the other compounds are ligands and are bound to the central Cr^{3+} ion. The Cl^- ions in $[Cr(NH_3)_6]Cl_3$ are counter ions needed to produce a neutral compound while the NH_3 molecules are the ligands bound to Cr^{3+}.

33. To determine the oxidation state of the metal, you must know the charges of the various common ligands (see Table 20.13 of the text).

a. pentaamminechlororuthenium(III) ion

b. hexacyanoferrate(II) ion

c. tris(ethylenediamine)manganese(II) ion

d. pentaamminenitrocobalt(III) ion

35. a. hexaamminecobalt(II) chloride

b. hexaaquacobalt(III) iodide

c. potassium tetrachloroplatinate(II)

d. potassium hexachloroplatinate(II)

e. pentaamminechlorocobalt(III) chloride

f. triamminetrinitrocobalt(III)

37. a. $[Co(C_5H_5N)_6]Cl_3$

b. $[Cr(NH_3)_5I]I_2$

c. $[Ni(NH_2CH_2CH_2NH_2)_3]Br_2$

d. $K_2[Ni(CN)_4]$

e. $[Pt(NH_3)_4Cl_2][PtCl_4]$

39. a.

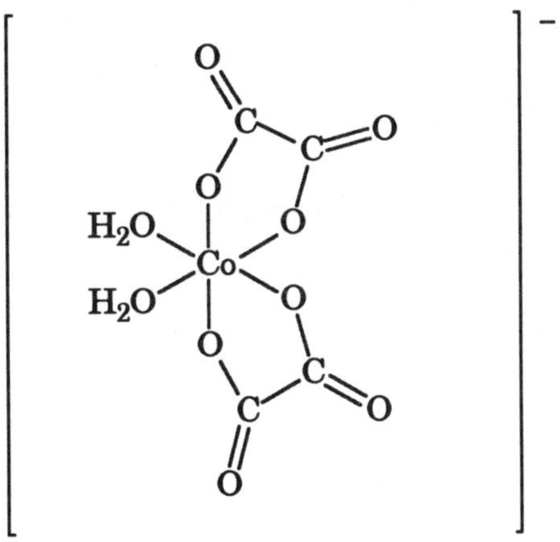

H₂O ligands, Co, oxalate structures:

$[Co(H_2O)_2(C_2O_4)_2]^-$

cis

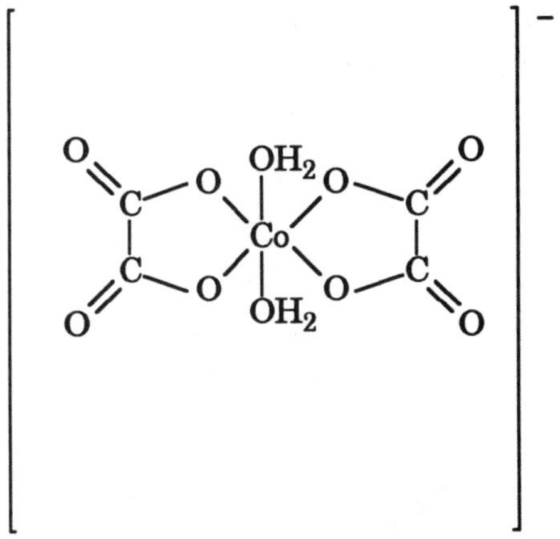

trans

Note: $C_2O_4^{2-}$ is a bidentate ligand. Bidentate ligands bond to the metal at two positions which are 90° apart from each other in octahedral complexes. Bidentate ligands do not bond to the metal at positions 180° apart.

b.

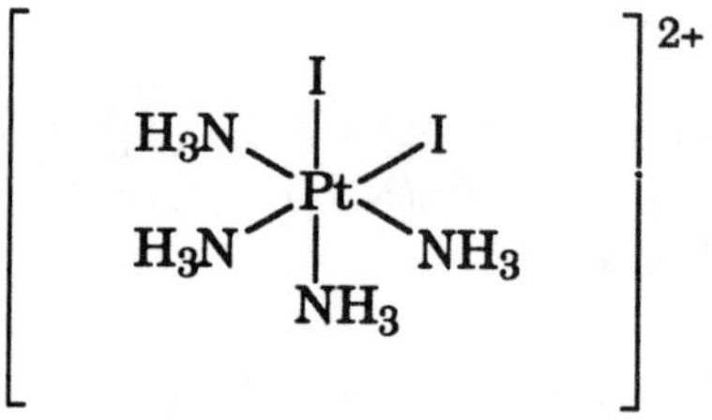

$[Pt(NH_3)_4I_2]^{2+}$

cis 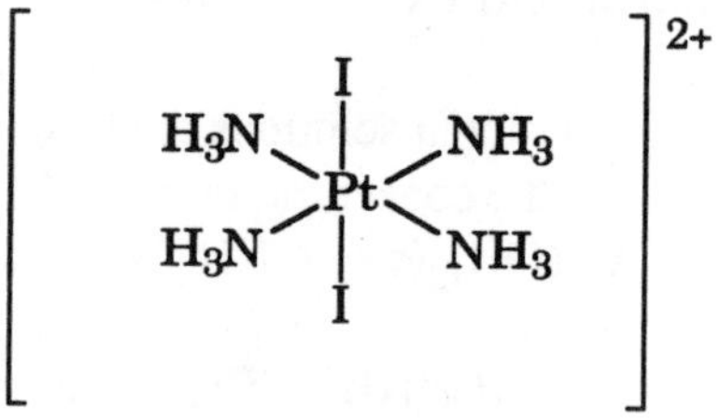 trans

c.

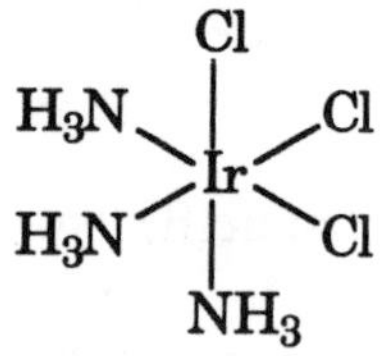

$Ir(NH_3)_3Cl_3$

cis 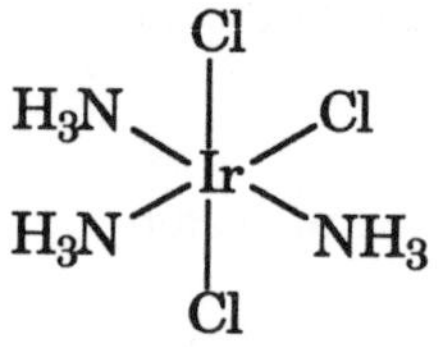 trans

d.

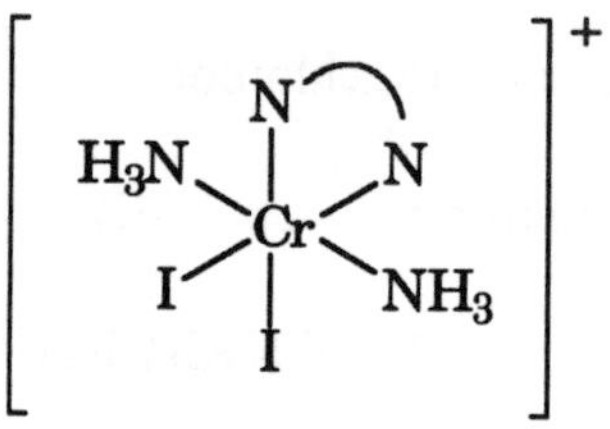

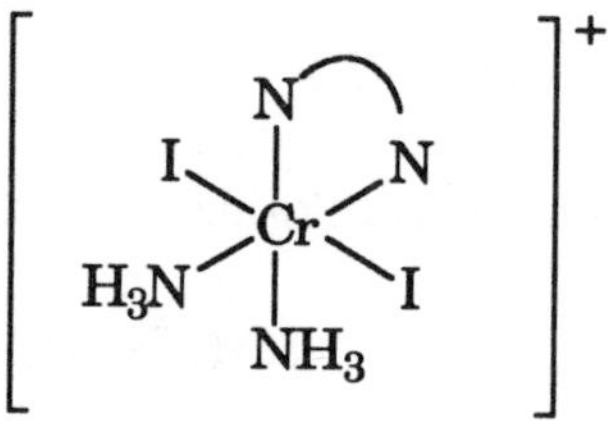

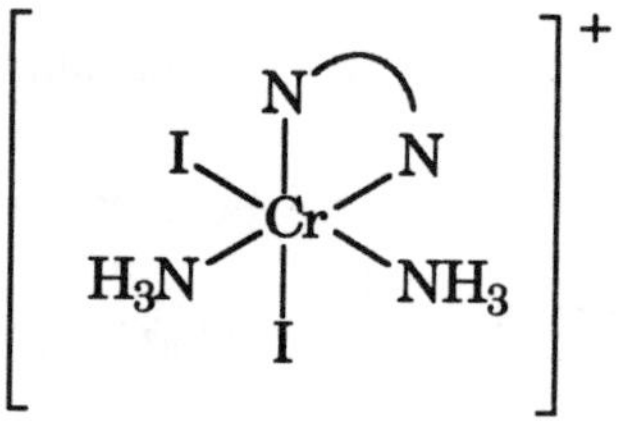

$[Cr(en)(NH_3)_2I_2]^{+}$

Note: N⌒N is an abbreviation for the bidentate ligand ethylenediamine ($H_2NCH_2CH_2NH_2$).

41.

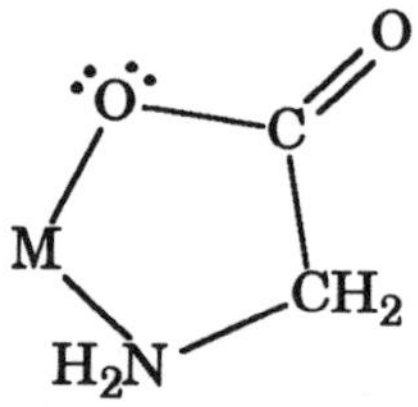

M = transition metal ion

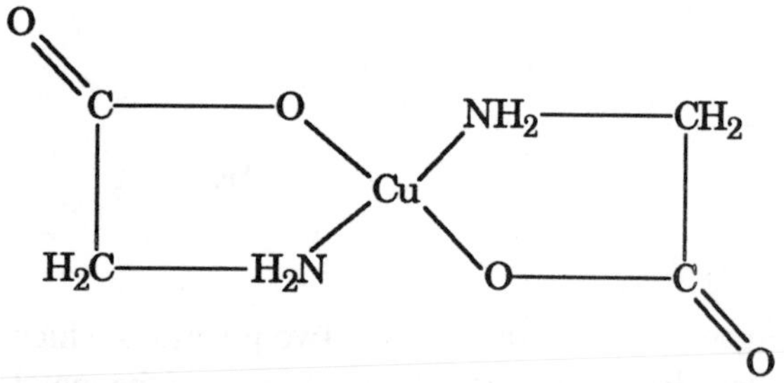

and

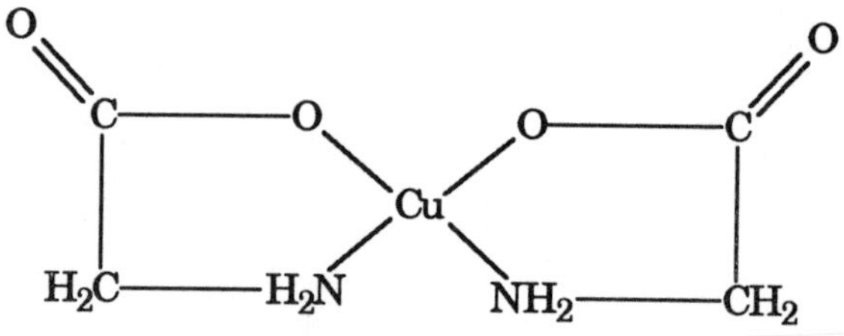

43. Linkage isomers differ in the way the ligand bonds to the metal. SCN^- can bond through the sulfur or through the nitrogen atom. NO_2^- can bond through the nitrogen or through the oxygen atom. OCN^- can bond through the oxygen or through the nitrogen atom. N_3^-, $NH_2CH_2CH_2NH_2$ and I^- are not capable of linkage isomerism.

45. There are four geometrical isomers (labeled i-iv). Isomers iii and iv are optically active and the nonsuperimposable mirror images are shown.

i. $[Cr(NH_3)_2BrCl(NH_2CH_2CH_2NH_2)]^+$ (structure: Br and Cl trans; H_3N, H_3N, NH_2–CH_2–CH_2–NH_2 in plane)

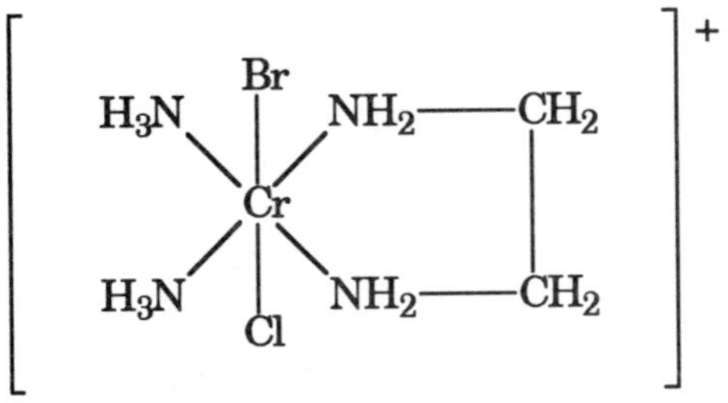

ii. $[Cr(NH_3)_2BrCl(NH_2CH_2CH_2NH_2)]^+$ (structure: NH_3 and NH_3 trans; Br, Cl, NH_2–CH_2–CH_2–NH_2 in plane)

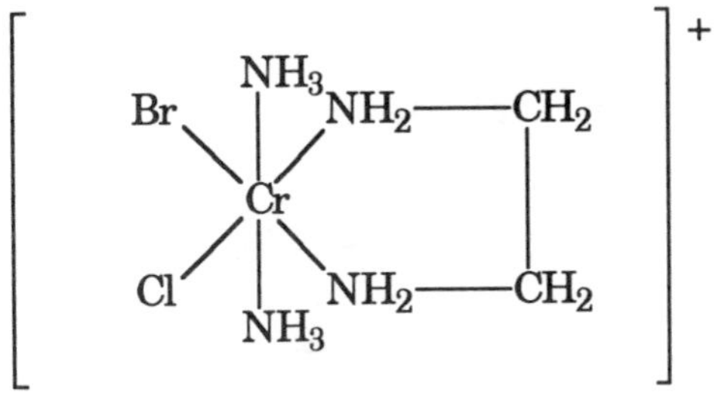

iii. $[Cr(NH_3)_2BrCl(NH_2CH_2CH_2NH_2)]^+$ (structure: Br and NH_3 trans; Cl, H_3N, NH_2–CH_2–CH_2–NH_2 in plane)

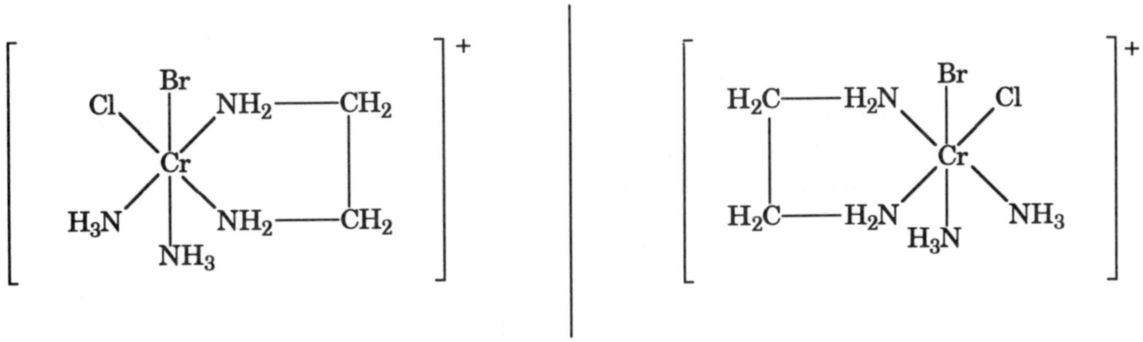

optically active | mirror | mirror image of iii (nonsuperimposable)

iv. $[Cr(NH_3)_2BrCl(NH_2CH_2CH_2NH_2)]^+$ (structure: Cl and NH_3 trans; Br, H_3N, NH_2–CH_2–CH_2–NH_2 in plane)

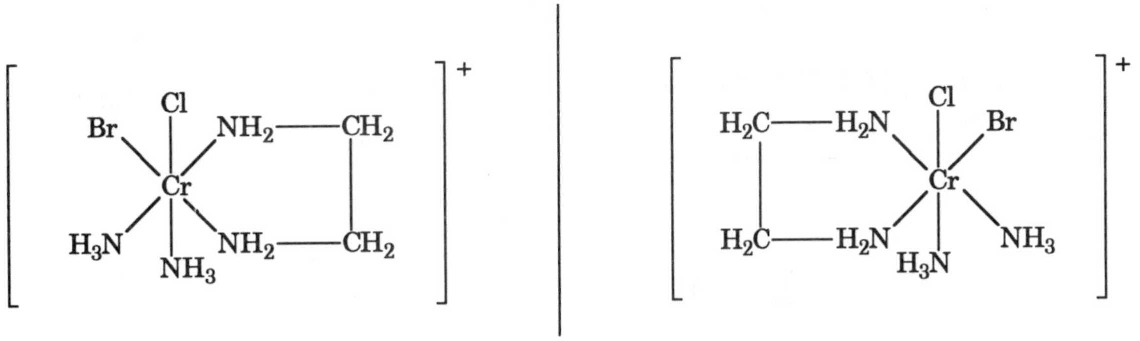

optically active | mirror | mirror image of iv (nonsuperimposable)

Bonding, Color, and Magnetism in Coordination Compounds

47. a. Fe^{2+}: [Ar]$3d^6$

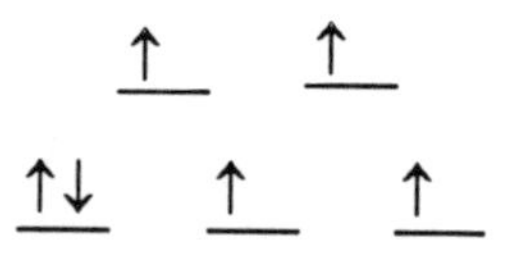

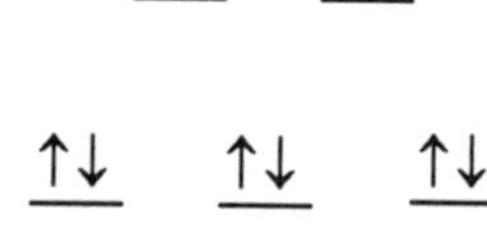

High spin, small Δ

Low spin, large Δ

b. Fe^{3+}: [Ar]$3d^5$

↑ ↑

↑ ↑ ↑

High spin, small Δ

c. Ni^{2+}: [Ar]$3d^8$

↑ ↑

↑↓ ↑↓ ↑↓

49. Since fluorine has a -1 charge as a ligand, then chromium has a +2 oxidation state in CrF_6^{4-}. The electron configuration of Cr^{2+} is: [Ar]$3d^4$. For four unpaired electrons, this must be a weak-field (high-spin) case where the splitting of the d-orbitals is small and the number of unpaired electrons is maximized. The crystal field diagram for this ion is:

↑ —

↑ ↑ ↑ ↕ small Δ

51. To determine the crystal field diagrams, you need to determine the oxidation state of the transition metal which can only be determined if you know the charges of the ligands (see Table 20.13). The electron configurations and the crystal field diagrams follow.

a. Ru^{2+}: [Kr]$4d^6$, no unpaired e^-

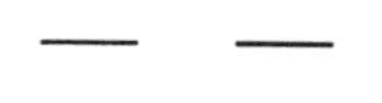

↑↓ ↑↓ ↑↓

Low spin, large Δ

b. Ni^{2+}: [Ar]$3d^8$, 2 unpaired e^-

↑ ↑

↑↓ ↑↓ ↑↓

c. V^{3+}: $[Ar]3d^2$, 2 unpaired e^-

___ ___

↑ ↑ ___

Note: Ni^{2+} must have 2 unpaired electrons, whether high-spin or low-spin, and V^{3+} must have 2 unpaired electrons, whether high-spin or low-spin.

53. Transition compounds exhibit the color complementary to that absorbed. From Table 20.16, $Ni(H_2O)_6Cl_2$ absorbs red light and $Ni(NH_3)Cl_2$ absorbs yellow-green light. $Ni(NH_3)_6Cl_2$ absorbs the shorter wavelength light which is the higher energy light ($E = hc/\lambda$). Therefore, Δ is larger for $Ni(NH_3)_6Cl_2$ which means that NH_3 is a stronger field ligand than H_2O. This is consistent with the spectrochemical series.

55. $CoCl_4^{2-}$; $CoCl_6^{4-}$ has an octahedral molecular structure and $CoCl_4^{2-}$ has a tetrahedral molecular structure (as are most Co^{2+} complexes with four ligands). Coordination complexes absorb light of energy equal to the energy difference between the split d-orbitals. Since the tetrahedral d-orbital splitting is less than one-half of the octahedral d-orbital splitting, then tetrahedral complexes will absorb lower energy light which corresponds to light with longer wavelengths ($E = hc/\lambda$).

57. Since the ligands are Cl^-, then iron is in the +3 oxidation state. Fe^{3+}: $[Ar]3d^5$

↑ ↑ ↑

↑ ↑

Since all tetrahedral complexes are high spin, then there are 5 unpaired electrons in $FeCl_4^-$.

Metallurgy

59. a. $\Delta H° = 2(-1117) + (-393.5) - [3(-826) + (-110.5)] = -39$ kJ

$\Delta S° = 2(146) + 214 - [3(90.) + 198] = 38$ J/K

b. $\Delta G° = \Delta H° - T\Delta S°$; $T = 800. + 273 = 1073$ K

$\Delta G° = -39$ kJ $- 1073$ K$(0.038$ kJ/K$) = -39$ kJ $- 41$ kJ $= -80.$ kJ

Additional Exercises

61. a. 4 O atoms on faces × 1/2 O/face = 2 O atoms, 2 O atoms inside body, Total: 4 O atoms

8 Ti atoms on corners × 1/8 Ti/corner + 1 Ti atom/body center = 2 Ti atoms

Formula of the unit cell is Ti_2O_4. The empirical formula is TiO_2.

b. $\overset{+4\ -2}{2\ TiO_2} + \overset{0}{3\ C} + \overset{0}{4\ Cl_2} \rightarrow \overset{+4\ -1}{2\ TiCl_4} + \overset{+4\ -2}{CO_2} + \overset{+2\ -2}{2\ CO}$

Cl is reduced and C is oxidized. Cl_2 is the oxidizing agent and C is the reducing agent.

$\overset{+4\ -1}{TiCl_4} + \overset{0}{O_2} \rightarrow \overset{+4\ -2}{TiO_2} + \overset{0}{2\ Cl_2}$

O is reduced and Cl is oxidized. O_2 is the oxidizing agent and $TiCl_4$ is the reducing agent.

63.

$$(Au(CN)_2^- + e^- \rightarrow Au + 2\ CN^-) \times 2 \qquad E° = -0.60\ V$$
$$Zn + 4\ CN^- \rightarrow Zn(CN)_4^{2-} + 2\ e^- \qquad -E° = 1.26\ V$$

$$2\ Au(CN)_2^-(aq) + Zn(s) \rightarrow 2\ Au(s) + Zn(CN)_4^{2-}(aq) \qquad E°_{cell} = 0.66\ V$$

$$\Delta G° = -nFE°_{cell} = -(2\ mol\ e^-)(96{,}485\ C/mol\ e^-)(0.66\ J/C) = -1.3 \times 10^5\ J = -130\ kJ$$

$$E° = \frac{0.0592}{n}\log K,\ \log K = \frac{nE°}{0.0592} = \frac{2(0.66)}{0.0592} = 22.30,\ K = 10^{22.30} = 2.0 \times 10^{22}$$

Note: We carried extra significant figures to determine K.

65. a. 2; Forms bonds through the lone pairs on the two oxygen atoms.

b. 3; Forms bonds through the lone pairs on the three nitrogen atoms.

c. 4; Forms bonds through the two nitrogen atoms and the two oxygen atoms.

d. 4; Forms bonds through the four nitrogen atoms.

67. a. $Ru(phen)_3^{2+}$ exhibits optical isomerism [similar to $Co(en)_3^{3+}$ in Figure 20.17 of the text].

b. Ru^{2+}: $[Kr]4d^6$; Since there are no unpaired electrons, then Ru^{2+} is a strong-field (low-spin) case.

— —

large Δ

↑↓ ↑↓ ↑↓

Challenge Problems

69. i. $0.0203\ g\ CrO_3 \times \frac{52.00\ g\ Cr}{100.0\ g\ CrO_3} = 0.0106\ g\ Cr;\ \%\ Cr = \frac{0.0106}{0.105} \times 100 = 10.1\%\ Cr$

ii. $32.93\ mL\ HCl \times \frac{0.100\ mmol\ HCl}{mL} \times \frac{1\ mmol\ NH_3}{mmol\ HCl} \times \frac{17.03\ mg\ NH_3}{mmol} = 56.1\ mg\ NH_3$

$$\% \text{ NH}_3 = \frac{56.1 \text{ mg}}{341 \text{ mg}} \times 100 = 16.5\% \text{ NH}_3$$

iii. 73.53% + 16.5% + 10.1% = 100.1%; The compound must be composed of only Cr, NH_3, and I.

Out of 100.00 of compound:

$$10.1 \text{ g Cr} \times \frac{1 \text{ mol}}{52.00 \text{ g}} = 0.194 \qquad \frac{0.194}{0.194} = 1.00$$

$$16.5 \text{ g NH}_3 \times \frac{1 \text{ mol}}{17.03 \text{ g}} = 0.969 \qquad \frac{0.969}{0.194} = 4.99$$

$$73.53 \text{ g I} \times \frac{1 \text{ mol}}{126.9 \text{ g}} = 0.5794 \qquad \frac{0.5794}{0.194} = 2.99$$

$Cr(NH_3)_5I_3$ is the empirical formula. Cr(III) forms octahedral complexes. So compound A is made of the octahedral $[Cr(NH_3)_5I]^{2+}$ complex ion and two I^- ions or $[Cr(NH_3)_5I]I_2$. Lets check this proposed formula using the freezing point data.

iv. $\Delta T_f = iK_f m$; For $[Cr(NH_3)_5I]I_2$, i = 3.0 (assuming complete dissociation).

$$m = \frac{0.601 \text{ g complex}}{1.000 \times 10^{-2} \text{ kg H}_2\text{O}} \times \frac{1 \text{ mol complex}}{517.9 \text{ g complex}} = 0.116 \text{ molal}$$

$$\Delta T_f = 3.0 \times 1.86°\text{C/molal} \times 0.116 \text{ molal} = 0.65°\text{C}$$

Since ΔT_f is close to the measured value, then this is consistent with the formula $[Cr(NH_3)_5I]I_2$.

71. No; In all three cases, six bonds are formed between Ni^{2+} and nitrogen, so ΔH values should be similar. $\Delta S°$ for formation of the complex ion is most negative for 6 NH_3 molecules reacting with a metal ion (7 independent species become 1). For penten reacting with a metal ion, 2 independent species become 1, so $\Delta S°$ is least negative of all three of the reactions. Thus, the chelate effect occurs because the more bonds a chelating agent can form to the metal, the more favorable $\Delta S°$ is for the formation of the complex ion and the larger the formation constant.

73.

The d_{z^2} orbital will be destabilized much more than in the trigonal planar case (see Exercise 20.72). The d_{z^2} orbital has electron density on the z-axis directed at the two axial ligands. The $d_{x^2-y^2}$ and d_{xy} orbitals are in the plane of the three trigonal planar ligands and should be destabilized a lesser amount as compared to the d_{z^2} orbital; only a portion of the electron density in the $d_{x^2-y^2}$ and d_{xy} orbitals is directed at the ligands. The d_{xz} and d_{yz} orbitals will be destabilized the least since the electron density is directed between the ligands.

CHAPTER TWENTY-ONE

THE NUCLEUS: A CHEMIST'S VIEW

Questions

1. Fission: Splitting of a heavy nucleus into two (or more) lighter nuclei.

   Fusion: Combining two light nuclei to form a heavier nucleus.

   Fusion is more likely for elements lighter than Fe; fission is more likely for elements heavier than Fe.

3. The assumptions are that the ^{14}C level in the atmosphere is constant or that the ^{14}C level at the time the plant died can be calculated. A constant ^{14}C level is a poor assumption and accounting for variation is complicated. Another problem is that some of the material must be destroyed to determine the ^{14}C level.

5. No, coal fired power plants also pose risks. A partial list of risks are:

Coal	Nuclear
Air pollution	Radiation exposure to workers
Coal mine accidents	Disposal of wastes
Health risks to miners (black lung disease)	Meltdown
	Terrorists
	Public fear

7. For fusion reactions, a collision of sufficient energy must occur between two positively charged particles to initiate the reaction. This requires high temperatures. In fission, an electrically neutral neutron collides with the positively charged nucleus. This has a much lower activation energy.

Exercises

Radioactive Decay and Nuclear Transformations

9. All nuclear reactions must be charge balanced and mass balanced. To charge balance, balance the sum of the atomic numbers on each side of the reaction and to mass balance, balance the sum of the mass numbers on each side of the reaction.

a. $^{51}_{24}Cr + ^{0}_{-1}e \rightarrow ^{51}_{23}V$

b. $^{131}_{53}I \rightarrow ^{0}_{-1}e + ^{131}_{54}Xe$

11. a. $^{3}_{1}H \rightarrow ^{3}_{2}He + ^{0}_{-1}e$

c. $^{7}_{4}Be + ^{0}_{-1}e \rightarrow ^{7}_{3}Li$

b. $^{8}_{3}Li \rightarrow ^{8}_{4}Be + ^{0}_{-1}e$

$^{8}_{4}Be \rightarrow 2\,^{4}_{2}He$

$^{8}_{3}Li \rightarrow 2\,^{4}_{2}He + ^{0}_{-1}e$

d. $^{8}_{5}B \rightarrow ^{8}_{4}Be + ^{0}_{+1}e$

e. $^{32}_{15}P \rightarrow ^{32}_{16}S + ^{0}_{-1}e$

13. a. $^{207}_{82}Pb$; Complete decay is 7α and 4β particles: $^{235}_{92}U \rightarrow ^{207}_{82}Pb + 7\,^{4}_{2}He + 4\,^{0}_{-1}e$

b. $^{235}_{92}U \rightarrow ^{231}_{90}Th + ^{4}_{2}He \rightarrow ^{231}_{91}Pa + ^{0}_{-1}e \rightarrow ^{227}_{89}Ac + ^{4}_{2}He$

$^{215}_{84}Po + ^{4}_{2}He \leftarrow ^{219}_{86}Rn + ^{4}_{2}He \leftarrow ^{223}_{88}Ra + ^{4}_{2}He \leftarrow ^{227}_{90}Th + ^{0}_{-1}e$

$^{4}_{2}He + ^{211}_{82}Pb \rightarrow ^{211}_{83}Bi + ^{0}_{-1}e \rightarrow ^{207}_{81}Tl + ^{4}_{2}He \rightarrow ^{207}_{82}Pb + ^{0}_{-1}e$

The intermediate nucludes are:

$^{231}_{90}Th$, $^{231}_{91}Pa$, $^{227}_{89}Ac$, $^{227}_{90}Th$, $^{223}_{88}Ra$, $^{219}_{86}Rn$, $^{215}_{84}Po$, $^{211}_{82}Pb$, $^{211}_{83}Bi$ and $^{207}_{81}Tl$.

15. Reference Table 21.2 of the text for potential radioactive decay processes. ^{8}B and ^{9}B contain too many protons or too few neutrons. Electron capture or positron production are both possible decay mechanisms that increase the neutron to proton ratio. Alpha particle production also increases the neutron to proton ratio, but it is not likely for these light nuclei. ^{12}B and ^{13}B contain too many neutrons or too few protons. Beta production lowers the neutron to proton ratio, so we expect ^{12}B and ^{13}B to be β-emitters.

17. a. $^{1}_{1}H + ^{14}_{7}N \rightarrow ^{11}_{6}C + ^{4}_{2}He$

b. $2\,^{3}_{2}He \rightarrow ^{4}_{2}He + 2\,^{1}_{1}H$

c. $^{1}_{1}H + ^{1}_{1}H \rightarrow ^{2}_{1}H + ^{0}_{+1}e$ (positron)

d. $^{1}_{1}H + ^{12}_{6}C \rightarrow ^{13}_{7}N$

Kinetics of Radioactive Decay

19. All radioactive decay follows first-order kinetics where $t_{1/2} = (\ln 2)/k$.

$$t_{1/2} = \frac{\ln 2}{k} = \frac{0.693}{1.0 \times 10^{-3}\ h^{-1}} = 690\ h$$

21. Half-life for first-order kinetics is independent of concentration (or number of nuclides). To go from 1.00×10^{20} to 2.50×10^{19} nuclides represents 2 half-lives.

$$1.00 \times 10^{20} \xrightarrow[t_{1/2}]{} 5.00 \times 10^{19} \xrightarrow[t_{1/2}]{} 2.50 \times 10^{19};\quad 2 \times t_{1/2} = 10.0\ \text{min},\quad t_{1/2} = 5.00\ \text{min}$$

23. Units for N and N_o are usually number of nuclei but can also be grams if the units are the same for both N and N_o. In this problem m_o = the initial mass of $^{47}Ca^{2+}$ to be ordered.

$$k = \frac{\ln 2}{t_{1/2}};\ \ln\left(\frac{N}{N_o}\right) = -kt = \frac{-0.693\ t}{t_{1/2}},\ \ln\left(\frac{5.0\ \mu g\ Ca^{2+}}{m_o}\right) = \frac{-0.693\ (2.0\ d)}{4.5\ d} = -0.31$$

$$\frac{5.0}{m_o} = e^{-0.31} = 0.73,\ m_o = 6.8\ \mu g \text{ of } {}^{47}Ca^{2+} \text{ needed initially}$$

$$6.8\ \mu g\ {}^{47}Ca^{2+} \times \frac{107.0\ \mu g\ {}^{47}CaCO_3}{47.0\ \mu g\ {}^{47}Ca^{2+}} = 15\ \mu g\ {}^{47}CaCO_3 \text{ should be ordered at the minimum.}$$

25. $t = 52.0$ yr; $k = \frac{\ln 2}{t_{1/2}};\ \ln\left(\frac{N}{N_o}\right) = -kt = \frac{-0.6931 \times 52.0\ yr}{28.8\ yr} = -1.25,\ \left(\frac{N}{N_o}\right) = e^{-1.25} = 0.287$

28.7% of the ^{90}Sr remains as of July 16, 1997.

27. $t_{1/2} = 5730$ yr; $k = (\ln 2)/t_{1/2}$; $\ln (N/N_o) = -kt$

$$\ln\left(\frac{N}{N_o}\right) = \frac{-0.6931\ t}{t_{1/2}} = \frac{-0.6931 \times 2200\ yr}{5730\ yr} = -0.27,\ \frac{N}{N_o} = e^{-0.27} = 0.76 = 76\% \text{ of } {}^{14}C \text{ remains}$$

29. Since 4.5×10^9 years is equal to the half-life, then one-half of the ^{238}U atoms will have been converted to ^{206}Pb. The numbers of atoms of ^{206}Pb and ^{238}U will be equal. Thus, the mass ratio is equal to the molar mass ratio:

$$\frac{206}{238} = 0.866$$

Energy Changes in Nuclear Reactions

31. $\Delta E = \Delta mc^2,\ \Delta m = \frac{\Delta E}{c^2} = \frac{3.9 \times 10^{23}\ kg\ m^2/s^2}{(3.00 \times 10^8\ m/s)^2} = 4.3 \times 10^6\ kg$

The sun loses 4.3×10^6 kg of mass each second. Note: $1\ J = 1\ kg\ m^2/s^2$

33. We need to determine the mass defect, Δm, between the mass of the nucleus and the mass of the individual parts that make up the nucleus. Once Δm is known, we can then calculate ΔE (the binding energy) using $E = mc^2$. Note: $1\ J = 1\ kg\ m^2/s^2$.

For $^{232}_{94}$Pu (94 e, 94 p, 138 n):

mass of ^{232}Pu nucleus = 3.85285×10^{-22} g - mass of 94 electrons

mass of ^{232}Pu nucleus = 3.85285×10^{-22} g - $94(9.10939 \times 10^{-28})$ g = 3.85199×10^{-22} g

$\Delta m = 3.85199 \times 10^{-22}$ g - (mass of 94 protons + mass of 138 neutrons)

$\Delta m = 3.85199 \times 10^{-22}$ g - $[94(1.67262 \times 10^{-24}) + 138(1.67493 \times 10^{-24})]$ g = -3.168×10^{-24} g

For 1 mol of nuclei: $\Delta m = -3.168 \times 10^{-24}$ g/nuclei $\times$ 6.0221×10^{23} nuclei/mol = -1.908 g/mol

$\Delta E = \Delta mc^2 = (-1.908 \times 10^{-3}\ \text{kg/mol})(2.9979 \times 10^8\ \text{m/s})^2 = -1.715 \times 10^{14} = \text{J/mol}$

For $^{231}_{91}$Pa (91 e, 91 p, 140 n):

mass of ^{231}P nucleus = 3.83616×10^{-22} g - $91(9.10939 \times 10^{-28})$ g = 3.83533×10^{-22} g

$\Delta m = 3.83533 \times 10^{-22}$ g - $[91(1.67262 \times 10^{-24}) + 140(1.67493 \times 10^{-24})]$ g = -3.166×10^{-24} g

$$\Delta E = \Delta mc^2 = \frac{-3.166 \times 10^{-27}\ \text{kg}}{\text{nuclei}} \times \frac{6.0221 \times 10^{23}\ \text{nuclei}}{\text{mol}} \times \left(\frac{2.9979 \times 10^8\ \text{m}}{\text{s}}\right)^2$$

$$= -1.714 \times 10^{14}\ \text{J/mol}$$

35. 1.0078 amu = mass of $^1_1$H atom; 1.0087 amu = mass of neutron; m_e = mass of electron

mass defect = Δm = mass of ^{24}Mg nucleus - [mass of 12 protons + mass of 12 neutrons]

$\Delta m = 23.9850\ \text{amu} - 12\ m_e - [12(1.0078 - m_e) + 12(1.0087)]$; Mass of electrons cancel.

$\Delta m = 23.9850 - [12(1.0078) + 12(1.0087)] = -0.2130$ amu

$$\Delta E = \Delta mc^2 = -0.2130\ \text{amu} \times \frac{1.6606 \times 10^{-27}\ \text{kg}}{\text{amu}} \times (2.9979 \times 10^8\ \text{m/s})^2 = -3.179 \times 10^{-11}\ \text{J}$$

$$\frac{\text{BE}}{\text{nucleon}} = \frac{-3.179 \times 10^{-11}\ \text{J}}{24\ \text{nucleons}} = -1.325 \times 10^{-12}\ \text{J/nucleon}$$

For ^{27}Mg (12 e, 12 p, 15 n):

$\Delta m = 26.9843 - 12\ m_e - [12(1.0078 - m_e) + 15(1.0087)] = -0.2398$ amu

$$\Delta E = \Delta mc^2 = -0.2398\ \text{amu} \times \frac{1.6606 \times 10^{-27}\ \text{kg}}{\text{amu}} \times (2.9979 \times 10^8\ \text{m/s})^2 = -3.579 \times 10^{-11}\ \text{J}$$

$$\frac{\text{BE}}{\text{nucleon}} = \frac{-3.579 \times 10^{-11}\ \text{J}}{27\ \text{nucleons}} = -1.326 \times 10^{-12}\ \text{J/nucleon}$$

37. $^{1}_{1}H + ^{1}_{0}n \rightarrow 2\,^{1}_{1}H + ^{1}_{0}n + ^{1}_{-1}H$; mass $^{1}_{-1}H$= mass $^{1}_{1}H$= 1.00728 amu = mass of proton = m_p

$\Delta m = 3\ m_p + m_n - (m_p + m_n) = 2\ m_p = 2(1.00728) = 2.01456$ amu

$$\Delta E = \Delta mc^2 = 2.01456 \text{ amu} \times \frac{1.66056 \times 10^{-27}\text{ kg}}{\text{amu}} \times (2.997925 \times 10^8\text{ m/s})^2$$

$\Delta E = 3.00660 \times 10^{-10}$ J of energy is absorbed per nuclei or 1.81062×10^{14} J/mol nuclei.

The source of energy is the kinetic energy of the proton and the neutron in the particle accelerator.

Detection, Uses, and Health Effects of Radiation

39. The Geiger-Müller tube has a certain response time. After the gas in the tube ionizes to produce a "count," some time must elapse for the gas to return to an electrically neutral state. The response of the tube levels because at high activities, radioactive particles are entering the tube faster than the tube can respond to them.

41. All evolved oxygen in O_2 comes from water and not from carbon dioxide.

43. Release of Sr is probably more harmful. Xe is chemically unreactive. Strontium is in the same family as calcium and could be absorbed and concentrated in the body in a fashion similar to Ca. This puts the radioactive Sr in the bones: red blood cells are produced in bone marrow. Xe would not be readily incorporated in the body.

The chemical properties determine where a radioactive material may be concentrated in the body or how easily it may be excreted. The length of time of exposure and what is exposed to radiation significantly affects the health hazard. (See exercise 21.44 for a specific example.)

Additional Exercises

45. The most abundant isotope is generally the most stable isotope. The periodic table predicts that the most stable isotopes for exercises a - d are ^{39}K, ^{56}Fe, ^{23}Na and ^{204}Tl. (Reference Table 21.2 of the text for potential decay processes.)

a. Unstable; ^{45}K has too many neutrons and will undergo beta particle production.

b. Stable

c. Unstable; ^{20}Na has too few neutrons and will most likely undergo electron capture or positron production. Alpha particle production makes too severe of a change to be a likely decay process for the relatively light ^{20}Na nuclei. Alpha particle production usually occurs for heavy nuclei.

d. Unstable; ^{194}Tl has too few neutrons and will undergo electron capture, positron production and/or alpha particle production.

47. $k = \frac{\ln 2}{t_{1/2}}$; $\ln\left(\frac{N}{N_o}\right) = -kt = \frac{-0.6931\ t}{t_{1/2}}$, $\ln\left(\frac{N}{15.3}\right) = \frac{-0.693\ (15{,}000\text{ yr})}{5730\text{ yr}} = -1.8$

$\frac{N}{15.3} = e^{-1.8} = 0.17$, $N = 15.3 \times 0.17 = 2.6$ counts per minute per g of C

If we had 10. mg C, we would see:

$$10.\text{ mg} \times \frac{1\text{ g}}{1000\text{ mg}} \times \frac{2.6\text{ counts}}{\text{min g}} = \frac{0.026\text{ counts}}{\text{min}}$$

It would take roughly 40 min to see a single disintegration. This is too long to wait and the background radiation would probably be much greater than the ^{14}C activity. Thus, ^{14}C dating is not practical for very small samples.

49. $20{,}000\text{ ton TNT} \times \frac{4 \times 10^9\text{ J}}{\text{ton TNT}} \times \frac{1\text{ mol }^{235}U}{2 \times 10^{13}\text{ J}} \times \frac{235\text{ g }^{235}U}{\text{mol }^{235}U} = 940\text{ g }^{235}U \approx 900\text{ g }^{235}U$

This assumes that all of the ^{235}U undergoes fission.

Challenge Problems

51. Assuming that the radionuclide is long lived enough such that no significant decay occurs during the time of the experiment, the total counts of radioactivity injected are:

$$0.10\text{ mL} \times \frac{5.0 \times 10^3\text{ cpm}}{\text{mL}} = 5.0 \times 10^2\text{ cpm}$$

Assuming that the total activity is uniformly distributed only in the rats blood, the blood volume is:

$$V \times \frac{48\text{ cpm}}{\text{mL}} = 5.0 \times 10^2\text{ cpm},\ V = 10.4\text{ mL} = 10.\text{ mL}$$

53. a. ^{12}C; It takes part in the first step of the reaction but is regenerated in the last step. ^{12}C is not consumed so it is not a reactant.

b. ^{13}N, ^{13}C, ^{14}N, ^{15}O, and ^{15}N are the intermediates.

c. $4\,{}^{1}_{1}H \rightarrow {}^{4}_{2}He + 2\,{}^{0}_{+1}e$; $\Delta m = 4.00260\text{ amu} - 2\,m_e + 2\,m_e - [4(1.00782\text{ amu} - m_e)]$

$\Delta m = 4.00260 - 4(1.00782) + 4(0.000549) = -0.02648$ amu for 4 protons reacting

For 4 mol of protons, $\Delta m = -0.02648$ g and ΔE for the reaction is:

$$\Delta E = \Delta mc^2 = -2.648 \times 10^{-5}\text{ kg} \times (2.9979 \times 10^8\text{ m/s})^2 = -2.380 \times 10^{12}\text{ J}$$

For 1 mol of protons reacting: $\frac{-2.380 \times 10^{12}\text{ J}}{4\text{ mol }^1H} = -5.950 \times 10^{11}\text{ J/mol }^1H$

CHAPTER TWENTY-TWO

ORGANIC CHEMISTRY

Questions

1. There is only one consecutive chain of C-atoms in the molecule. They are not all in a true straight line since the bond angles at each carbon are the tetrahedral angles of 109.5°.

3. Substitution: An atom or group is replaced by another atom or group.

   e.g., H in benzene is replaced by Cl. $C_6H_6 + Cl_2 \xrightarrow{\text{catalyst}} C_6H_5Cl + HCl$

   Addition: Atoms or groups are added to a molecule.

   e.g., Cl_2 adds to ethene. $CH_2=CH_2 + Cl_2 \rightarrow CH_2Cl-CH_2Cl$

5. A thermoplastic polymer can be remelted; a thermoset polymer cannot be softened once it is formed.

7. Plasticizer compounds make a polymer more flexible by inserting themselves between adjacent polymer chains which weakens the intermolecular forces. Crosslinking makes a polymer more rigid by bonding adjacent polymer chains together.

9. a. Replacement of hydrogens with halogens results in fewer H and OH radicals in the flame. In addition, as halogen atoms are released in a flame, they react readily with H radicals by forming hydrogen halides. Thus, halogens act as free radical scavengers.

   b. With aromatic groups present, it is more difficult to "chip off" pieces of the polymer in the pyrolysis zone due to the increased strength of the bonds in aromatic rings. In addition, polymers with aromatic rings tend to char naturally, which inhibits combustion.

   c. It is more difficult to "chip off" fragments of the polymer in the pyrolysis zone. The crosslinked polymer chars instead of burns.

11. Polyvinyl chloride contains some polar C–Cl bonds as compared to only nonpolar C–H bonds in polyethylene. The stronger interparticle forces would be found in polyvinyl chloride since there are dipole-dipole forces present in PVC that are not present in polyethylene.

Exercises

Hydrocarbons

13. i.

$CH_3\text{–}CH_2\text{–}CH_2\text{–}CH_2\text{–}CH_2\text{–}CH_3$

ii.

$$CH_3\text{—}CH(CH_3)\text{—}CH_2\text{—}CH_2\text{—}CH_3$$

iii.

$$CH_3\text{—}CH_2\text{—}CH(CH_3)\text{—}CH_2\text{—}CH_3$$

iv.

$$CH_3\text{—}C(CH_3)_2\text{—}CH_2\text{—}CH_3$$

v.

$$CH_3\text{—}CH(CH_3)\text{—}CH(CH_3)\text{—}CH_3$$

All other possibilities are identical to one of these five compounds.

15. See Exercise 22.13 for the structures. The names of structure i - v respectively, are: hexane (or n-hexane), 2-methylpentane, 3-methylpentane, 2,2-dimethylbutane and 2,3-dimethylbutane.

Note: A difficult task in Exercise 22.13 is recognizing different compounds from compounds that differ by rotations about one or more C–C bonds (called conformations). The best way to distinguish different compounds is to name them. Different name = different compound; same name = same compound so it is not an isomer.

17. a.

$$CH_3\text{—}CH(CH_3)\text{—}CH_2\text{—}CH_2CH_3$$

b.

$$CH_3\text{—}C(CH_3)_2\text{—}CH_2\text{—}CH(CH_3)\text{—}CH_3$$

c.

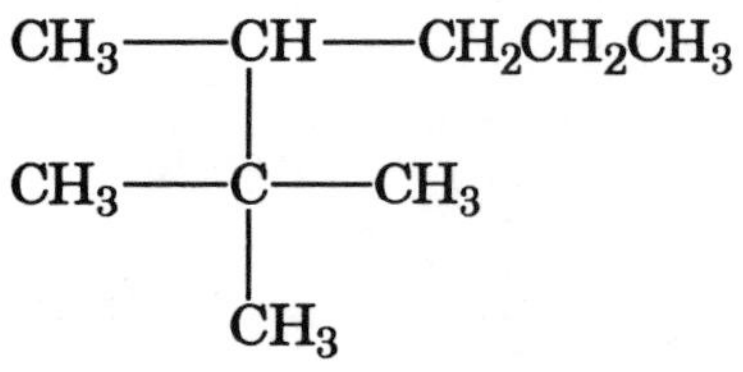

$$\begin{array}{l} CH_3-CH-CH_2CH_2CH_3 \\ \quad\quad\ \ | \\ CH_3-C-CH_3 \\ \quad\quad\ \ | \\ \quad\quad CH_3 \end{array}$$

d. The longest chain is 6 carbons long.

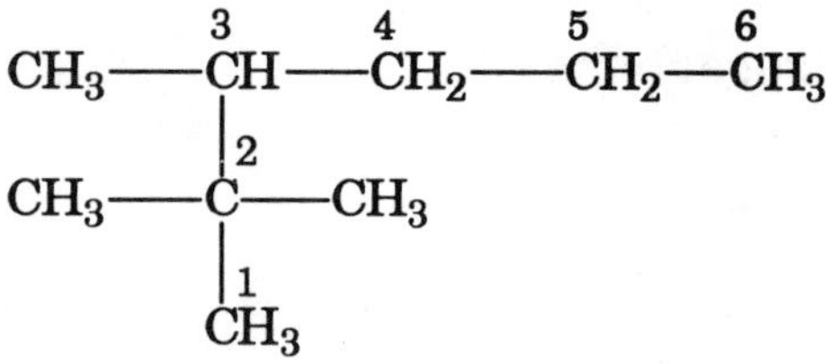

$$\begin{array}{l} CH_3-\overset{3}{C}H-\overset{4}{C}H_2-\overset{5}{C}H_2-\overset{6}{C}H_3 \\ \quad\quad\ \ |\,2 \\ CH_3-C-CH_3 \\ \quad\quad\ \ |\,1 \\ \quad\quad CH_3 \end{array}$$

2,2,3-trimethylhexane

19. a. 2,3,3-trimethylhexane b. 8-ethyl-2,5,5-trimethyldecane c. 3-methylhexane

Note: For alkanes always identify the longest carbon chain for the base name first, then number the carbons to give the lowest overall numbers for the substituent groups.

21. a. 1-butene b. 2-methyl-2-butene c. 2,5-dimethyl-3-heptene

Note: The multiple bond is assigned the lowest number possible.

23. a. $CH_3\text{–}CH_2\text{–}CH{=}CH\text{–}CH_2\text{–}CH_3$

b. $CH_3\text{–}CH{=}CH\text{–}CH{=}CH\text{–}CH_2CH_3$

c.

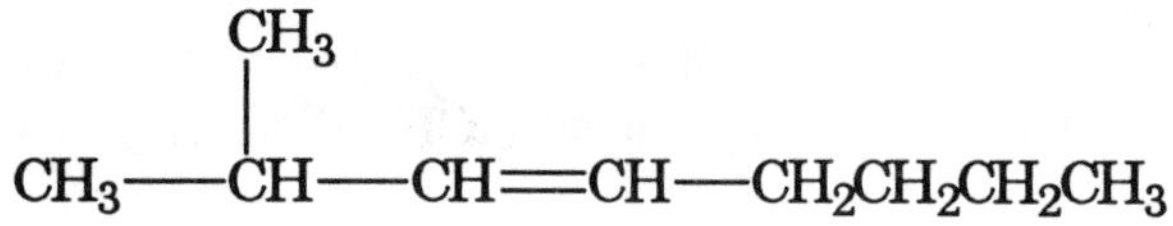

$$\begin{array}{l} \quad\quad CH_3 \\ \quad\quad\ \ | \\ CH_3-CH-CH{=}CH-CH_2CH_2CH_2CH_3 \end{array}$$

25. a.

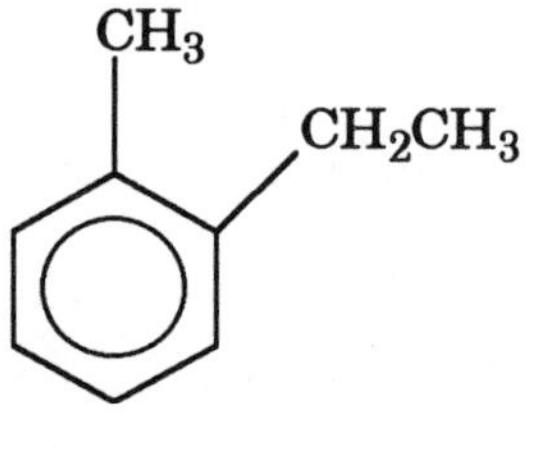

CH_3, CH_2CH_3 (benzene ring with adjacent CH_3 and CH_2CH_3 substituents)

b.

$$\begin{array}{l} \quad\quad CH_3 \quad\quad\quad\quad\quad\quad CH_3 \\ \quad\quad\ | \quad\quad\quad\quad\quad\quad\quad | \\ H_3C-C-C_6H_4-C-CH_3 \\ \quad\quad\ | \quad\quad\quad\quad\quad\quad\quad | \\ \quad\quad CH_3 \quad\quad\quad\quad\quad\quad CH_3 \end{array}$$

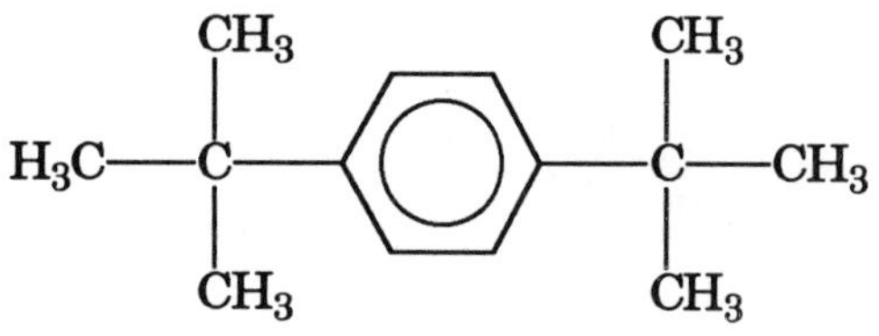

c.

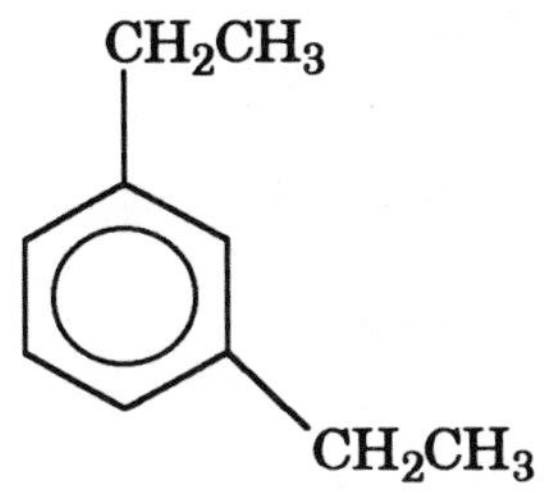

(Structure: benzene ring with CH_2CH_3 and CH_2CH_3 substituents in meta positions)

d.

(Structure: benzene ring attached to $CH_2—CH{=}CH—CH_3$)

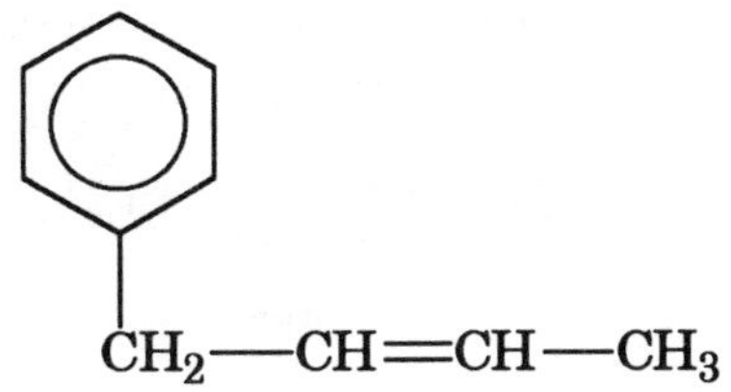

27. a. 1,3-dichlorobutane

b. 1,1,1-trichlorobutane

c. 2,3-dichloro-2,4-dimethylhexane

d. 1,2-difluoroethane

Isomerism

29. To exhibit cis-trans isomerism, each carbon in the double bond must have two structurally different groups bonded to it. In Exercise 22.21, this only occurs for compound c. The cis and trans isomers for 21c are:

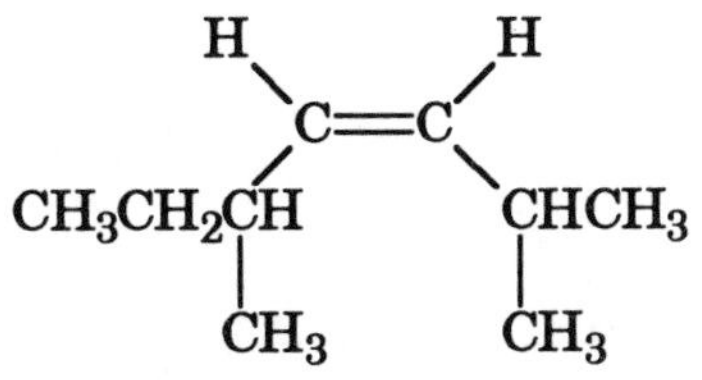

cis: H and H on the same side of C=C; $CH_3CH_2CH(CH_3)$ and $CH(CH_3)CH_3$ (CHCH$_3$ with CH_3) on the same side.

trans: H on one carbon and $CHCH_3$ (with CH_3) on the other carbon across; $CH_3CH_2CH(CH_3)$ and H.

cis 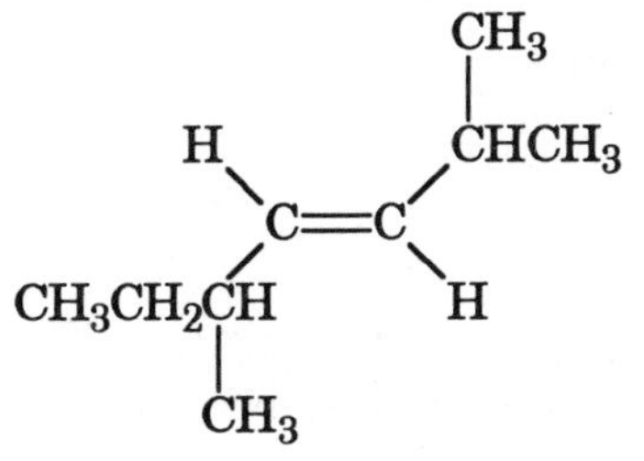 trans

Similarly, all the compounds in Exercise 23.23 can also exhibit cis-trans isomerism.

In the other compounds in Exercise 23.21, each carbon in the double bond does not contain two different groups. In 21a, the first carbon in the double bond contains two H atoms and in 21b, the first carbon in the double bond contains 2 CH_3 groups. To illustrate that these compounds do not exhibit cis-trans isomerism, lets look at the potential cis-trans isomers for the compound in Exercise 23.21a.

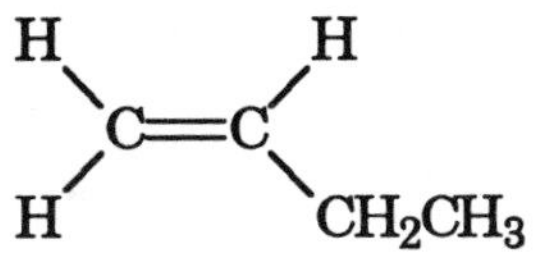

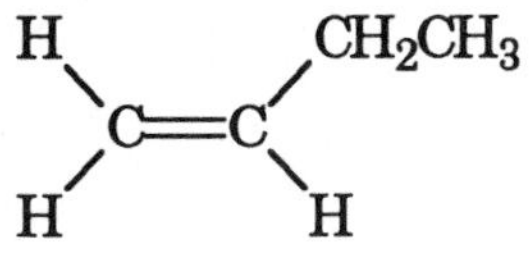

$H_2C{=}CH(CH_2CH_3)$ with H on top right and CH_2CH_3 on bottom right; $H_2C{=}CH(CH_2CH_3)$ with CH_2CH_3 on top right and H on bottom right.

These are the same compounds; they only differ by a simple rotation of the molecule. Therefore, they are not isomers of each other.

31. C_5H_{10} has the general formula for alkenes, C_nH_{2n}. To distinguish the different isomers from each other, we will name them. Each isomer must have a different name.

$CH_2{=}CHCH_2CH_2CH_3$

1-pentene

$CH_3CH{=}CHCH_2CH_3$

2-pentene

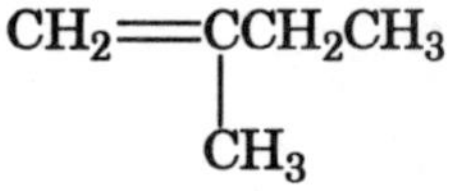

$CH_2{=}C(CH_3)CH_2CH_3$

2-methyl-1-butene

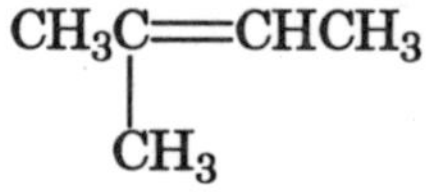

$CH_3C(CH_3){=}CHCH_3$

2-methyl-2-butene

$CH_3CH(CH_3)CH{=}CH_2$

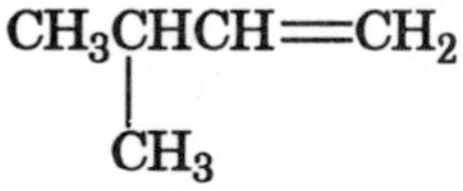

3-methyl-1-butene

33. To help distinguish the different isomers, we will name them.

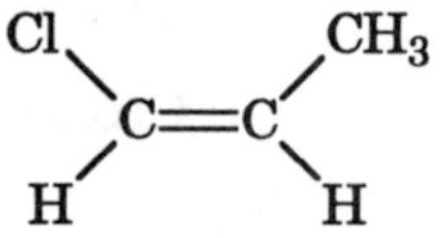

(Cl)(H)C=C(CH₃)(H), Cl and CH₃ on the same side

cis-1-chloro-1-propene

(H)(Cl)C=C(CH₃)(H), Cl and CH₃ on opposite sides

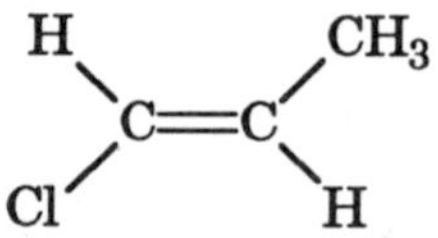

trans-1-chloro-1-propene

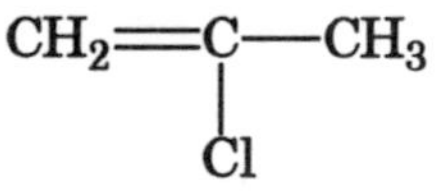

$CH_2{=}C(Cl){-}CH_3$

2-chloro-1-propene

$CH_2{=}CH{-}CH_2Cl$

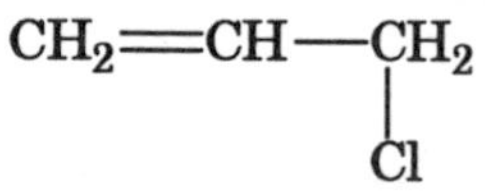

3-chloro-1-propene

35.

(F)(F)C=C(H)(H)

(F)(H)C=C(F)(H), both F on the same side

(F)(H)C=C(H)(F), F atoms on opposite sides

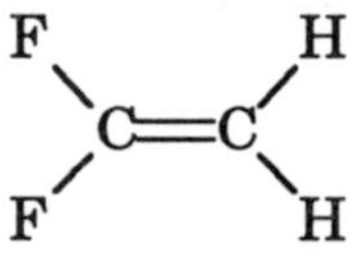

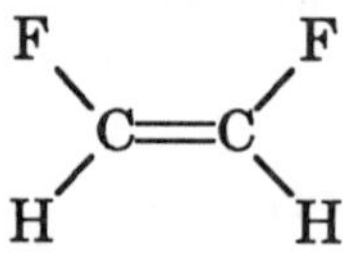

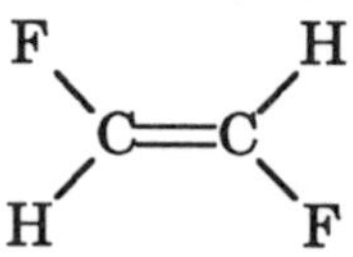

37.

a. H_3C, H on one carbon; $CH_2CH_2CH_3$, H on the other: $\mathrm{H_3C(H)C{=}C(H)CH_2CH_2CH_3}$

b. H_3C, H on one carbon; H, CH_3 on the other: $\mathrm{H_3C(H)C{=}C(H)CH_3}$

c. H_3C, Cl on one carbon; CH_2CH_3, Cl on the other: $\mathrm{H_3C(Cl)C{=}C(Cl)CH_2CH_3}$

39. a. $CH_3^*–CH_2^*–CH_2–CH_3$; There are two "types" of hydrogens in n-butane (see asterisks). Thus they are two monochloro isomers of n-butane (1-chlorobutane and 2-chlorobutane).

b.

$$CH_3^*–\overset{\displaystyle CH_3}{\underset{\displaystyle H^*}{C}}–CH_3$$

There are two types of hydrogens in 2-methylpropane (see *). Thus, there are two monochloro isomers of 2-methylpropane (1-chloro-2-methylpropane and 2-chloro-2-methylpropane).

c.

$$CH_3–\overset{\displaystyle CH_3}{\underset{\displaystyle CH_3}{C}}–CH_3$$

There is only one type of hydrogen in this compound so there is only one monochloro isomer (1-chloro-2,2-dimethylpropane).

d.

$$CH_3^*–CH_2^*–\overset{\displaystyle CH_3^*}{\underset{\displaystyle H^*}{C}}–CH_2–CH_3$$

There are four types of hydrogens in this compound (see *) so they are four monochloro isomers possible.

Functional Groups

41. Reference Table 22.5 for the common functional groups.

a. ketone b. aldehyde c. carboxylic acid d. amine

43. a.

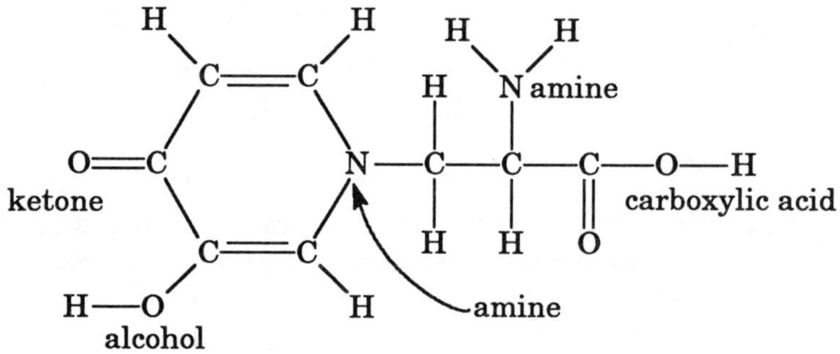

```
          H           H    H    H
           \         /      \  /
            C=======C    H   N amine
           /         \   |   |
   O===C              N--C---C---C---O---H
   ketone            /   |   |   ||  carboxylic acid
           \        /    H   H   O
            C=======C
           /         \
     H---O            H      amine
     alcohol
```

b. 5 carbons in ring and the carbon in $-CO_2H$: sp^2; the other two carbons: sp^3

c. 24 sigma bonds; 4 pi bonds

45. a. 2-methyl-2-propanol; Since there are 3 R groups attached to the carbon containing the OH group, then this is a tertiary alcohol.

b. cyclobutanol; Since the carbon containing the OH group is bonded to two other carbons (2 R groups), then this is a secondary alcohol.

c. 2,2-dimethyl-1-propanol; primary alcohol (1 R group bonded to carbon containing the OH group)

47.

$HO—CH_2—CH_2—CH_2—CH_3$

1-butanol

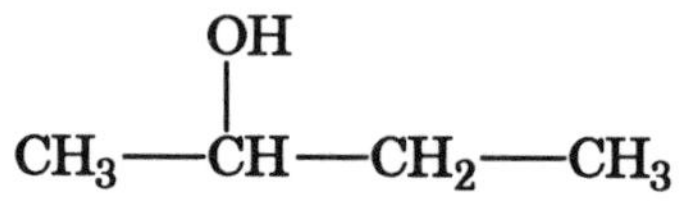

$CH_3—CH(OH)—CH_2—CH_3$

2-butanol

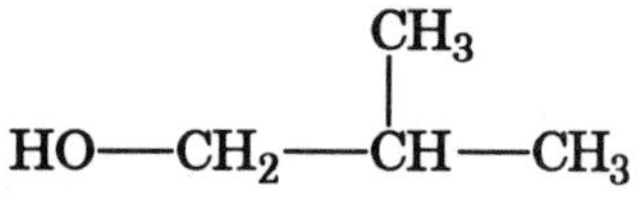

$HO—CH_2—CH(CH_3)—CH_3$

2-methyl-1-propanol

$CH_3—C(CH_3)(OH)—CH_3$

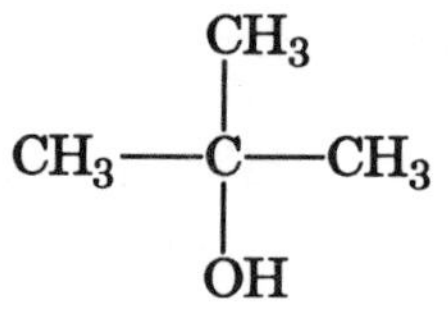

2-methyl-2-propanol

There are three possible ethers with the formula $C_4H_{10}O$. They are:

$CH_3CH_2—O—CH_2CH_3$

diethyl ether

$CH_3—O—CH_2CH_2CH_3$

methylpropyl ether

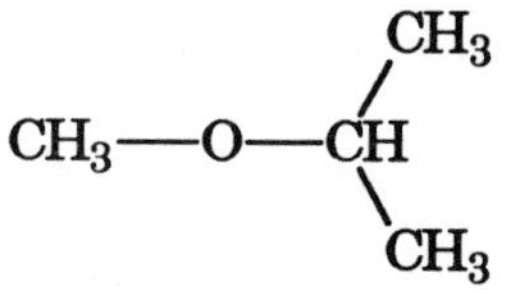

$CH_3—O—CH(CH_3)_2$

isopropylmethyl ether

49. a. 1-bromo-2-propanone b. 2-methylbutanal

Reactions of Organic Compounds

51. a. $CH_2=CH_2 + Br_2 \rightarrow CH_2Br-CH_2Br$

b. $C_6H_6 + Br_2 \xrightarrow{FeBr_3} C_6H_5Br + HBr$

c. $CH_2=CH-CH_2-CH=CH_2 + 2\,H_2 \xrightarrow{Pt} CH_3-CH_2-CH_2-CH_2-CH_3$

53.

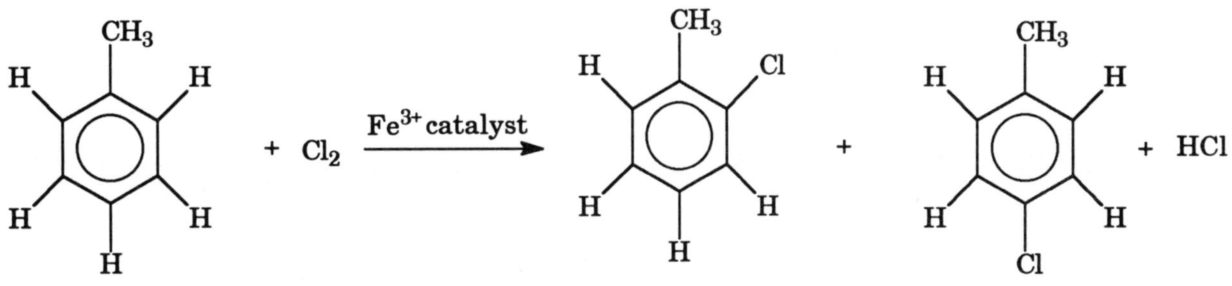

$C_6H_5CH_3 + Cl_2 \xrightarrow{Fe^{3+}\text{ catalyst}}$ ortho-chlorotoluene + para-chlorotoluene + HCl

ortho para

$C_6H_5CH_3 + Cl_2 \xrightarrow{\text{light}} C_6H_5CH_2Cl + HCl$

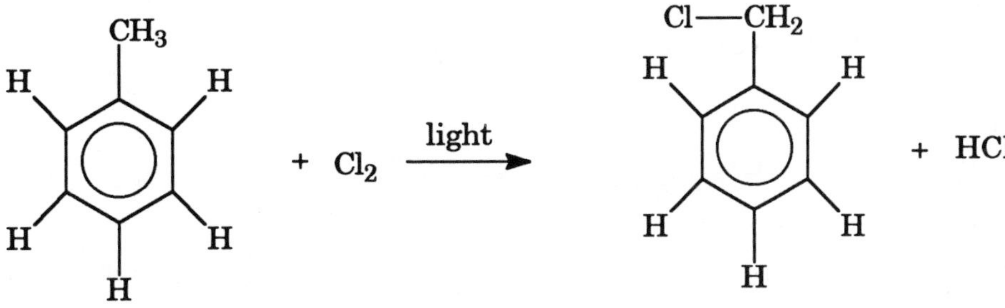

To substitute for the benzene ring hydrogens, an iron(III) catalyst must be present. Without this special iron catalyst, the benzene ring hydrogens are unreactive. To substitute for an alkane hydrogen, light must be present. For toluene, the light catalyzed reaction substitutes a chlorine for a hydrogen in the methyl group attached to the benzene ring.

55. Primary alcohols (a and e) are oxidized to aldehydes which can be oxidized further to carboxylic acids. Secondary alcohols (b and c) are oxidized to ketones and tertiary alcohols (d) do not undergo this type of oxidation reaction. For the primary alcohols (a and e), we listed both the aldehyde and the carboxylic acid as possible products.

a. $CH_3-C(=O)-H + CH_3-C(=O)-OH$

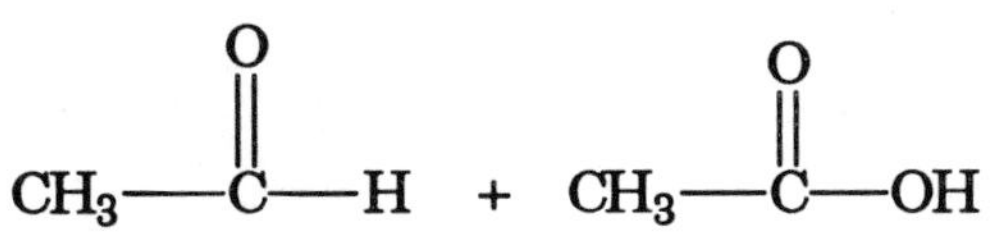

b. $CH_3-C(=O)-CH_3$

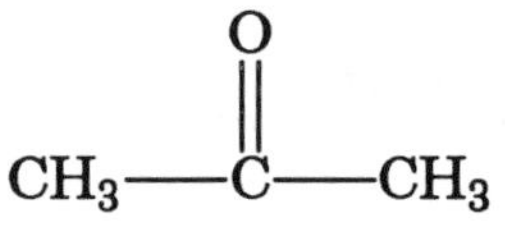

c.

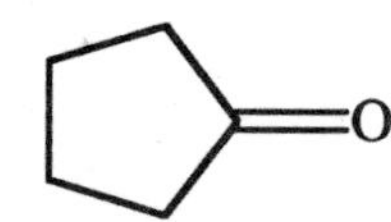

d.

No reaction occurs

e.

$$CH_3-\underset{CH_3}{\overset{CH_3}{C}}-\overset{O}{\overset{\|}{C}}-H \;+\; CH_3-\underset{CH_3}{\overset{CH_3}{C}}-\overset{O}{\overset{\|}{C}}-OH$$

57. a. $CH_3CH=CH_2 + Br_2 \rightarrow CH_3CHBrCH_2Br$ (Addition reaction of Br_2 with propene)

b. $CH_3C \equiv CH + H_2 \xrightarrow{\text{catalyst}} CH_3CH=CH_2 + Br_2 \rightarrow CH_3CHBrCH_2Br$

Hydrogenation of propyne followed by the addition reaction of Br_2 could yield the desired product.

59. When an alcohol is reacted with a carboxylic acid, an ester is produced.

a. $$CH_3\overset{O}{\overset{\|}{C}}-OH + HO-CH_3 \longrightarrow CH_3\overset{O}{\overset{\|}{C}}-O-CH_3 + H_2O$$

b. $$H-\overset{O}{\overset{\|}{C}}-OH + HO-CH_2CH_2CH_3 \longrightarrow H-\overset{O}{\overset{\|}{C}}-O-CH_2CH_2CH_3 + H_2O$$

Polymers

61. The backbone of the polymer contains only carbon atoms which indicates that Kel-F is an addition polymer. The smallest repeating unit of the polymer and the monomer used to produce this polymer are:

$$\left(\underset{Cl}{\overset{F}{C}}-\underset{F}{\overset{F}{C}} \right)_n \qquad \underset{Cl}{\overset{F}{C}}=\underset{F}{\overset{F}{C}}$$

Note: Condensation polymers generally have O or N atoms in the backbone of the polymer.

63.

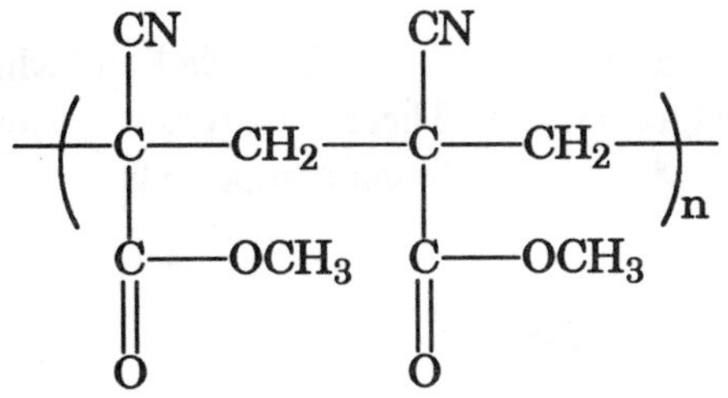

CN CN

$-\!\!\left(\!-C-CH_2-C-CH_2-\!\right)_n$

C—OCH_3 C—OCH_3

O O

Super glue is an addition polymer formed by reaction of the C=C bond in methyl cyanoacrylate.

65. H_2O is eliminated when Kevlar forms. Two repeating units of Kevlar are:

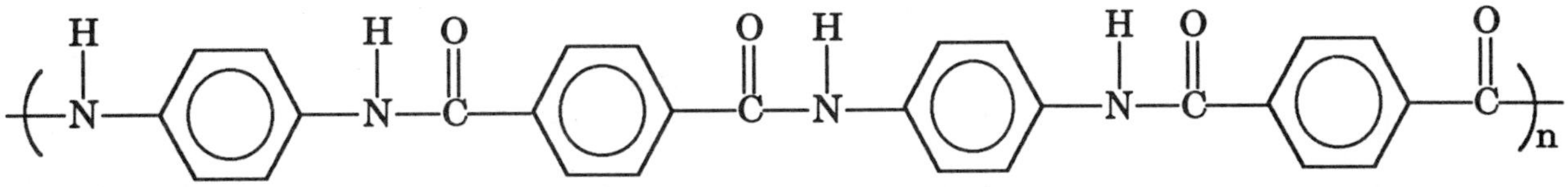

H H O O H H O O

$-\!\!\left(\!-N-C_6H_4-N-C-C_6H_4-C-N-C_6H_4-N-C-C_6H_4-C-\!\right)_n$

67. This is a condensation polymer where two molecules of H_2O form when the monomers link together.

$H_2N-C_6H_4-NH_2$ and HO_2C, CO_2H, HO_2C, CO_2H (on benzene ring)

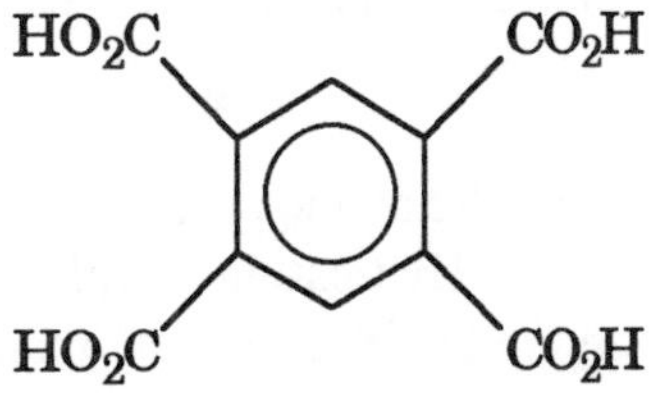

69. Divinylbenzene has two reactive double bonds which are both used when divinylbenzene inserts itself into two adjacent polymer chains. The chains cannot move past each other because the crosslinks bond adjacent polymer chains together making the polymer more rigid.

71. a. The polymer formed using 1,2-diaminoethane will exhibit relatively strong hydrogen bonding interactions between adjacent polymer chains. Since hydrogen-bonding is not present in the ethylene glycol polymer (a polyester polymer forms), then the 1,2-diaminoethane polymer will be stronger.

b. The presence of rigid groups (benzene rings or multiple bonds) makes the polymer stiffer. Hence, the monomer with the benzene ring will produce the more rigid polymer.

c. Polyacetylene will have a double bond in the carbon backbone of the polymer.

$$n\ HC\equiv CH \longrightarrow -\!\!\left(\!-CH=CH-\!\right)_n$$

The presence of the double bond in polyacetylene will make polyacetylene a more rigid polymer than polyethylene. Polyethylene doesn't have C=C bonds in the backbone of the polymer (the double bonds in the monomers react to form the polymer).

Additional Exercises

73. $CH_2Cl–CH_2Cl$, 1-2-dichloroethane; There is free rotation about the C–C single bond which doesn't lead to different compounds. CHCl=CHCl, 1,2-dichloroethene; There is no rotation about the C=C double bond. This creates the cis and trans isomers which are different compounds.

75. The cis isomer has the CH_3 groups on the same

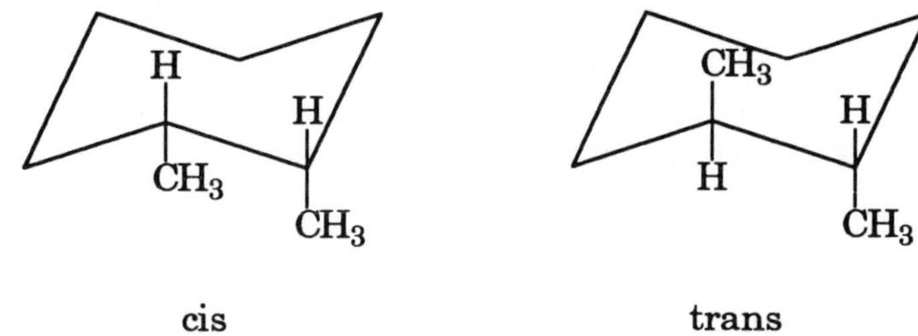

77. Alcohols consist of two parts, the polar OH group and the nonpolar hydrocarbon chain attached to the OH group. As the length of the nonpolar hydrocarbon chain increases, the solubility of the alcohol decreases in water, a very polar solvent. In methyl alcohol (methanol), the polar OH group can override the effect of the nonpolar CH_3 group and methyl alcohol is soluble in water. In stearyl alcohol, the molecule consists mostly of the long nonpolar hydrocarbon chain so it is insoluble in water.

79. In the presence of H^+, the amine group is protonated creating a positive charge on morphine ($R_3N + H^+ \rightarrow R_3\overset{+}{N}H$). By treating morphine with HCl, an ionic compound results which is more soluble in water and in the blood stream than the neutral covalent form of morphine.

81. a.

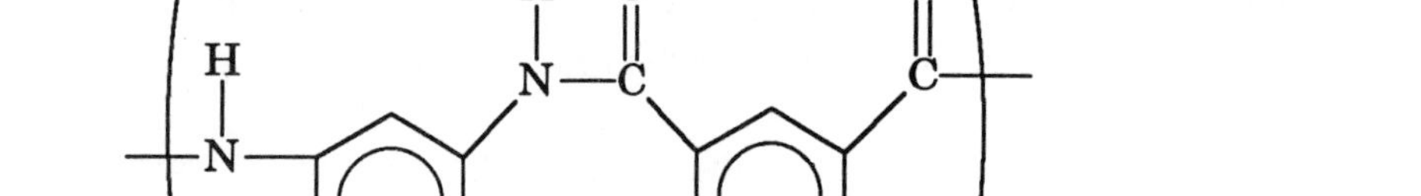

H_2N— —NH_2 and HO_2C— —CO_2H

b. Repeating unit:

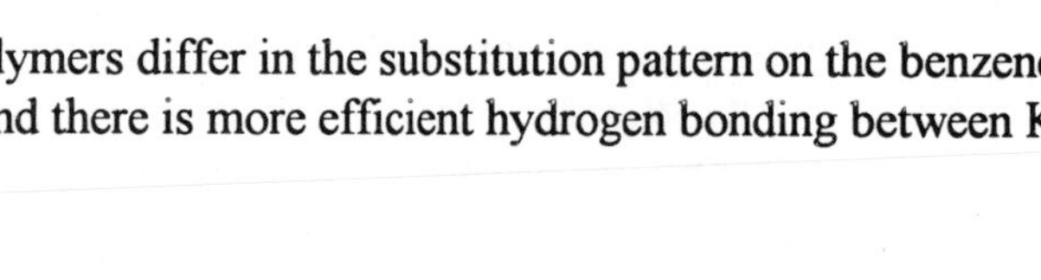

H O O
H N—C C
N
n

The two polymers differ in the substitution pattern on the benzene rings. The Kevlar chain is straighter and there is more efficient hydrogen bonding between Kevlar chains than between Nomex chains.

83. As the car ages the plasticizers escape from the seat covers: i) The waxy coating is the escaped plasticizer; ii) The new car smell is the smell of the plasticizers (di-octylphthalate); iii) Loss of plasticizer causes and vinyl to become brittle.

Challenge Problems

85.

$$\underset{13}{HC}\equiv\underset{12}{C}-\underset{11}{C}\equiv\underset{10}{C}-\underset{9}{CH}=\underset{8}{C}=\underset{7}{CH}-\underset{6}{CH}=\underset{5}{CH}-\underset{4}{CH}=\underset{3}{CH}-\underset{2}{CH_2}-\underset{1}{\overset{\overset{\displaystyle O}{\|}}{C}}-OH$$

87. Out of 100.00 g:

$$71.89\text{ g C} \times \frac{1\text{ mol C}}{12.01\text{ g C}} = 5.986\text{ mol} \approx 6\text{ mol C}$$

$$12.13\text{ g H} \times \frac{1\text{ mol H}}{1.008\text{ g H}} = 12.03\text{ mol} \approx 12\text{ mol H}$$

The empirical formula is $C_6H_{12}O$.

$$15.98\text{ g O} \times \frac{1\text{ mol O}}{16.00\text{ g O}} = 0.9988\text{ mol} \approx 1\text{ mol O}$$

$$R_1-\overset{\overset{\displaystyle O}{\|}}{C}-O-R_2 + H_2O \longrightarrow R_1-\overset{\overset{\displaystyle O}{\|}}{C}-OH + HOCH_2CH_3$$

R_2 must be CH_3CH_2 since CH_3CH_2OH is one of the products. The molar mass of $-CO_2H$ is $\approx$ 45 g/mol, so the mass of R_1 is 172 - 45 = 127. If R_1 is $CH_3-(CH_2)_n-$, then 15 + n(14) = 127 and n = 112/14 = 8. Ethyl caprate is:

$$CH_3(CH_2)_8\overset{\overset{\displaystyle O}{\|}}{C}-OCH_2CH_3$$

The molecular formula of $C_{12}H_{24}O_2$ agrees with the empirical formula.

89.

$$\left[-OCH_2CH_2O\overset{\overset{\displaystyle O}{\|}}{C}\underset{\underset{\displaystyle H}{|}}{N}-C_6H_4-\underset{\underset{\displaystyle H}{|}}{N}\overset{\overset{\displaystyle O}{\|}}{C}OCH_2CH_2O\overset{\overset{\displaystyle O}{\|}}{C}\underset{\underset{\displaystyle H}{|}}{N}-C_6H_4-\underset{\underset{\displaystyle H}{|}}{N}\overset{\overset{\displaystyle O}{\|}}{C}-\right]_n$$

91. a. The temperature of the rubber band increases when it is stretched.

b. Exothermic since heat is released.

c. As the chains are stretched, they line up more closely together resulting in stronger London dispersion forces between the chains. Heat is released as the strength of the intermolecular forces increases.

d. Stretching is not spontaneous so ΔG is positive. $\Delta G = \Delta H - T\Delta S$; Since ΔH is negative then ΔS must be negative in order to give a positive ΔG.

e.

unstretched

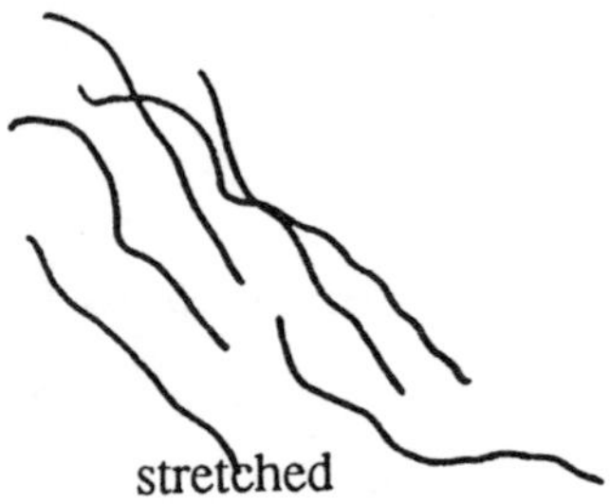

stretched

The structure of the stretched polymer is more ordered (lower S).

CHAPTER TWENTY THREE

BIOCHEMISTRY

Questions

1. Primary: The amino acid sequence in the protein. Covalent bonds (peptide linkages) are the forces that link the various amino acids together in the primary structure.

   Secondary: Includes structural features known as α-helix or pleated sheet. Both are maintained mostly through hydrogen bonding interactions.

   Tertiary: The overall shape of a protein, long and narrow or globular. Maintained by hydrophobic and hydrophilic interactions, such as salt linkages, hydrogen bonds, disulfide linkages and dispersion forces.

3. Both denaturation and inhibition reduce the catalytic activity of an enzyme. Denaturation changes the three-dimensional structure of an enzyme. Inhibition involves the attachment of an incorrect molecule at the active site, preventing the substrate from interacting with the enzyme.

5. Hydrogen bonding occurs between the -OH groups of starch and water molecules.

7. They all contain nitrogen atoms with lone pairs of electrons.

9. A deletion may change the entire code for a protein, thus giving an entirely different sequence of amino acids. A substitution will change only one single amino acid in a protein.

11. A polyunsaturated fat contains more than one carbon-carbon double bond.

Exercises

Proteins and Amino Acids

13. They are both hydrophilic amino acids because both contain highly polar R groups.

15. a. Aspartic acid and phenylalanine make up aspartame.

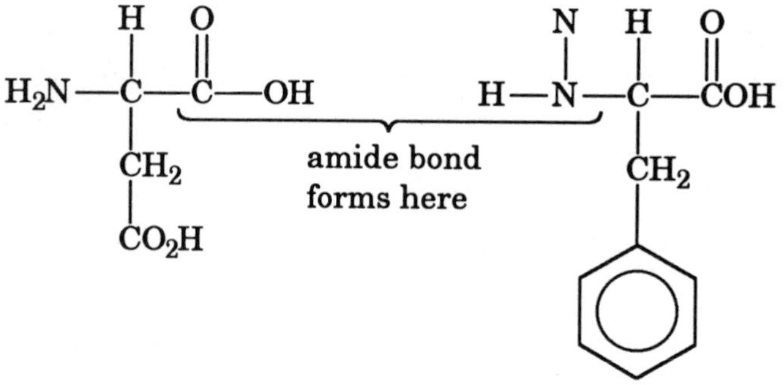

$$H_2N{-}\underset{\substack{|\\CH_2\\|\\CO_2H}}{\overset{\substack{H\\|}}{C}}{-}\overset{\substack{O\\||}}{C}{-}OH \qquad H{-}\overset{\substack{N\\|}}{N}{-}\underset{\substack{|\\CH_2\\|\\C_6H_5}}{\overset{\substack{H\\|}}{C}}{-}\overset{\substack{O\\||}}{C}OH$$

amide bond forms here

b. Aspartame contains the methyl ester of phenylalanine. This ester can hydrolyze to form methanol:

$$R{-}CO_2CH_3 + H_2O \rightleftharpoons RCO_2H + HOCH_3$$

17. a. Six tripeptides are possible. They are (from NH_2 to CO_2H end):

gly-phe-ala, gly-ala-phe, phe-gly-ala, phe-ala-gly, ala-phe-gly, ala-gly-phe

b. Three tripeptides are possible. They are (from NH_2 to CO_2H end):

ala-ala-gly, ala-gly-ala, gly-ala-ala

19. There are 5 possibilities for the first amino acid, 4 possibilities for the second amino acid, 3 possibilities for the third amino acid, 2 possibilities for the fourth amino acid and 1 possibility for the last amino acid. The number of possible sequences is:

$5 \times 4 \times 3 \times 2 \times 1 = 5! = 120$ different pentapeptides

21. a. Ionic: Need NH_2 on side chain of one amino acid with CO_2H on side chain of the other amino acid. The possibilities are:.

NH_2 on side chain = His, Lys or Arg; CO_2H on side chain = Asp or Glu

b. Hydrogen bonding: Need N–H or O–H bond in side chain. The hydrogen bonding interaction occurs between the X– H bond and a carbonyl group from any amino acid.

X–H $\cdots\cdots\cdot$ O = C (carbonyl group)

Ser	Asn	Any amino acid
Glu	Thr	
Tyr	Asp	
His	Gln	
Arg	Lys	

c. Covalent: Cys – Cys (forms a disulfide linkage)

d. London dispersion: All amino acids with nonpolar R groups. They are:

Gly, Ala, Pro, Phe, Ile, Trp, Met, Leu and Val

e. Dipole-dipole: Need side chain with OH group. Tyr, Thr and Ser all could form this specific dipole-dipole force with each other since all contain an OH group in the side chain.

23. Glutamic acid: R = $-CH_2CH_2CO_2H$; Valine: R = $-CH(CH_3)_2$; A polar side chain is replaced by a nonpolar side chain. This could affect the tertiary structure of hemoglobin and the ability of hemoglobin to bind oxygen.

Carbohydrates

25. Of the pentoses, D-ribulose has only two chiral carbon atoms while the other pentoses have three chiral carbon atoms (see the following structure where the asterisks denotes the various chiral carbon atoms). Of the hexoses, D-fructose has only three chiral carbon atoms versus four chiral carbon atoms in the other hexoses. Note: A chiral carbon atom has four different substituent groups attached.

```
       CH2OH                    CH2OH
        |                        |
        C=O                      C=O
        |                        |
    H--*C--OH               HO--*C--H
        |                        |
    H--*C--OH                H--*C--OH
        |                        |
       CH2OH                 H--*C--OH
                                 |
                                CH2OH
```

D-Ribulose

D-Fructose

27. See Figures 23.17 and 23.18 of the text for examples of the cyclization process.

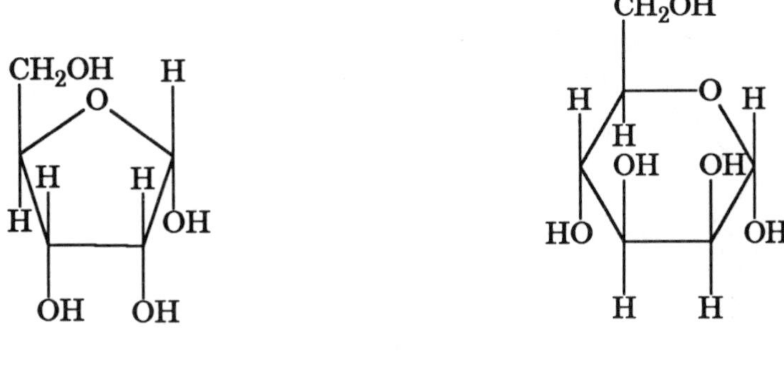

D-Ribose

D-Mannose

29. The α and β forms of glucose differ in the orientation of a hydroxy group on one specific carbon in the cyclic forms (see Figure 23.18 of the text). Starch is a polymer composed of only α-D-glucose and cellulose is a polymer composed of only β-D-glucose.

Optical Isomerism and Chiral Carbon Atoms

31. A chiral carbon has four different groups attached to it. A compound with a chiral carbon is optically active. Isoleucine and threonine contain more than the one chiral carbon atom (see asterisks).

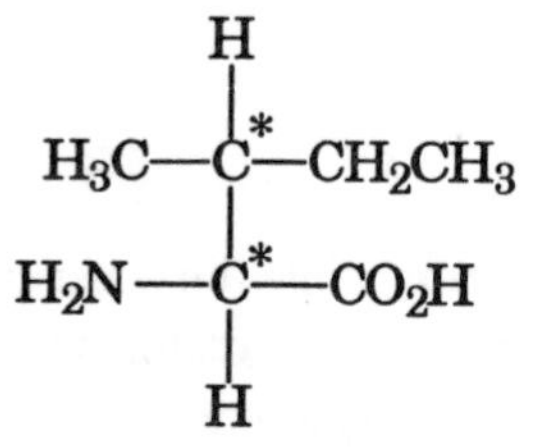

H
H_3C—C*—CH_2CH_3
H_2N—C*—CO_2H
H

isoleucine

H
H_3C—C*—OH
H_2N—C*—CO_2H
H

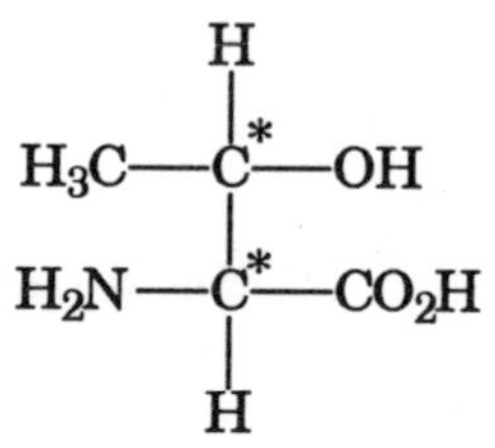

threonine

33. Only one of the isomers is optically active. The chiral carbon in this optically active isomer is marked with an asterisk.

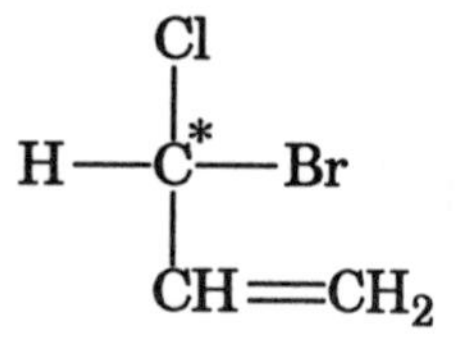

Cl
H—C*—Br
CH=CH_2

Nucleic Acids

35. The complimentary base pairs in DNA are cytosine (C) and guanine (G), and thymine (T) and adenine (A). The complimentary sequence is: T-A-C-G-C-C-G-T-A

37. Uracil will H-bond to adenine.

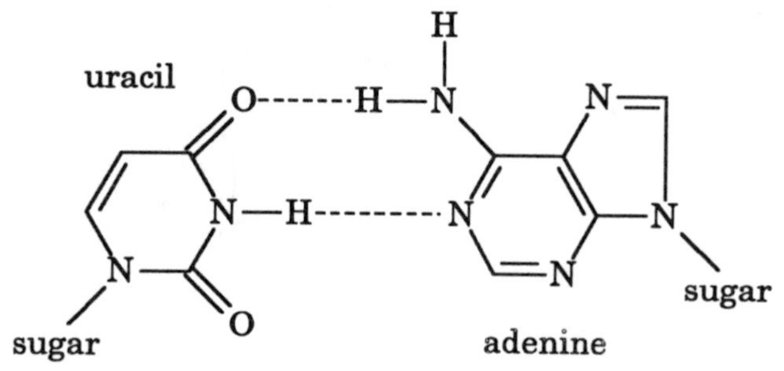

uracil O------H—N (H)
N—H--------N
sugar adenine sugar

39. Base pair:

RNA	DNA
A	T
G	C
C	G
U	A

a. Glu: CTT, CTC

Val: CAA, CAG, CAT, CAC

Met: TAC

Trp: ACC

Phe: AAA, AAG

Asp: CTA, CTG

b. DNA sequence for Met - Met - Phe - Asp - Trp:

TAC - TAC - AAA - CTA - ACC
or or
AAG CTG

c. Due to phe and asp, there is a possibility of four different DNA sequences.

d.

C—T—T—A—C—C—A—A—A

Glu - Trp - Phe

e. C - T - C - A - C - C - A - A - A

C - T - T - A - C - C - A - A - G

C - T - C - A - C - C - A - A - G

Lipids and Steroids

41.

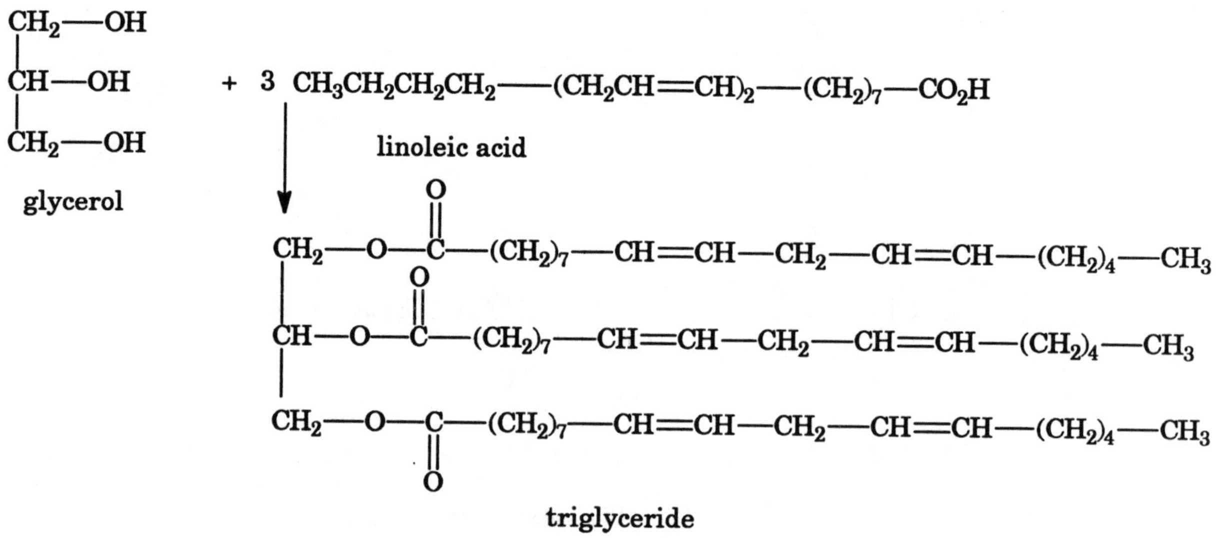

$$\begin{array}{l} CH_2{-}OH \\ | \\ CH{-}OH \\ | \\ CH_2{-}OH \end{array} \quad + \; 3\; CH_3CH_2CH_2CH_2{-}(CH_2CH{=}CH)_2{-}(CH_2)_7{-}CO_2H$$

glycerol — linoleic acid

$$\downarrow$$

$$\begin{array}{l} CH_2{-}O{-}C(=O){-}(CH_2)_7{-}CH{=}CH{-}CH_2{-}CH{=}CH{-}(CH_2)_4{-}CH_3 \\ | \\ CH{-}O{-}C(=O){-}(CH_2)_7{-}CH{=}CH{-}CH_2{-}CH{=}CH{-}(CH_2)_4{-}CH_3 \\ | \\ CH_2{-}O{-}C(=O){-}(CH_2)_7{-}CH{=}CH{-}CH_2{-}CH{=}CH{-}(CH_2)_4{-}CH_3 \end{array}$$

triglyceride

43. Hydrogenation converts unsaturated fats (double bonds present) to saturated fats (no double bonds present). Unsaturated fats occur as oily liquids and saturated fats occur as solids. Therefore, hydrogenation will solidify fats. The triglyceride in Exercise 23.41 contains 6 carbon-carbon double bonds, hence, 6 mol of H_2 are required to completely hydrogenate 1 mol of this triglyceride.

45. A detergent molecule has a polar (or ionic) head group and a long nonpolar hydrocarbon tail.

47. No; Chloresterol compounds are naturally occuring compounds in all humans and are essential for human life. Some important compounds produced from cholesterol are bile acids, steroid hormones and vitamin D.

49. See Figure 23.33b of the text for the structure. Since vitamin D_3 is composed almost entirely of carbon and hydrogen, then it is essentially nonpolar and will be soluble in nonpolar fat. Fat soluble vitamins will accumulate in the body more readily than water soluble vitamins. Excess water soluble vitamins will pass through the body.

Additional Exercises

51. Glutamic acid:

$$H_2N—CH(CH_2CH_2CO_2H)—CO_2H$$

One of the two acidic protons in the carboxylic acid groups is lost to form MSG. Which proton is lost is impossible for you to predict.

Monosodium glutamate:

$$H_2N—CH(CH_2CH_2CO_2^-Na^+)—CO_2H$$

In MSG, the acidic proton from the carboxylic acid in the R group is lost, allowing formation of the ionic compound.

53. $\Delta G = \Delta H - T\Delta S$; For the reaction, we break a P–O and O–H bond and form a P–O and O–H bond. Thus, $\Delta H \approx 0$. $\Delta S < 0$, since 2 molecules are going to form one molecule (order increases). Thus, $\Delta G > 0$ and the reaction is not spontaneous.

55. Glycine can be thought of as a diprotic acid. The first proton to leave comes from the carboxylic acid end with $K_a = 4.3 \times 10^{-3}$. The second proton to leave comes from the protonated amine end (K_a for $R–NH_3^+ = K_w/K_b = 1.0 \times 10^{-14}/6.0 \times 10^{-3} = 1.7 \times 10^{-12}$).

In 1.0 M H^+, both the carboxylic acid and the amine end will be protonated since H^+ is in excess. The protonated form of glycine is $^+H_3NCH_2COOH$. In 1.0 M OH^-, the dibasic form of glycine will be present since the excess OH^- will remove all acidic protons from glycine. The dibasic form of glycine is $H_2NCH_2COO^-$.

57. $5 \times 10^9 \text{ pairs} \times \dfrac{340 \times 10^{-12}\text{ m}}{\text{pair}} = 1.7\text{ m} \approx 2\text{m}$; 1.7 m corresponds to 5' 7".

Challenge Problems

59. The initial increase in rate is a result of the effect of temperature on the rate constant. At higher temperatures the enzyme begins to denature, losing its activity and the rate decreases.

61. a. Even though this form of tartaric acid contains 2 chiral carbon atoms (see asterisks in the following structure), the mirror image of this form of tartaric acid is superimposable. Therefore, it is not optically active. An easier way to identify optical activity in molecules with two or more chiral carbon atoms is to look for a plane of symmetry in the molecule. If a molecule has a plane of symmetry, then it is never optically active. A plane of symmetry is a plane that bisects the molecule where one side exactly reflects on the other side.

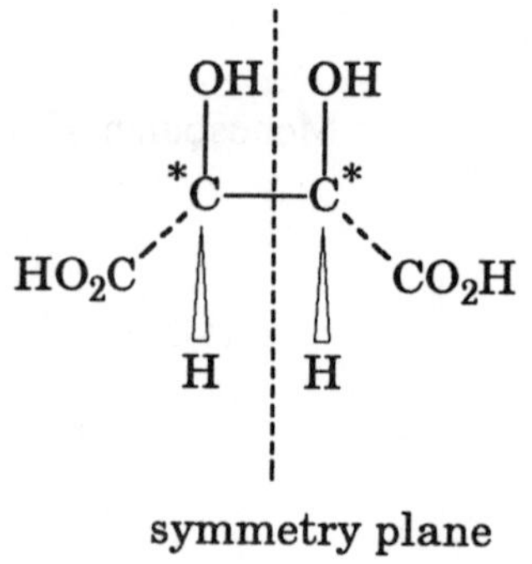

OH OH

C—C

HO_2C CO_2H

H H

symmetry plane

b. The optically active forms of tartaric acid have no plane of symmetry. The structures of the optically active forms of tartaric acid are:

OH OH OH OH

C—C C—C

H CO_2H HO_2C H

CO_2H H H CO_2H

mirror

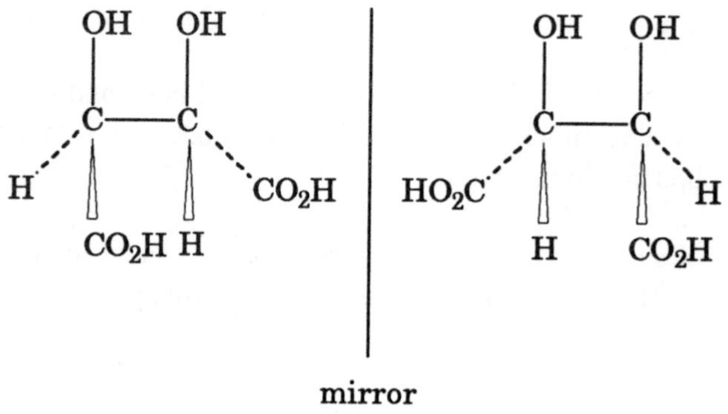

These two forms of tartaric acid are nonsuperimposable.